THE CATHOLIC BIBLICAL QUARTERLY
MONOGRAPH SERIES

II
THE HIDDEN KINGDOM

A REDACTION-CRITICAL STUDY OF THE REFERENCES TO THE KINGDOM OF GOD IN MARK'S GOSPEL

EDITORIAL BOARD

Joseph A. Fitzmyer, S.J., Chairman

Raymond E. Brown, S.S.
Frank M. Cross, Jr.
J. Louis Martyn

Dennis J. McCarthy, S.J.
Roland E. Murphy, O.Carm.
Bruce Vawter, C.M.

THE HIDDEN KINGDOM

A REDACTION-CRITICAL STUDY OF THE
REFERENCES TO THE KINGDOM OF
GOD IN MARK'S GOSPEL

BY

ALOYSIUS M. AMBROZIC

The Catholic Biblical Association of America
Washington, D. C. 20017
1972

THE HIDDEN KINGDOM: A REDACTION-CRITICAL STUDY OF THE REFERENCES TO THE KINGDOM OF GOD IN MARK'S GOSPEL

by Aloysius M. Ambrozic.

© 1972 The Catholic Biblical Association of America
Washington, D. C.

All rights reserved. No part of this book
may be reproduced in any form or by
any means without permission in writing
from the chairman of the editorial board.

PRINTED IN THE UNITED STATES
by the Heffernan Press Inc.
35 New Street, Worcester, Mass. 01605

LIBRARY OF CONGRESS CATALOGUE CARD NUMBER: 72-89100

MATERI IN POKOJNEMU OČETU V ZAHVALO

TO MY MOTHER AND TO MY LATE FATHER IN GRATITUDE

Table of Contents

Author's Preface	xi
Introduction	1
Chapter I THE COMING KINGDOM	3
(1) Mark 1:14-15	3
(A) The Origin and Structure of Mark 1:14-15	4
(a) *The Redactional Character of Mark 1:14-15*	4
(b) *The Structure of Mark 1:14-15*	6
(B) Mark's euangelion	8
(a) *Principles Guiding This Discussion*	8
(b) *The Content of Mark's euangelion*	13
(1) A Brief Survey of the Passages in Which the Term Occurs	13
(2) The Kingdom of God Is at Hand	15
(c) *The Function of Mark's euangelion*	27
(2) Mark 11:10	32
(A) Redactional or Redactionally Retouched Verses in Mark 11:1-10	32
(B) Marcan Source	35
(C) Mark's Understanding of the Pericope	40
(3) Concluding Remarks on Chapter I	44
Chapter II THE HIDDEN KINGDOM	46
(1) Mark 4:10-12	46
(A) The Origin of the Logion and Mark's Redactional Treatment of It	47
(B) Those Outside	53
(C) In Parables	72
(D) The Mystery of the Kingdom	92
(2) Mark 4:26-29	106
(A) The Text and the Original Form of the Parable	106

  (B) Interpretation of Jesus' Parable 108
  (C) Mark's Understanding of the Parable 120

 (3) Mark 4:30-32 122
  (A) The Text of Mark 4:30-32 and Luke 13:18-19 122
  (B) Discovering the Meaning of Jesus' Parable 123
  (C) Mark's Understanding of the Parable 132

 (4) Concluding Remarks 134

Chapter III THE ETHICAL DEMANDS OF THE KINGDOM 136

 (1) Mark 10:14-15 136
  (A) The Original Isolated Character of Mk 10:15 and the Question of Its Genuinity 136
  (B) The Meaning of Jesus' Logion 139
   (a) *"To Enter the Kingdom"* 139
   (b) *"To Receive the Kingdom"* 143
   (c) *"Like a Child"* 148
  (C) Mark's Message in 10:15 154

 (2) Mark 10:23-25 158
  (A) Text and Composition of Mark 10:23-25 158
  (B) Mark 10:23-25 on the Lips of Jesus 165
  (C) Mark's Understanding of Vss. 23-25 170

 (3) Mark 9:47 171
  (A) The Context of the Saying 172
  (B) The Message of Mark 9:33-50 174
  (C) Mark 9:47 175

 (4) Mark 12:34 177
  (A) Redaction and Tradition in Mark 12:28-34 177
  (B) The Meaning of Mark 12:34a 179

 (5) Concluding Remarks 181

Chapter IV THE THEME OF THE KINGDOM IN THE LITURGY OF THE COMMUNITY 183

Mark 14:25 183
  (A) The Text, Original Context, and Form of the Logion .. 183

(B) The Message of the Logion		189
(C) The Logion in the Gospel of Mark		197
Chapter V THE FUTURE KINGDOM		203
(1) Mark 9:1		203
(A) The Text and the Possible Redactional Elements of the Logion		203
(B) The Message and the Authenticity of the Logion		209
(C) Mark 13		222
(D) The Function of the Logion within the Gospel		231
(2) A Note on Mark 15:43		240
Chapter VI CONCLUSIONS		244
Bibliography		248
Index of Modern Authors		267
Index of Subjects		271
Biblical Index		272
Extrabiblical Literature Index		280

AUTHOR'S PREFACE

The following work is based on a dissertation accepted by the theological faculty of the Bayerische Julius-Maximilians-Universität, Würzburg, in the summer of 1970 in fulfilment of the requirement for the degree of Doctor of Theology. The title of the dissertation was *St. Mark's Concept of the Kingdom of God: A Redaction-Critical Study of the References to the Kingdom of God in the Second Gospel*. It has been reworked for publication.

I would like to thank all those who in various ways have made this work possible and contributed to its growth: the Most Rev. Philip F. Pocock, Archbishop of Toronto, for releasing me from other duties that I might write it; and the Rev. Basil F. Courtemanche for suggesting this to him. My most sincere gratitude is owed to all my German and non-German friends who made my stay in Germany so pleasant and profitable. From among them I would mention in particular my fellow-students in the Neutestamentliches Oberseminar in Würzburg whose friendship I appreciated and enjoyed more than I can say, and the Slovene priests in Munich whose home was my home whenever I needed a holiday from exegesis. I thank Prof. DDr. Joseph Reuss for acting as the second reader of the dissertation, the Rev. William R. Crummer for correcting my English, and the Rev. Joseph A. Fitzmyer, S.J., for his patient kindness in preparing the book for publication in the CBQMS.

The greatest debt of gratitude, however, I owe to Prof. Dr. Dr. h. c. Rudolf Schnackenburg. In him I found not only an outstanding *didaskalos* who guided and accompanied my work with attentive and painstaking care, but also, in the fullest sense, a *presbyteros tēs ekklēsias tou theou*.

I should also like to thank the following publishers for the kind permission granted to quote from their works:
SCM Press, London, and Charles Scribner's Sons, New York: J. Jeremias, *The Parables of Jesus*; Clarendon Press, Oxford: R. H. Charles, *Apocrypha and Pseudepigrapha of the Old Testament*; Cambridge University Press: C. E. B. Cranfield, *The Gospel according to St Mark*; Macmillan, London: V. Taylor, *The Gospel according to St. Mark*; Penguin Books Ltd., Harmondsworth: G. Vermes, *The Dead Sea Scrolls in English*; Universitetsforlaget, Oslo: N. A. Dahl, "The Parables of Growth," *Studia Theologica* 5 (1951) 132-66.

ALOYSIUS M. AMBROZIC

INTRODUCTION

At first sight there seems to be little order and method in the manner in which Mark arranges his material. He gives the impression of having been strongly dominated by the form which his pericopes had assumed in the process of oral transmission and, sometimes at least, by the order imposed upon them by earlier collectors. Many attempts have been made to find a biographical progression of events or an orderly logical division of matter in the Gospel; these attempts have either come to grief or carry little conviction. It is quite impossible to reconstruct a life of Jesus or to extract a theological treatise from the work of the traditional Second Evangelist. Yet Mark is no mere collector who has only gathered traditional material, stringing it together by frequent use of *kai* and *euthys*. The redaction-critical study of the Gospel which has been going on for some decades has indicated the unity of concept and purpose underlying his work. We say "decades," because the analyses of the Gospel made by such scholars as E. Lohmeyer and R. H. Lightfoot deserve to be called redaction-criticism, even though the term had not yet been invented and some of the methods in use today had not yet been fully developed.

However, the recognition and affirmation of the unity of concept and purpose of the Gospel does not seem to help much in arriving at a clear picture of Mark's notion of the kingdom. The passages that speak of the kingdom seem to be scattered; his motives for placing them redactionally where they are now found, or for accepting them as part and parcel of already formed units of tradition, seem to have been, in many cases, other than that of conveying his thought on the kingdom. But in this respect again we must not judge too hastily from appearances alone. While it would be quite wrong to attribute a disproportionate role to the theme of the kingdom in the Gospel, it is significant that 1:15, the verse which summarizes the message of the entire work, proclaims that "the kingdom of God is at hand." A good number of other references to the kingdom are clustered in two clearly distinguishable sections of the Gospel. Three of them occur in the parable chapter, and six of them in chs. 9 and 10. A certain systematization seems to be observable. The redactionally inserted 4:11 speaks of the mystery of the kingdom, and the seeming insignificance of the kingdom, invisible to man's eye, is brought out with particular clarity in the two parables which expressly speak of it, viz., that of the Seed Growing Secretly and that of the Mustard Seed. Sayings on entry into the kingdom are found in the section of the Gospel in which Jesus is devoting almost all his time to the instruction of the disciples, calling them to follow him. Other references to the kingdom are dispersed. We shall attempt to see whether they found their way into the Gospel by the express intention of the redactor or by the pull exerted upon him by the tradition at his disposal; and so we shall consider the light they throw on Mark's thought about the kingdom of God.

The purpose of this study is to examine all Mark's references to the kingdom, and thus to arrive at an understanding of the idea that he has of it. Since this is primarily a redaction-critical inquiry, we shall restrict it to those passages in which the evangelist himself sees such a reference; the parable of the Servants of the Absent Householder (13:34), for example, which may originally have been a kingdom parable, will not be considered, since in the present context the absent householder has been allegorized to represent the Son of Man. Obviously we will learn more about Mark's concept of the kingdom, as well as about his entire Gospel, from the sayings and the passages which have been given their present place by the deliberate intention of the redactor. But the other references also have a function in the Gospel; not to give them due consideration would entail the risk of disregarding potentially important aspects of its message.

Our inquiry is divided into five chapters. In the first we consider the passages which proclaim the coming of the kingdom and its hidden presence in Jesus' word and work (1:15; 11:10). The second chapter is devoted to its mysterious activity among men (4:10-12, 21-25, 26-29, 30-32). The third chapter deals with the ethical demands of the kingdom (10:13-16, 21-27; 9:47; 12:34). The subject matter of the fourth chapter is the theme of the kingdom in the liturgical celebration of the community (14:25). In the last chapter we examine the passages which speak of the kingdom as a future reality (9:1; 15:43); in this connection a consideration of the eschatological discourse and, in particular, of the saying in 13:30 is evidently imperative.

Biblical quotations in English are taken from the *New American Bible* (the term "reign," however, is replaced by "kingdom"). Intertestamental literature is quoted from R. H. Charles, *The Apocrypha and Pseudepigrapha of the Old Testament in English*. G. Vermes, *The Dead Sea Scrolls in English,* is used when Qumran literature is quoted in English.

Chapter I

THE COMING KINGDOM

The theme of the coming kingdom in Mark's Gospel is found in two main passages, Mk 1:14-15 and 11:10. These passages will be discussed individually. The origin and structure of Mk 1:14-15 are considered first; there follows a discussion of the content and function of the term *euangelion* in the Gospel. Mk 11:10 is studied within its immediate context, 11:1-11. An examination of Mark's redactional interventions in the entire passage throws light on his message in 11:10.

(1) Mark 1:14-15

After John's arrest, Jesus appeared in Galilee proclaiming the good news of God: ¹⁵"This is the time of fulfilment. The kingdom of God is at hand! Reform your lives and believe in the gospel!"

Mark 1:14-15 is the first and the most important of the passages dealing with the kingdom in the Second Gospel. Its origin and structure, its function in the Gospel, and its message must be considered. A brief look at its structure suffices to show that the term *euangelion* plays a paramount role; and the verb *ēngiken* is significant for our understanding of its message. Our foremost attention will thus center on these two words.

As to the text of the passage, there is a certain amount of disagreement among critical editions. The first disagreement concerns the first two words of vs. 14. Westcott-Hort, Nestle, and Tasker prefer the reading *kai meta*, found in the Codices Vaticanus and Bezae and in some versions. Merk ascribes to this reading the same, or almost the same, validity as the reading which he has in the text. The United Bible Societies edition, Merk (with the reservation just mentioned), Tischendorf, von Soden and Vogels, on the other hand, favor the reading *meta de*, which is found in other uncials and in all minuscules. The second of the two readings should be accepted. One reason in its favor is the number of MSS which contain it; another is the greater likelihood that in the textual transmission of the Second Gospel, in which *kai* abounds and *de* is rare, *kai* would replace *de* than vice versa.

The second disagreement concerns the first two words of vs. 15. Tischendorf and Tasker omit *kai legōn*; Nestle has doubts about it though he allows it to stand, while von Soden doubts the authenticity of *kai* alone. The United Bible Societies edition, Merk, Westcott-Hort, and Vogels accept the phrase. The reason for doubt comes from the *prima manus* of the Codex

Sinaiticus, some versions, and Greek Origen. We feel that the omission of the phrase is explained by its apparent superfluousness.

(A) THE ORIGIN AND STRUCTURE OF MARK 1:14-15

(a) *The Redactional Character of Mark 1:14-15*

We can assert with practical certainty that vs. 14 is redactional.[1] In it we find such typically Marcan terms as *euangelion*,[2] Galilee,[3] and *kēryssein*.[4] We hope to show below that *meta de to paradothēnai* also plays an important role in Mark's concept of the Baptist's function.[5]

Vs. 15 also should be regarded as redactional.[6] To begin at the end of the

---

[1] See W. Marxsen, *Mark the Evangelist: Studies on the Redaction History of the Gospel* (Nashville: Abingdon, 1969) 38-40; F. Mussner, "Die Bedeutung von Mk 1,14f. für die Reichsgottesverkündigung," *TTZ* 66 (1957) 258; "Gottesherrschaft und Sendung Jesu nach Mk 1,14f," *Praesentia Salutis: Gesammelte Studien zu Fragen und Themen des Neuen Testaments* (Düsseldorf: Patmos, 1967), 82 (since "Gottesherrschaft" is a strongly reworked rendition of "Bedeutung" they are referred to separately); J. Sundwall, *Die Zusammensetzung des Markusevangeliums* (Acta Academiae Aboensis, Humaniora IX:2; Åbo:Åbo Akademi, 1934) 8; K. L. Schmidt, *Der Rahmen der Geschichte Jesu: Literarische Untersuchungen zur ältesten Jesusüberlieferung* (Darmstadt: Wissenschaftliche Buchgesellschaft, 1964) 32-33; W. Trilling, *Christusverkündigung in den synoptischen Evangelien: Beispiele gattungsgemässer Auslegung* (Biblische Handbibliothek IV; München: Kösel, 1969) 49.

[2] That *euangelion* is a redactional term introduced into the synoptic material by Mark has been shown by W. Marxsen (*Mark,* 119-26). In this he has found wide agreement: see G. Bornkamm, *RGG*³ 2, 760; L. E. Keck, "The Introduction to Mark's Gospel," *NTS* 12 (1965-66) 357; S. Schulz, "Markus und das Alte Testament," *ZTK* 58 (1961) 189; J. Lambrecht, *Die Redaktion der Markus-Apokalypse: Literarische Analyse und Strukturuntersuchung* (Rom: Päpstliches Bibelinstitut, 1967) 126-30 (particularly p. 128, n. 1); F. Hahn (*Mission in the New Testament* [*SBT* 47; London: SCM, 1965] 70-73) questions W. Marxsen's opinion; his brief remarks, however, fail to cast serious doubt on the arguments and conclusions of Marxsen. Hahn's simple denial of the redactional character of Mk 1:14-15 hardly amounts to a convincing argument. With regard to ch. 13, one would like to ask why Hahn looks upon vss. 7c and 33-37 as redactional and vs. 10 as traditional.

[3] That Galilee has been introduced by Mark into preexisting material has been shown by W. Marxsen (*Mark,* 57-61).

[4] The redactional character of *kēryssein* in Mark has been pointed out by E. Schweizer ("Anmerkungen zur Theologie des Markus," *Neotestamentica* [Zürich: Zwingli, 1963] 93-94).

[5] R. Pesch ("Anfang des Evangeliums Jesu Christi," *Die Zeit Jesu: Festschrift für Heinrich Schlier* [eds. G. Bornkamm and K. Rahner; Freiburg: Herder, 1970] 115) contends that vss. 14-15 are, with the exception of vs. 15c (*pisteuete en tō euangeliō*), to be ascribed to a pre-Marcan redactor. However, one should not multiply redactors in the case of a text in which there occur so many Marcan redactional terms.

[6] See R. Bultmann, *History of the Synoptic Tradition* (New York: Harper & Row,

verse, the phrase *pisteuete en tō euangeliō* should in all probability be ascribed to Mark himself. It is true that, as E. Lohmeyer observes,[7] the language of vs. 15 is strongly colored by the missionary terminology of the primitive Christian community; parallels to vs. 15b are found in Acts 5:31; 11:18; 20:21. The fact, however, that the Marcan phrase occurs nowhere else in the NT hardly speaks in favor of his suggestion that Mark is here quoting a traditional formula. It should be evident, of course, that neither *pisteuein* nor *euangelion* had been coined by Mark; since the latter term belongs to his redactional vocabulary, however, the phrase as such should be considered redactional.

The term *metanoia* was also much used in the missionary activity of the primitive Church.[8] This usage likely had its roots in Jesus' own preaching.[9] The fact that *metanoia* and *pistis* are frequently found in apposition in early Christian preaching (Acts 11:17-18; 20:21) probably had some influence on Mark's composition of vs. 15b. Yet there is merit in J. Sundwall's suggestion[10] that Mark placed *metanoeite* here as a refrain to the Baptist's proclamation of the *baptisma metanoias* in 1:4. It is surely significant that the ministry of Jesus is introduced by terminology very similar to that employed to present John's proclamation in the desert. K. Romaniuk[11] has argued that the parallelism which exists between the themes of fulfilment and conversion as well as between those of nearness and faith in the logion entitles us to assume the existence of an Aramaic saying which might be traced to Jesus himself. There is little to quarrel with in his observation on parallelism in the verse. But in view of the fact that the term *pisteuein* most likely became a technical term in Christian missionary vocabulary only in the course of the Christian mission,[12] and in view of the above observations on the redactional features of the verse, his opinion fails to convince. There

---

1963) 118; K. L. Schmidt, *Rahmen*, 32-34; L. E. Keck, "Introduction," 358, n. 7; E. Haenchen, *Der Weg Jesu: Eine Erklärung des Markus–Evangeliums und der kanonischen Parallelen* (Berlin: de Gruyter, 1968) 73; W. Trilling, *Christusverkündigung*, 50-51.

[7] *Das Evangelium des Markus* (*Meyer* 1/2; Göttingen: Vandenhoeck & Ruprecht, 1963) 30.

[8] See E. Lohmeyer, *Markus*, 30; F. Mussner, "Bedeutung," 261; J. Behm, *TDNT* 4, 1003-5.

[9] See J. Behm, *TDNT* 4, 1001-3.

[10] *Zusammensetzung*, 8.

[11] "Repentez-vous, car le royaume des cieux est tout proche (Matt. iv. 17 par.)," *NTS* 12 (1965-66) 265. N. Perrin (*The Kingdom of God in the Teaching of Jesus* [*NT* Library; London: SCM, 1963] 199-201) points out the improbability of such an opinion on the grounds of content alone.

[12] See E. Lohmeyer, *Markus*, 29; D. E. Nineham, *The Gospel of St. Mark* (*Pelican Gospel Commentaries*; Harmondsworth: 1963) 68-70; R. Bultmann, *TDNT* 6, 208-9.

may well be echoes of Jesus' own words in the saying, but to affirm that the saying as such derives from Jesus or from pre-Marcan tradition is quite another matter.

The *hoti recitativum*, the particle introducing the logion, is again typically Marcan.[13] Are we, on the other hand, to consider the statement placed on the lips of Jesus in vs. 15a as a traditional formula? Echoes of early Christian preaching are undoubtedly as strong in the first as they are in the second part of the verse. The thought that with Jesus there came the fulfilment of time is frequently repeated in the NT.[14] The nearness of the kingdom, likewise, is proclaimed often in the Gospels. Yet there is reason to doubt the proposition that Mark is reproducing a ready-made formula here. Although the two themes, that of fulfilment and that of the kingdom, are closely related,[15] there is no evidence in the NT that Mk 1:15a was a set formula in wide circulation among Christian missionaries. What seems to be better attested is the connection between the fulfilment and the coming of Jesus,[16] a link which may have been operative when Mark wrote *ēlthen* in vs. 14 and *peplērōtai* in vs. 15a.

We would conclude, therefore, that in vs. 15a, as well as in 15b, Mark is composing freely, using vocabulary of the primitive Christian mission. Although in vs. 15 this vocabulary is particularly in evidence,[17] and though some of the terms should be traced back to the preaching practice of Jesus himself (kingdom, conversion), nevertheless Mark himself is responsible for the formulation of vss. 14-15.[18]

(b) *The Structure of Mark 1:14-15*

The structure of the passage is not difficult to discern: vs. 14 forms the introduction, vs. 15 contains the message. Vs. 15 is neatly subdivided into

---

[13] See F. Mussner, "Bedeutung," 259-60; "Gottesherrschaft," 83. *Hoti recitativum* occurs in Mark 50 times; see M. Zerwick, *Untersuchungen zum Markus-Stil: Ein Beitrag zur stilistischen Durcharbeitung des Neuen Testaments* (Romae: Institutum Biblicum, 1937) 39, 42.

[14] See Acts 3:18; Gal 4:4; Eph 1:10. The same idea lies at the base of Matthew's references to individual OT passages which find their fulfilment in Christ's words or actions.

[15] See F. Mussner, "Gottesherrschaft," 83-84.

[16] See note 14.

[17] See R. Bultmann, *History*, 118, 341; D. E. Nineham, *Mark*, 68; E. Lohmeyer, *Markus*, 29-30; W. G. Kümmel, "Die Naherwartung in der Verkündigung Jesu," *Heilsgeschehen und Geschichte: Gesammelte Aufsätze 1933-1964* (Marburger theologische Studien 3; Marburg: Elwert, 1965) 459.

[18] See W. Trilling, *Christusverkündigung*, 49-56.

the message proper in 15a, and exhortation in 15b.[19] In vs. 15a we find two indicatives, and their position at the beginning of the two phrases shows what the evangelist wishes to emphasize: not the "time" and the "kingdom," but the fact that the time is fulfilled and that the kingdom is near.[20] Vs. 15b enunciates the response which the fact announced in 15a should evoke: conversion and faith. There is a further parallelism between the announced fulfilment and the conversion required, as well as between the nearness of the kingdom and faith.[21] Fulfilment and conversion look backwards, expressing the end of the time of waiting and a turning away from the manner of life which is understood to be part and parcel of it. Faith, on the other hand, is man's proper response to the definitive saving act of God, the establishing of his kingdom.

Another feature of these verses is of utmost importance for our discussion: the twice repeated Marcan redactional term "gospel." Nowhere else in his Gospel does Mark employ this term twice in such close succession. For vs. 15a he has thus created a setting by means of his chief redactional term.[22] A further consideration confirms this view. F. Mussner remarks that *kai legōn* is an exceptional construction for Mark,[23] and suggests that the "and" should be looked upon as epexegetical. Vs. 15a is, therefore, an explication of the "gospel of God" of vs. 14.

Hence for Mark, whose chief concern is to present the "good news of Jesus Christ" (1:1), this consists fundamentally in the fact that "this is the time of fulfilment; the kingdom of God is at hand."[24] Having arrived at the

---

[19] See F. Mussner, "Bedeutung," 257; "Gottesherrschaft," 81, 86; R. Asting, *Die Verkündigung des Wortes Gottes im Urchristentum* (Stuttgart: Kohlhammer, 1939) 316. W. Trilling (*Christusverkündigung*, 47-49) speaks of the indicative in vs. 15a and the imperative in vs. 15b.

[20] See M. Zerwick, *Markusstil*, 102; E. Lohmeyer, *Markus*, 30; F. Mussner, "Gottesherrschaft," 90.

[21] See K. Romaniuk, "Repentez-vous," 265; M.-J. Lagrange, *Evangile selon Saint Marc* (Paris: Gabalda, 1920) 17.

[22] See J. Sundwall, *Zusammensetzung*, 8; F. Mussner, "Gottesherrschaft," 86; K.-G. Reploh, *Markus—Lehrer der Gemeinde: Eine redaktionsgeschichtliche Studie zu den Jüngerperikopen des Markus-Evangeliums* (Stuttgarter biblische Monographien; Stuttgart: Katholisches Bibelwerk, 1969) 15.

[23] "Bedeutung," 259, and "Gottesherrschaft," 83; against the omission of *legōn* in some critical editions, see C. H. Turner, "Marcan Usage: Notes, Critical and Exegetical, on the Second Gospel," *JTS* 28 (1926-27) 10, 14.

[24] See F. Mussner, "Gottesherrschaft," 85; J. M. Robinson, *The Problem of History in Mark* (*SBT* 21; London: SCM, 1962) 23, n. 2; J. Schniewind, *Das Evangelium nach Markus* (München: Siebenstern Taschenbuch Verlag, 1968) 43; N. Q. Hamilton, "Resurrection Tradition and the Composition of Mark," *JBL* 84 (1965) 419.

primary content of Mark's "gospel," we must examine the term a little more closely.

(B) MARK'S EUANGELION

(a) *Principles Guiding This Discussion*

Before we begin the study of the content of Mark's "gospel," something should be said about the method to be employed in it. This is important, because some of the investigations made thus far, though contributing greatly to our understanding of this Marcan redactional term, lead to false conclusions and introduce foreign material into Marcan theology because of preconceptions linked with the term. These preconceptions are derived from various sources and may be fully legitimate within certain areas of NT study. Indiscriminate attribution of them to Mark, however, may be inappropriate. Some principles which seem to be important should, therefore, be mentioned.

(1) The primary source for discovering the meaning of Mark's "gospel" should be the Gospel of Mark. This principle should be self-evident. Yet it has not been followed by such exegetes as J. Wellhausen,[25] W. Marxsen,[26] and J. Schreiber,[27] who deduce the meaning of the term from Paul and pre-Pauline hellenistic Christianity. It would be difficult to doubt that Mark was influenced by the use made of the term in the hellenistic church, but we must keep in mind that "gospel" is by no means as univocal in Paul

---

[25] *Das Evangelium Marci* (Berlin: Reimer, 1909) 3, 67; see J. M. Robinson's criticism in *Problem*, 23, n. 2.

[26] See *Mark*, 126-38, particularly pp. 127-29, where W. Marxsen, to determine the meaning of the term in the Second Gospel, chooses the two passages which say least about its content, viz., 8:35 and 10:29. He can draw such conclusions at the price of having already made up his mind about the matter on the basis of Pauline writings, and not on the basis of Mark.

[27] "Die Christologie des Markusevangeliums: Beobachtungen zur Theologie und Komposition des zweiten Evangeliums," *ZTK* 58 (1961) 154-83. For a criticism, see L. E. Keck, "Mark 3, 7-12 and Mark's Christology," *JBL* 84 (1965) 355-57. For a criticism of Bultmann's suggestion that Mark is a conglomerate of the hellenistic Christ-myth and the story about Jesus, the suggestion taken for granted by J. Schreiber in his "Christologie" as well as in his *Theologie des Vertrauens: Eine redaktionsgeschichtliche Untersuchung des Markusevangeliums* (Hamburg: Furche, 1967), see P. Vielhauer, "Erwägungen zur Christologie des Markusevangeliums," *Zeit und Geschichte: Dankesgabe an R. Bultmann* (Tübingen: Mohr, 1964) 155-69; *Aufsätze zum Neuen Testament* (*Theol. Bücherei*, 31; München: Kaiser, 1965) 199-214. See also E. Best, *The Temptation and the Passion: The Markan Soteriology* (*NTSMS* 2; Cambridge: Univ. Press, 1965) 125-33; J. Roloff, "Das Markusevangelium als Geschichtsdarstellung," *EvT* 27 (1969) 76-77, particularly p. 77, n. 19.

as is sometimes presumed,[28] that it remains to be proven that pre-Pauline elements in Pauline epistles give us a full picture of the meaning of the term in the hellenistic community, and that Mark could very well have endowed a term taken from the tradition with a content which it had not possessed before him. It is exegetically more correct to follow the example of W. Wrede[29] who, in attempting to discover the content of the "mystery" in 4:11, felt obliged to begin his investigation with the evidence in the Second Gospel.

(2) In an examination of the Marcan passages in which the term "gospel" occurs, preference should be given to those passages which say something about it and which are primarily concerned with it. Thus one does not begin by discussing Mk 8:35 and 10:29,[30] since one cannot determine the relationship existing between Jesus and the "gospel" from these verses alone. The passages which must be consulted first are 1:14-15; 13:9-11; 14:9; after them 1:1; and only then 8:35 and 10:29.

(3) We should not presume that the evangelist shared our insight into the growth of the synoptic tradition; even less should we attribute to him the mental constructs which we employ to unify the phenomena of this growth. Marxsen's comparison of Mark's actualization of the traditions about the earthly Jesus with the actualization which took place in the Israelite cult[31] may in itself be enlightening and helpful, but, in view of some recent studies, the conclusions he draws from the comparison are oversimplified. G. von Rad[32] has made it clear that the concept and the feeling of real contemporaneity in Israelite cult were no longer possible by the time that Deuteronomy came to be written. H. Zirker[33] would date this impossibility much earlier than von Rad; he is of the opinion that the consciousness of the lapse of time between the founding-event and its commemoration in the cult was already present when the sacred calendar of the Canaanites was historicized. There is no doubt that Israel's cult turned to the past because

---

[28] That the Pauline concept of the "gospel" is not as univocal as imagined by W. Marxsen (*Mark*, 130) who bases his view on G. Friedrich (*TDNT* 2, 731) is made clear by Friedrich himself and by F. Mussner, "Evangelium und Mitte des Evangeliums: Ein Beitrag zur Kontroverstheologie," *Praesentia Salutis*, 160-67.

[29] *Das Messiasgeheimnis in den Evangelien: Zugleich ein Beitrag zum Verständnis des Markusevangeliums* (Göttingen: Vandenhoeck & Ruprecht, 1963) 58.

[30] See L. E. Keck, "Introduction," 357, who accuses W. Marxsen of pushing into the background the main point enunciated by Mk 8:35; see also note 26 above. Marxsen offers no proof whatsoever that the *kai* joining *emou* and *euangeliou* is epexegetical.

[31] *Mark*, 113, n. 170.

[32] *Old Testament Theology* (Edinburgh: Oliver and Boyd, 1962-65) 2, 108-11.

[33] *Die kultische Vergegenwärtigung der Vergangenheit in den Psalmen* (*BBB* 20; Bonn: Hanstein, 1964) 115-16.

of its meaning for the present. We must, however, avoid two extremes: that of denying all theological relevance to the past event as such and reducing it to a mere address to the present generation, and that of attributing to the Israelite cult a concept of time which would enable the worshipper to forget the distinctions between past, present, and future and really participate in the past event commemorated by the cult.[34] Israel's cult was, according to Zirker, always aware of the chronological succession of generations.[35]

Perhaps the most blatant example of attributing our mental constructs to the evangelist is supplied by A. Suhl.[36] In his summary of Marxsen's views on Mark's "gospel" as an address to the community of Mark's time,[37] he agrees that Mark had no real interest in the earthly Jesus. Though he does not say so expressly, he gives every reason to suppose that he also agrees with Marxsen's assertion that it is the risen Jesus who speaks in the Second Gospel.[38] Suhl then proceeds to compare Mark with the Book of Deuteronomy. He refers to G. von Rad, according to whom Deuteronomy extinguishes some seven centuries of disobedience and places Israel in the desert again to be instructed by Moses.[39] We feel that the comparison with Deuteronomy is quite legitimate. Yet von Rad's statement asserts the very opposite of what Suhl understands it to say. It is not Moses who is raised up and transported from Mt. Nebo to Jerusalem, but the community of Israel which is taken back into the desert. Deuteronomic traditions grew at a time when the cultures of the Near East felt the approaching end and attempted, in order to counteract their inner insecurity, to recapture the foundations on which they rested.[40] Deuteronomy itself was "a conscious effort to recapture both the letter and the spirit of Mosaism."[41] It may be

---

[34] Ibid., 110-18.

[35] His remark (*Vergegenwärtigung*, 117) on the cultic "today": "Das 'Heute' will nie die Vergangenheit selbst kultisch gegenwärtig setzen oder in die Vergangenheit eintreten, sondern die sich im Gedächtnis versammelnde Gegenwart neu unter die in der Geschichte gestiftete Bedeutsamkeit des Bundes stellen." For a similar view of OT actualization, see C. Westermann, "Vergegenwärtigung der Geschichte in den Psalmen," *Forschung am Alten Testament* (Theol. Bücherei 24; München: Kaiser, 1964) 315. On the gospels as consciously presenting a series of past events, see pp. 317-18, and B. S. Childs, *Memory and Tradition in Israel* (SBT 37; London: SCM, 1962) 81-85.

[36] *Die Funktion der alttestamentlichen Zitate und Anspielungen im Markusevangelium* (Gütersloh: Mohn, 1965) 168.

[37] Ibid., 166-68.

[38] *Mark*, 131.

[39] *Theology* 1, 231.

[40] See W. F. Albright, *From the Stone Age to Christianity: Monotheism and the Historical Process* (Garden City, N.Y.: Doubleday, 1957) 315-21.

[41] Ibid., 319.

significant that the book, besides moving the community backwards into the presence of Moses in the desert, is looking forward to a new Moses.⁴²

To return to Mark, there is no doubt that he wishes to address his contemporaries; we would also agree with Marxsen and Suhl and, for that matter, with the vast majority of modern exegetes, that Mark is not writing a historically reconstructed life of Jesus. However, Marxsen's view of the Second Gospel tends to force a dilemma upon us: either we place Jesus in the situation of Mark's community or Mark's community in the situation of the earthly Jesus. This dilemma is quite inadmissible, since it involves an acceptance of certain presuppositions which were foreign to Mark. Yet for the sake of argument, it might at least be as correct to claim that Mark places his community in the company of the earthly Jesus, the Son of God,⁴³ on his way to the cross as to assert with Marxsen that it is the risen Jesus who is speaking to the community through the medium of Mark's Gospel. There is no doubt that Mark is updating the Christian message; but do we have any reason to think that he looked upon his undertaking as updating, rather than as a return to the origins? The procedure which we meet in Deuteronomy is by no means limited to that book alone. We encounter it frequently in the OT, which attributes the Torah to Moses, psalms to David, wisdom to Solomon, an apocalypse to Daniel. We may further ask whether the phenomenon of pseudonymy in the intertestamental literature is not merely that. This type of procedure may seem to us to be fiction; but it is a different matter to assert that it was fiction to biblical authors.

(4) The next principle flows, to a degree, from the previous one. It should be clear even from a cursory examination of the Second Gospel that Jesus is the primary proclaimer of the "gospel." One cannot deny, of course, that the needs, preoccupations, and experiences, past and present, of the community for which Mark wrote influenced the evangelist's image of Jesus and of his teaching. But to imagine that Mark reasoned in Marxsen's manner,⁴⁴ viz., that the Jesus who proclaims must be the risen Lord, since the

---

⁴² See G. von Rad, *Theology* 2, 260-62; *Deuteronomy* (*OT Library*; Philadelphia: Westminster, 1966) 123-25.

⁴³ Mark's Jesus is the Son of God not in virtue of his resurrection, as he is in the traditional formula quoted in Rom 1:3-4, but already at his baptism. Here we meet a process parallel to that undergone by the term "gospel": as Mark's "gospel" is made to announce not only the risen but the earthly Jesus, so is divine sonship predicated of Jesus not only in his risen state but also in his earthly existence. See also C. Burger, *Jesus als Davidssohn: Eine traditionsgeschichtliche Untersuchung* (*FRLANT* 98; Göttingen: Vandenhoeck & Ruprecht, 1970) 68.

⁴⁴ *Mark*, 131-38. L. E. Keck's statement ("Introduction," 358) that "Marxsen has

"gospel" of the community is being proclaimed in the present, is an unwarranted conclusion. In Mark's view it is Jesus' way which determines the way of the community.[45] It is not Jesus who suffers the fate of the community, but the community which suffers the fate of Jesus; it is not Jesus who is made to proclaim the "gospel" of the community, but the community which must proclaim the "gospel of God" brought by Jesus.[46] In Mark 1:1 the governing word is not "the gospel" but "Jesus Christ." If we are searching for phrases which express the matter succinctly, we would choose R. Asting's, "Jesus embodies the gospel,"[47] as an appropriate way of stating the relationship between Jesus and the "gospel." This phrase avoids the impression that, in Mark's view, the "gospel" of the hellenistic church determines the earthly Jesus. We know, and Mark may well have known, that the hellenistic community associated Jesus' divine sonship with his resurrection. But for Mark the heavenly voice at the baptism was a simple statement of fact, not a literary device whereby he succeeded in projecting this sonship into Jesus' earthly existence.[48]

It is not the risen Lord who addresses us in the Second Gospel, but the earthly Jesus, the Son of God. There is no doubt that Mark believes in Jesus' resurrection; yet his placing of the climax of the Gospel at the death of Jesus in 15:39[49] must surely have a significance. Marxsen's statement that the Second Gospel is a commentary on the "gospel" as found in Paul[50] can be accused of two methodical blunders: first, of assuming, with no proof whatever, that Mark assigns the same meaning to the term as Paul; secondly, of forcing the material collected and shaped by Mark into a mold which it clearly resists, instead of using the material itself as a guide in discovering the meaning of the term "gospel."

(5) We should avoid coarsening the evangelist's thought and message by identifications which are either unwarranted or too literal or exaggerated. Statements like that of D. Bosch,[51] "the person of Jesus is to be neither

simply modernized Mark's theology into Marxsen's" may sound unduly harsh, but it is hardly incorrect.

[45] See J. Delorme, "Aspects doctrinaux du second évangile," *ETL* 43 (1967) 93-94, 98.

[46] See E. Käsemann, "Blind Alleys in the 'Jesus of History' Controversy," *New Testament Questions of Today* (*NT Library*; London: SCM, 1969) 49-50, 62-63.

[47] *Verkündigung*, 320.

[48] See E. Schweizer, "Mark's Contribution to the Quest of the Historical Jesus," *NTS* 10 (1963-64) 431.

[49] See P. Vielhauer, "Christologie," 164-65; E. Schweizer, *The Good News According to Mark* (Richmond, Va.: John Knox, 1970) 358; *TWNT* 8, 381; H. F. Peacock, "The Theology of the Gospel of Mark," *RevExp* 55 (1958) 396.

[50] *Mark*, 138.

[51] *Die Heidenmission in der Zukunftsschau Jesu: Eine Untersuchung zur Eschatolo-*

separated nor distinguished from his word, i.e., the gospel of Jesus Christ," can be understood correctly if they are taken as a piece of tolerable rhetorical exaggeration. But understood as sober scientific exegesis, it vulgarizes the evangelist's thought by contaminating it with modern phenomenology. To say that "to believe in the gospel" means "to believe in Jesus Christ"[52] is, in a fashion, correct; but, in Marxsen's hands, it leads to an objectification of the "gospel"—a thought quite foreign to Mark for whom the "gospel" is that which is being proclaimed by Jesus and the community (see 1:14-15; 13:9-11). It is precisely this coarsening and objectification which enables Marxsen to suggest that the Jesus who speaks to us in Mark's Gospel is the risen Lord. Once he has identified the "gospel" with Jesus the Christ and observed that "gospel" is a present reality, no other conclusion is possible.[53] The "gospel" thus not only represents but becomes a representative of Jesus Christ. That we should proceed with much greater care in these matters has been shown by Asting,[54] who, with his customary finesse, perceives the nuances in such realities as the "gospel," the proclamation, the written Gospel, and Jesus Christ.[55]

The principles enunciated above amount to this: we must avoid at all costs importing into Mark's thought and message themes which stem from other NT books unless we can show, with some degree of probability, that Mark had made them his own. Above all, however, we must shun attributing to Mark our own preconceptions and our mental constructs.

(b) *The Content of Mark's euangelion*

(1) A Brief Survey of the Passages in Which the Term Occurs

We have already discussed Mk 1:14-15; it is the only passage in which the content of the gospel is directly indicated. Since it can be considered as a summary and programmatic statement of the entire written work,[56] it is

gie der synoptischen Evangelien (*ATANT* 36; Zürich: Zwingli, 1959) 58 (my translation).

[52] W. Marxsen, *Mark*, 135.

[53] In *The Resurrection of Jesus of Nazareth* (London: SCM, 1970, pp. 77-78), Marxsen seems to defend the opposite thesis: what the evangelists intend to show by means of their resurrection narratives is the fact that Jesus' work continues even after his death. It is not up to us to determine whether the two theses really or only seemingly contradict each other.

[54] *Verkündigung*, 320-23, 355-59.

[55] See *Verkündigung*, pp. 352-53, where he rejects the identification of Jesus with the "gospel" in Rom 1:3-4.

[56] This will be discussed below.

of paramount importance for this study. The most obvious similarity between 1:14-15; 13:10; and 14:9 consists in the fact that in them the gospel is said to be proclaimed (*kēryssein*). Though the gospel should not be reduced to a mere function and must not be looked upon simply as an act of proclamation,[57] there is nonetheless a close link between these two Marcan redactional terms: the "gospel" is there to be "proclaimed," ultimately to the entire world (13:10). Since gospel is a redactional term, introduced into the synoptic material by Mark, we should presume that it has the same content in 13:10 and 14:9 as in the programmatic statement in 1:14-15. Whoever disagrees with this assumption must shoulder the burden of providing evidence to the contrary.[58] A note of caution should be sounded with regard to 14:9. This passage, taken in isolation, does not necessarily indicate that Mark thought of the gospel as a narrative;[59] for the word which he uses to describe the handing on of the tradition reported in 14:3-8 is not *kēryssein*, but *lalein*. In Mk 1:1 the verb "to proclaim" does not appear; we feel, however, that it is implied. Mark is writing the book in order to proclaim the good news.

Thus in 1:14-15; 13:10 and 14:9 the good news and its proclamation are expressly linked, while in 1:1 this connection is implied. Is this connection also implied in 8:35 and 10:29? There seem to be indications in favor of this view. Mk 10:29 is introduced by the question placed on Peter's lips by the redactor: "We have left everything and followed you." In the Second Gospel the disciples alone are said to have "left to follow." In 1:16-20, where we meet the disciples for the first time, Jesus calls them in order to "become fishers of men"; the disciples immediately leave behind their nets and their families. In 3:14, a redactional verse, Jesus calls the Twelve "to be with him" and "to be sent to preach" (*kēryssein*). Mark thus defines the disciples by the words "to leave," "to follow," "to preach." In 10:28, moreover, there is a conscious contrast between the disciples who have left everything and followed him and the rich man who has refused to leave all he possessed to follow Jesus. It would seem, then, that *heneken tou euangeliou* in 10:29 suggests more than faith in the good news.

---

[57] See R. Asting, *Verkündigung*, 321.

[58] M. Werner (*Der Einfluss paulinischer Theologie im Markusevangelium* [*BZNW* 1; Giessen: Töpelmann, 1923] 98) agrees with J. Wellhausen that in Mk 13:10 and 14:9 "gospel" has the meaning of "the message about Jesus Christ"; he sees, however, that Wellhausen could not be correct in assigning the same meaning to the term in 1:14-15 (p. 100); his refusal to follow Wellhausen's lead in this case is welcome, but he does not give reasons for assigning a different sense to the term elsewhere.

[59] This view is held by, among others, R. Asting, *Verkündigung*, 322-23; W. Marxsen, *Mark*, 129; F. Mussner, "Evangelium," 168.

Participation in the work of Jesus, i.e., in the proclamation of the good news, seems to be implied.[60] The proclamation of the good news is also present in 8:35; this seems to be indicated by its place between the confession of Jesus' messiahship by Peter in the name of other disciples (8:27-29) and Jesus' affirmation that, at his coming in glory, the Son of Man will be ashamed of those who are ashamed of him and his words in the world (8:38).[61]

(2)   The Kingdom of God Is at Hand

Since 1935, when C. H. Dodd proposed that in Mk 1:15 *ēngiken* should be translated not "has come near" but "has arrived" or "has come upon you,"[62] the discussion of this translation has revived periodically. Although the concept of the kingdom in the thought of Jesus which Dodd presents[63] has not found acceptance,[64] his suggestion that *ēngiken* has the meaning of "has arrived" has fared better. There is still no unanimity on the subject. To mention but a few names: on the side of Dodd, one finds R. H. Lightfoot, M. Black, and W. R. Hutton,[65] whereas among his opponents we find J. Y. Campbell, W. G. Kümmel, J. M. Creed, and K. W. Clark.[66] It is not necessary to go over the battleground again; having read through the minute discussions in which the same text can be used to argue for both sides,[67] and where, on occasion, it seems to be forgotten that the sacred

---

[60] See J. Delorme, "Aspects," 80-81; F. Hahn (*Mission*, 71) remarks that the term "gospel" is primarily a *nomen actionis*.

[61] See J. Delorme, "Aspects," 80-81. A suggestive parallel is Rom 1:16, where Paul speaks of his apostolic work; see also 2 Tim 1:8,12.

[62] *The Parables of the Kingdom* (rev. ed.; London: Collins, 1967) 35-41, 43, n. 23; "'The Kingdom of God Has Come,'" *ExpT* 48 (1936-37) 138-42.

[63] *Parables*, 35, 44.

[64] See R. Schnackenburg, *God's Rule and Kingdom* (Montreal: Palm, 1963) 129-43; N. Perrin, *Kingdom*, 64-78; G. Lundström, *The Kingdom of God in the Teaching of Jesus* (Edinburgh: Oliver and Boyd, 1963) 113-24.

[65] R. H. Lightfoot, *History and Interpretation in the Gospels* (as quoted by K. W. Clark, "Realized Eschatology," *JBL* 59 [1940] 382); M. Black, "The Kingdom of God Has Come," *ExpT* 63 (1951-52) 289-90; W. R. Hutton, "The Kingdom of God Has Come," *ExpT* 64 (1952-53) 89-91.

[66] J. Y. Campbell, "'The Kingdom of God Has Come,'" *ExpT* 48 (1936-37) 91-94; W. G. Kümmel, *Promise and Fulfilment: The Eschatological Message of Jesus* (SBT 23; London: SCM, 1961) 19-25; K. W. Clark, "Realized Eschatology," *JBL* 59 (1940) 367-83; J. M. Creed ("'The kingdom of God Has Come,'" *ExpT* 48 [1936-37] 184-85) limits himself to refuting C. H. Dodd's translation of Mk 9:1.

[67] On Dan 4:11,22 (LXX), see J. Y. Campbell, "The Kingdom," 92; C. H. Dodd, "The Kingdom," 140-41; K. W. Clark, "Realized Eschatology," 369.

writers were not hair-splitting theologians and quibbling philologists,[68] one is only too ready to agree with R. F. Berkey's conclusion that philology alone cannot decide the issue,[69] and with R. Schnackenburg's observation that *ēngiken* could have been used by the early church to express the provisional presence of the kingdom.[70] K. W. Clark's conclusion that *engizein* means "to draw near, even to the very point of contact," but that "the experience which draws near is still sequential"[71] would, one suspects, bring an indulgent smile to the lips of biblical authors. Here we wish to consider only one of Clark's suggestions in somewhat greater detail since it seems not to have been dealt with by others. It is his suggestion that in Heb 7:19 and Jas 4:8 reverence for the transcendent and holy God dictated the use of the verb *engizein*.[72] This proposal could well be countered by another one, viz., that the verb in these texts derives from the sacrificial language of the OT, for sacrificial imagery is found in the context of both passages. Thus the verb is very likely a more or less direct translation of the Hebrew or Aramaic *qrb*, a technical term whose hiphil is translated as "bring," "present," "offer" (see Lev 1:2,3,5,10,13),[73] and whose qal is translated by "draw near" in the RSV (cf. Lev 9:5,7,8; 10:4,5; 16:1; 21:17),[74] but in the LXX by various forms of *proserchesthai* —a sure sign that the LXX translators were not haunted by our philological scruples. In this connection, we could well ask with Berkey[75] where we are to draw the line between drawing near and communion.

In Mark the verb *engizein* occurs three times only (1:15; 11:1; 14:42) and thus does not allow a firm conclusion in its regard. Its meaning in 1:15 will have to be decided on other grounds. Anticipating a little, we would agree with Berkey that the verb in 1:15 does not suggest nearness at the expense of arrival,[76] and that "one cannot eliminate the ministry as *one* of

---

[68] See W. R. Hutton's remark ("The Kingdom of God," 89) on the CCD and R. Knox translations of Acts 23:15: they "state in plain English exactly what is meant."

[69] "*EGGIZEIN, PHTHANEIN*, and Realized Eschatology," *JBL* 82 (1963) 177-87; see also F. W. Beare, *The Earliest Records of Jesus* (Oxford: Blackwell, 1962) 44.

[70] *God's Rule*, 141.

[71] "Realized Eschatology," 381.

[72] Ibid., 370.

[73] Cf. the RSV; the LXX translates it by various forms of *prosagō* and *prospherō*.

[74] In all these passages the Targum Onkelos has *qrb*, rendering the Hebrew *qrb*. W. G. Kümmel's remark (*Promise*, 23 n. 13), that the Aramaic *qrb* almost always denotes nearness, clearly does not apply to these texts.

[75] "*EGGIZEIN*," 183.

[76] Ibid., 187.

the decisive events in the total eschatological drama, and thus a 'realized' eschatological event."[77]

It is commonly agreed that Mk 1:14-15 serves as a programmatic statement which introduces, summarizes, and initiates the ministry of Jesus.[78] There are some exegetes who wish to limit this summary, to cover only a part of the Second Gospel.[79] Against such restrictions F. Mussner[80] points out that these exegetes do not prove their assertions; that, further, this is the only Marcan summary recapitulating Jesus' proclamation, while all the other summaries recount his activity; and that, since the statement of 1:15 is found but once, and only on the lips of Jesus, it should be looked upon as the opening logion in the light of which the rest of the book should be read. Mussner's arguments are good, but they need further substantiation. This we shall attempt to provide.

The common opinion which holds that 1:15 summarizes the entire proclamation of Jesus finds strong corroboration if we collate 1:1 with 1:14-15. Both passages are redactional,[81] both contain the term gospel, and their proximity to each other justifies our relating them. Even greater justification is provided by the tension observable between them: vs. 1 speaks of the "beginning" of the gospel, while vs. 14 introduces the "gospel of God"; vs. 1 is the heading of the entire work, to be sure, but more directly and immediately it is a preamble to the ministry of John the Baptist, while vs. 14 introduces the ministry of Jesus. Vs. 14 is set off from the preceding section by *de*, a relatively infrequent particle in Mark which

---

[77] Ibid., 185 (Berkey's italics).

[78] See J. M. Robinson, *Problem*, 21; R. Bultmann, *History*, 341; K. L. Schmidt, *Rahmen*, 33; F. Mussner, "Gottesherrschaft," 91; M.-J. Lagrange, *Marc*, 15; E. Schweizer, *Mark*, 44; W. Marxsen, *Mark*, 65-66; A. Kuby, "Zur Konzeption des Markus-Evangeliums," *ZNW* 49 (1958) 54; G. Schille, "Bemerkungen zur Formgeschichte des Evangeliums: Rahmen und Aufbau des Markus-Evangeliums," *NTS* 4 (1957-58) 6, 14; J. Schmid, *The Gospel according to Mark* (= *RNT* 1; Staten Island, N.Y.: Mercier Press, 1968) 29-30; D. E. Nineham, *Mark*, 67-68.

[79] V. Taylor, *The Gospel according to St. Mark* (London: Macmillan, 1966) 165; C. E. B. Cranfield, *The Gospel according to Saint Mark* (*The Cambridge Greek Testament Commentary*; Cambridge: University Press, 1963) 61 (the summary intends to cover 1:14-3:6); S. E. Johnson, *A Commentary on the Gospel according to St. Mark* (*Black's NT Commentaries*; London: Black, 1960) 24; R. H. Lightfoot, *The Gospel Message of St. Mark* (Oxford: Clarendon, 1952) 20.

[80] "Gottesherrschaft," 91 n. 50.

[81] For 1:1, see R. Bultmann, *History*, 245; T. A. Burkill, *Mysterious Revelation: An Examination of the Philosophy of St. Mark's Gospel* (Ithaca, N.Y.: Cornell, 1963) 9-10.

should, for that very reason, be looked upon as a strong link (by way of contradistinction) with what precedes.[82]

Before we compare the two passages, however, we must discuss the character and function of the first of them. Opinions on vs. 1 vary; a useful summary of a number of them can be found in C. E. B. Cranfield's commentary (pp. 34-35). For the purpose of this discussion we must decide on the connection, or lack of it, between vs. 1 and vss. 2-3.[83] Are we to consider Mk 1:1-3 as one sentence, or should we put a period at the end of vs. 1? Since, in this matter, various preconceptions can play havoc with the exegesis, we limit ourselves to philological observations. The decisive words are *kathōs gegraptai* introducing vs. 2. It is significant that Mark never begins a sentence with the word *kathōs*.[84] Moreover, "without exception *kathōs*-clauses follow their main clauses in Matthew and Mark (12 instances) . . . where *kathōs* introduces a following quotation in the New Testament it invariably follows its main clause . . . . Some editors break this rule in their punctuation of Mk 1, 1f, but there is no need to do this."[85] Lagrange's objection that the fusion of the first three verses of Mark into one sentence is a "construction . . . très dure"[86] need not disturb us when dealing with the Second Gospel. To his contention that on such a supposition *archē* should have the article,[87] we can reply that in the Second Gos-

---

[82] See M. Zerwick, *Markusstil*, 1-2, 6.

[83] For a discussion of various opinions, see E. Klostermann, *Das Markus-Evangelium* (*HNT* 3; Tübingen: Mohr, 1950) 3; M.-J. Lagrange, *Marc*, 1.

[84] E. Schweizer, *Mark*, 30.

[85] G. D. Kilpatrick, "The Punctuation of John VII. 37-38," *JTS* ns 11 (1960) 340-41; see also M.-J. Lagrange, *Marc*, 2; W. Grundmann, *Das Evangelium nach Markus* (*Theologischer Handkommentar zum Neuen Testament* 2; Berlin: Evangelische Verlagsanstalt, 1965) 26. V. Taylor (*Mark*, 153) voices the opinion that in Mk 1:2 "*kathōs* stands at the beginning of a new sentence," as in Lk 11:30; 17:26; Jn 3:14; 1 Cor 2:9. The point at issue, however, is not *kathōs* alone, but *kathōs gegraptai*; of the passages referred to by Taylor only 1 Cor 2:9 has this phrase; and it confirms G. D. Kilpatrick's assertation, for vs. 9 is obviously quoted in support of vs. 8a; vs. 10 introduces a new theme. These facts militate against the opinion of R. Pesch ("Anfang," 116), according to which Mk 1:2-3 had served, for a pre-Marcan redactor, as an anticipated commentary on the report which follows it.

Pesch claims that Mk 1:2-8 is to be ascribed in its entirety to a pre-Marcan redactor ("Anfang," 113-14). But this claim is rather dubious because of another fact: in Mk 1:4-5 we find a habitual procedure of Mark, viz., the including with practically every occurrence of the verb *kēryssein* a reference to, or a suggestion of, the wide or universal echo of the proclamation (explicit reference: 1:4-5, 38-39, 45; 5:20; 7:36-37; 13:10; 14:9; suggestion: 1:7,14; 6:12; the only exception to the rule is 3:14). See also note 5 above.

[86] *Marc*, 1.

[87] *Marc*, 1.

pel *archē* never has the article, and that the absence of it in 1:1 can be explained by the fact that the verse serves as a heading.[88] Hence Mk 1:1-3 should be looked upon as one sentence.

If this conclusion is correct, it should not be difficult to discover to what the "beginning" in 1:1 refers. For the OT quotations undoubtedly refer to John the Baptist. The "beginning of the gospel" is thus John's ministry.[89] Since, however, his ministry is merely the "beginning," it is evident that the gospel itself is a larger complex. Thus the heading of the written Gospel, by the very fact that it contains the word gospel, calls for a complement. This complement is provided by 1:14-15. Moreover, the ministry of John the Baptist, as presented by Mark 1:1-9 and by the rest of the Gospel, looks beyond itself. A study of the redactional composition of Mk 1:1-13, and in particular of vss. 1-9, would take us too far afield.[90] All we need to note is the strong selectivity of Mark in reporting John's work in the introductory verses of his Gospel. "The narrative is composed by a process of selection and emphasis. Mark could have reported more concerning John. He knows that he gathered round him a body of disciples . . . , that he was put to death by Antipas . . . , and that his great success was an embarrassment to the chief priests."[91] But Mark is "concentrating attention upon the advent of the Mightier One and on the baptism he will dispense."[92] *Erchetai* of 1:7 finds its echo and fulfilment in *ēlthen* of 1:14; *kēryssōn* of 1:4 finds its complement in the *kēryssōn* of 1:14.

The OT quotations in 1:2-3 likewise reach beyond John the Baptist, for they describe him as preparing someone else's way. There is always a danger of reading too much into a passage; yet it may not be entirely false to

---

[88] See E. P. Gould, *A Critical and Exegetical Commentary on the Gospel according to St. Mark* (*ICC*; Edinburgh: Clark, 1955) 2, n. 1; V. Taylor, *Mark*, 152.

[89] See J. M. Robinson, *Problem*, 23-24; E. Lohmeyer, *Markus*, 10; J. Delorme, "Aspects," 82; E. Schweizer, "Anmerkungen," 94; J. Schreiber, "Christologie," 159-60. We cannot agree with L. E. Keck ("Introduction," 359, 367), who thinks that Jesus is the "beginning"; or with F. Mussner ("Evangelium," 168), who is of the opinion that Jesus' baptism constitutes the "beginning."

[90] For some of the attempts, see W. Marxsen, *Mark*, 30-44; K. L. Schmidt, *Rahmen*, 18-22; J. Sundwall, *Zusammensetzung*, 6-7; U. W. Mauser, *Christ in the Wilderness* (*SBT* 39; London: SCM, 1963) 77-80; R. T. Simpson, "The Major Agreements of Matthew and Luke against Mark," *NTS* 12 (1965-66) 276-77; R. Pesch, "Anfang." For the scene of Jesus' baptism, see M. Sabbe, "Le baptême de Jésus," *De Jésus aux évangiles: Tradition et rédaction dans les évangiles synoptiques* (ed. I. de la Potterie; *Bibliotheca ETL*, 25; Gembloux: Duculot, 1967) 184-211.

[91] V. Taylor, *Mark*, 151.

[92] V. Taylor, *Mark*, 152; see also D. E. Nineham, *Mark*, 57; J. Delorme, "Aspects," 82.

see in the "way" of 1:2[93] an anticipation of the typically Marcan *en tē hodō* which occurs later in the Gospel, mostly in redactionally formed verses.[94] This impression is strengthened by other considerations. The fact that OT prophecies appear as such in 1:2-3 only gives them a programmatic force; everything reported later is to be seen as God's action carrying out the promises they contain; in the story which is about to unfold the *eschata* are present.[95] Further, John the Baptist prepares the way of Jesus not only by means of the words which form the climax of Mark's report about him (1:7-8) but also by his destiny. In other words, he not only prepares, but goes the way of Jesus.[96] Mark seems to be intent on presenting the analogy between the divinely decreed destinies of the two men. Both are described as having been handed over,[97] both die a violent death which is due, ultimately, to the will of God (9:12-13); and the power of one as well as of the other comes from God, a fact which the authorities refuse to admit (11:27-33). Finally, we may even observe a certain similarity in the structure of 1:1-3 and 1:14-15. Both are introduced by a verse containing the term gospel, then comes a statement announcing the definitive act of salvation which, in its turn, is followed by an exhortation.

To sum up, vss. 1 and 14-15 complement and presuppose each other. The two passages are related to each other; not as promise and fulfilment since John the Baptist forms part of the gospel,[98] but as the introduction and unfolding of the same divinely preordained plot. Vs. 1 is the heading of the entire written work: it is that not only because it happens to stand at its beginning, but because it introduces us to the divine decree in the light of which we are to understand all that is to follow, and to the man who, as an integral part of the gospel, anticipates Jesus' proclamation as well as his destiny. Vss. 14-15 form the climax of the section begun in vs. 1 as well as

[93] There are some who consider the first of the two quotations to be a later gloss; for names and arguments, see V. Taylor, *Mark*, 153. But there is no textual evidence against its genuineness; this should be a sufficient argument in its favor. U. W. Mauser's remark (*Wilderness*, 81, n. 1) should be quoted: "If it is an interpolation, it was done most ingeniously."

[94] 8:27; 9:33,34; 10:32,52.

[95] See W. Grundmann, *Markus*, 26; E. Schweizer, *Mark*, 29-31; F. Mussner, "Evangelium," 168.

[96] See E. Lohmeyer, *Markus*, 29; E. Schweizer, "Die theologische Leistung des Markus," *EvT* 24 (1964) 347; C. E. B. Cranfield, *Mark*, 62; W. Trilling, *Christusverkündigung*, 43-44.

[97] W. Marxsen (*Mark*, 39, n. 33) observes that the verb *paradidonai* is used absolutely only of Jesus and of the Baptist (Baptist: 1:14; Jesus: 3:19; 14:11,18,21, 42; 15:10), with qualifications of Jesus and others (others: 13:9,11,12; Jesus: 9:31; 10:33; 14:41; 15:1,15). See also V. Taylor, *Mark*, 165; E. Schweizer, *Mark*, 44-45.

[98] See J. Delorme, "Aspects," 82; E. Lohmeyer, *Markus*, 10; J. Schreiber, "Christologie," 159-60.

the summary of all that is to follow, for they present that of which vs. 1 announces the beginning.[99] Precisely because the passage is the climax of what precedes, it can serve not only as the beginning but as the foundation of what follows;[100] the particle *de* connects as well as distinguishes. What is presented as having been foretold in vss. 2-3 is, in vs. 15, said to be fulfilled;[101] what vs. 1 proclaims as beginning, vss. 14-15 present as arriving. Vss. 1-3 and 14-15 give Jesus' entire work and proclamation an eschatological character.

We now turn to the question of the Marcan meaning of *ēngiken*. Should the phrase be translated "the kingdom of God is near" or "the kingdom of God has come"? We opt for the second translation. The kingdom is present at the very moment of Jesus' proclamation of its approach; it is, in fact, this proclamation which makes it present. Now the reasons which have led to this conclusion.

The first reason is found in the synthetic parallelism of the two statements in vs. 15a: *peplērōtai ho kairos* and *ēngiken hē basileia*.[102] Grammatically and stylistically, the two statements are alike; verbs in the perfect tense precede their respective subjects, and the completed action is thus emphasized. The first clause enunciates clearly that the divinely decreed time of waiting has come to an end.[103] The decisive manifestation of the saving God, promised in the prophecies quoted in vss. 2-3, must, therefore, be taking place. The second member of the parallel can be seen as interpreting the first;[104] it states the same truth. The only difference between the members: the first looks backward, while the second looks to the present

---

[99] There is much to be said in favor of L. E. Keck's opinion ("Introduction") that the introductory part of the Gospel extends as far as vs. 15. Vss. 14-15 are, after all, a direct result of the baptism and temptation scenes; they are the proclamation of victory after the struggle with Satan. J. M. Robinson (*Problem*, 31) says that 1:15a "provides a first commentary on the baptism-temptation unit." We wonder, however, whether one should attempt to divide Mark's Gospel into such clearly separate units. The very abundance of divisions should warn us that the Second Gospel does not lend itself easily to neat and clear-cut divisions and subdivisions. For a summary of such divisions, see R. Pesch, *Naherwartungen: Tradition und Redaktion in Mk 13* (*Kommentare und Beiträge zum Alten und Neuen Testament*; Düsseldorf: Patmos, 1968) 50-53. Pesch's own division (pp. 55-73) is much too artificial and contrived to be convincing. For a discussion of his division, see below, p. 24, n. 112.

[100] See J. M. Robinson, *Problem*, 30-31.

[101] See D. Bosch, *Heidenmission*, 54.

[102] See K.-G. Reploh, *Lehrer*, 20-22. It will be noted that we owe much to his discussion of this question.

[103] See W. Grundmann, *Markus*, 37; E. P. Gould, *Mark*, 16; D. E. Nineham, *Mark*, 69; and particularly F. Mussner, "Gottesherrschaft," 86-88.

[104] See E. Lohmeyer, *Markus*, 10; W. Trilling, *Christusverkündigung*, 46-47, 53, 56-57, 60.

and future; the first announces the end of the old era, the second proclaims the beginning of the new.

The second reason is found in the programmatic statement of vss. 14-15. If this is a summary of Jesus' work, as presented in the Second Gospel, then for Mark, Jesus constitutes the end of the time of waiting and of promises. Precisely because he constitutes the end, he is also the one who introduces the new reality which is the kingdom of God. The OT prophecies quoted by the evangelist and intended by him to characterize the entire written work speak of the way of the one who has been promised. The suggestion is that, since the time of promises looked forward to the way of Jesus, it is his way to the cross which constitutes the kingdom. As the "beginning of the gospel" consists in the ministry of the man who by his proclamation and destiny prepares the way, so the "gospel of God" serves as the heading of the ministry of the man who goes the way. Yet this heading speaks not of the way but of the coming of the kingdom.

We find the third reason in the word *kēryssein*. E. Schweizer has shown[105] that for Mark Jesus is primarily one who proclaims and teaches. The miracles and above all the exorcisms which abound in the Gospel should not be taken as the primary object of Mark's thought but as auxiliary and subordinate to Jesus' work of proclamation and of teaching.[106] What the OT has promised is, as far as Mark is concerned, the time of salvation which consists in the proclamation and teaching of Jesus. It is thus false to separate his proclamation from his miraculous activity and from his gathering of disciples.[107] It is the proclamation which is the properly eschatological event; in the proclamation itself the kingdom is present. The gospel, as soon as it resounds, brings in the new reality of the kingdom.[108]

There is, finally, another reason, viz., the evidence of other statements about the kingdom in the Gospel. Mk 4:11, though taken from the tradition, has been inserted into a pre-existing unit of tradition in the interest of Marcan theology. Missing in Mark's version of this logion is the reference

---

[105] "Anmerkungen," 93-97; see also L. E. Keck, "Introduction," 360.

[106] L. E. Keck, "Introduction," 361: "If it is true that for Mark 'Jesus' word is action' . . . , it is also equally true that for Mark Jesus' action is 'word.'"

[107] As is done by F. Mussner, "Gottesherrschaft," 88-90; since, according to him, Mark is writing a theologically recast "vita Jesu," we are led to imagine a salvific vacuum in the admittedly short lapse of time between vss. 14-15, which announce the end of the period of waiting, and vss. 16-20, which introduce the kingdom.

[108] See R. Asting, *Verkündigung*, 317-20, 359; E. Schweizer, *Mark*, 30-32; J. Delorme, "Aspects," 82; H. Flender, "Lehren und Verkündigung in den synoptischen Evangelien," *EvT* 25 (1965) 702-4; J. Schniewind, *Markus*, 43, 50.

to the disciples' knowledge, which may have been present in an earlier form. We shall discuss this passage in ch. II; but it is of interest here in that Mark's logion, because of the absence of the reference to knowledge, makes the kingdom a present gift. We notice a similar phenomenon in 10:15, where a probable earlier reference to knowledge is omitted in Mark's version of the logion. The kingdom is again portrayed as a reality into which men may now enter. In 9:1 Mark has possibly—see below—added the phrase "with power" at the end of the logion, presumably to distinguish the future coming of the kingdom from its presence in the word and work of Jesus Christ, the Son of God. Another observation may be added: it seems that for Mark Jesus is the Son of God precisely because through his ministry he brings the kingdom to pass. It may be significant that we find redactionally constructed or redactionally placed references to the kingdom in close proximity to two of the three proclamations of Jesus' divine sonship which the Second Evangelist not only allows to stand and make his own, but which serve him as pillars on which his Gospel rests (1:11; 9:7; 15:39). There seems to be a causal relationship between 1:9-11 and 1:14-15: Jesus, having been proclaimed Son of God by the divine voice, goes in the power of the spirit to meet Satan in the desert; then he comes to Galilee proclaiming victory in the gospel of vs. 15. In 9:1 the introductory phrase shows that Mark himself added the logion to the preceding section; the phrase "with power" and the perfect tense of the verb "to come" serve to stress the difference between the future fulfilment of the kingdom and its presence in the Son of God who is on his way to the cross. The presence of the kingdom is also suggested by the two parables which expressly speak of it (4:26-29, 30-32). The strong contrast between the inexplicable growth and the certainty of the harvest, between the smallness of the seed and the size of the tree is, admittedly, intended to emphasize the final state of the kingdom. But the very contrast suggests that the final result is not confined entirely to the future. The same thought is implied in 10:30 where the present rewards for following Jesus are broadly described and distinguished from the future ones. It should be noted that, in their parallels to Mk 10:30, Matthew omits the distinction and Luke the breadth of description.

These reasons are sufficient to show that Mark considers the kingdom to be already present in the word and work of Jesus and that the phrase "the kingdom of God is at hand," should be understood in the sense of "has come."

Hence, for Mark the kingdom is a present reality. But why does he use the verb *engizein* to express this fact? One reason is undoubtedly to be sought in the missionary vocabulary of the church; another is the peculiar character of this presence. The present kingdom is hidden[109] and is still waiting to become manifest and unfold all its eschatological powers.[110] The tension between its present hiddenness and its future glory is in the very flesh and blood of the Second Gospel. We find it, therefore, impossible to agree with R. Pesch who maintains that the Gospel would be complete, and that it actually existed for a time, without the eschatological discourse[111] —and this quite apart from our failure to be convinced by his division of the Gospel.[112] His remark, that the reader would notice no defect should

---

[109] See E. Haenchen, *Weg*, 73, n. 1a.

[110] See M.-J. Lagrange, *Marc*, 16; T. A. Burkill, *Revelation*, 29-31; J. Schreiber, *Vertrauen*, 121, 136; R. Asting, *Verkündigung*, 317-18.

[111] *Naherwartungen*, 55, 65-66.

[112] There are many objections which can be levelled against his method of determining the breaks in the narrative. For instance, his reliance on stichometry and on the supposition that the Gospel is structured according to a threefold arrangement of subsections (pp. 54-56) is so wooden that one gets the impression that the entire structure of the Gospel is based on the counting of lines. His description of ancient methods of writing is quite inadequate to justify his procedure which is presumably founded on them. The lingering mistrust which hounds the reader who follows Pesch's dividing of the Gospel into sections, and these in turn into subsections, is compounded rather than removed when he comes to the joining of sections into parts, and again dividing these into three subsections which cut across earlier divisions. Are we to take seriously the suggestion that in the section 8:27—10:52 the caesura is to be placed at 9:30 (p. 62), while in the part 6:30—10:52 it is to be placed at 9:13 (p. 69)? The "axis" of the last part (11:1—16:8, without ch. 13) is to be found in "Jesus' testament" (14:12-25—p. 69), whereas the axis of the section 14:1—16:8 is found in 14:53—15:5 (p. 66). When we finally arrive at his division of the entire Gospel into three parts (p. 70: 1:2—8:26; 8:27-30; 8:31—16:8), we are reminded of H. Gardner's remark in her criticism of A. Farrer's treatment of the Second Gospel (*The Business of Criticism* [Oxford: University Press, 1959] p. 122): "If patterns are what we are interested in, and patterns are what we are looking for, patterns can certainly be found." It is somewhat difficult to imagine Mark counting lines and pericopes, arranging the shifting axes of the sections, the parts and the entire Gospel in such an artificial manner. Is such an elaborate superstructure necessary to account for Mark's flow of thought? Mark, after all, was not writing a sonnet, but a gospel. We should resist the temptation to divide it too neatly.

The change of place and/or time which Pesch uses as a means of dividing his sections is taken into account or disregarded as it happens to suit him. This change in 1:14 and 4:1 might be more important than in 1:35; 2:1 and 5:21 where he places the dividing line (pp. 57, 59). His division of the fifth section (11:1—12:44) is as follows (pp. 64-65): 11:1-26; 11:27—12:12; 12:13-44. We should think that 12:34 would be a much more natural conclusion of a subsection than 12:12, for vs. 34 closes a series of disputes between Jesus and Jewish representatives (11:27; 12:13,18,28), where it is they who take the initiative; in 12:35-36 Jesus is speaking to the crowd on his own initiative. The parable in 12:1-12 may well be the climax of the entire section, but must we prove it by forcing the entire Gospel into an arithmetic straightjacket?

ch. 13 be missing,[113] can be accepted only on the supposition that Pesch's division is as evident to the reader as it seems to be to Pesch. The strong contrasts in the parables and in such passages as 8:38—9:1 and 14:62 look forward to the full manifestation of the kingdom, and it would be rather surprising if the ultimate significance of the passion narrative were not given a longer and more explicit statement than that provided by 14:62.[114] The verb *engizein* itself in 1:15 seems to indicate that the kingdom, though already present, has not yet fully appeared.

The kingdom is being brought about by the word and work of Jesus Christ. The kingdom and Jesus are thus intimately related to each other in the present; so they will remain in the future. The redactional joining of 9:1 to 8:38 shows this clearly enough. This kingdom, present and future, brought about by Jesus now and to be consummated by his coming "in the glory of his Father," is the content of the gospel proclaimed by John the Baptist, by Jesus, and by the church.[115] In this manner we must understand the assertion that Jesus Christ is the content of the gospel. What is primarily affirmed by the gospel is the hidden presence of the kingdom;[116] this kingdom, however, is being brought about by Jesus Christ. Only at second remove, as it were, is Jesus Christ the gospel. This identification should not be understood univocally. As Jesus is not to be looked upon as the kingdom pure and simple, so he must not be the function of saving proclamation carried out by himself or by the church. Jesus Christ, the kingdom, and the gospel are to Mark different realities, even though intimately related to one another in God's saving purpose. It may not even be correct to conceive "the gospel of God" in 1:14 and "the gospel of Jesus Christ" as absolutely identical. The church and the evangelist proclaim "the gospel of Jesus Christ," and John's proclamation is its "beginning." But Jesus stands in a special relationship to the Father; he is the Son of God, the proclaimer *par excellence*; his relationship to the gospel is, therefore, more immediate and more fully stamped by the ultimate source of the good news than that of the church.

Perhaps in the phrase, "the gospel of God," we should look for an explanation of the singular construction *pisteuete en tō euangeliō*. Mk 1:15 is the only passage in the NT where the phrase *pisteuein en* occurs; only here, moreover, is its object not a person. The phrase has been variously trans-

---

[113] *Naherwartungen*, 65.

[114] See R. H. Lightfoot, *Message*, 48-59; C. H. Dodd, *The Apostolic Preaching and its Developments* (London: Hodder and Stoughton, 1963) 50-51; H. Conzelmann, "Geschichte und Eschaton nach Mc 13," *ZNW* 50 (1959) 211.

[115] See E. Schweizer, "Anmerkungen," 94-95.

[116] Thus we cannot agree with G. Bornkamm, *RGG*³ 2, 761, who claims that in Mark the message of the kingdom is pushed into the background.

lated.[117] Although A. Oepke remarks that it need not be a Semitism,[118] it seems that many exegetes take it as such.[119] But is the presence of *en* to be regarded as no more than a Semitism, or does Mark use it with a definite purpose in mind? That it is more than a simple preposition would seem to be indicated by his preference for *eis* in passages where one would, in accordance with the classical usage, expect *en*.[120] C. H. Turner finds only two passages in which the contrary occurs (1:16; 4:36), viz., where *en* has taken the place of the classical *eis*. The first of these two passages seems to demand no such conclusion; thus there remains only 4:36. Since, then, *eis* is the preposition used to introduce the object of *pisteuein* in the NT,[121] and Mark prefers *eis* to *en*, and given 1:15 as redactional, *en* in 1:15 is undoubtedly neither a Semitism nor a mere replacement for *eis* due to the confusion of the two prepositions in hellenistic Greek. *En* should rather be given its full force by being translated, "on the basis of";[122] it has the same meaning as in Jn 16:30; 1 Cor 4:4; Acts 24:16.[123] For Mark, the opposite of faith is fear (see 4:40; 5:36; 9:6,15,23; 10:24).[124] The "good news of God" proclaimed by Jesus Christ should infuse courage into the hearts of the disciples and enable them to overcome their terror even in the face of seemingly overwhelming dangers.[125]

[117] For translations, see W. F. Arndt and F. W. Gingrich, *A Greek-English Lexicon of the New Testament and Other Early Christian Literature: A Translation and Adaptation of Walter Bauer's Griechisch-Deutsches Wörterbuch zu den Schriften des Neuen Testaments und der übrigen urchristlichen Literatur* (Chicago: University of Chicago, 1952) 666.

[118] *TDNT* 2, 433-34.

[119] See R. Bultmann, *TDNT* 6, 203; M.-J. Lagrange, *Marc*, 16; V. Taylor, *Mark*, 167; C. E. B. Cranfield, *Mark*, 68; F. Mussner, "Gottesherrschaft," 85.

[120] See C. H. Turner, "Usage," *JTS* 26 (1924-25) 15-20. For a somewhat different view, see J. J. O'Rourke, "A Note Concerning the Use of *eis* and *en* in Mark," *JBL* 85 (1966) 349-51 (O'Rourke strains the evidence unduly, at least in regard to 1:9; 10:10; 13:3).

[121] Besides other possible constructions (accusative, *hoti*, etc.); see Bauer-Arndt-Gingrich, *Lexicon*, 666.

[122] See Hoffman, as quoted there, *Lexicon*, 666; also E. Lohmeyer, *Markus*, 30; W. Marxsen, *Mark*, 135. As indicated, we cannot agree with the first half of Marxsen's interpretation which presupposes a literal identity of the "gospel" with the "Lord who is coming."

[123] See Bauer-Arndt-Gingrich, *Lexicon*, 260 (*en* III, 3a).

[124] See J. M. Robinson, *Problem*, 68-73; H. Riesenfeld, *Jésus transfiguré: L'arrière-plan du récit évangélique de la transfiguration de Notre-Seigneur* (*Acta seminarii neotestamentici uppsaliensis XVI*; Copenhagen: Munksgaard, 1947) 285; J. C. Fenton, "Paul and Mark," *Studies in the Gospels: Essays in Memory of R. H. Lightfoot* (ed. D. E. Nineham: Oxford: Blackwell, 1955) 108; R. H. Lightfoot, *Message*, 91.

[125] W. Trilling (*Christusverkündigung*, 52) takes *en* as having an instrumental

(c) *The Function of Mark's euangelion*

The function of the redactional term gospel is to bestow unity on the disparate material contained in the Second Gospel and to serve as the arch spanning the gap between Jesus and the community. Its unifying role has been shown in the discussion of the programmatic character of 1:14-15. Here we wish to consider the nature of the relationship which the gospel establishes between Jesus and the community. Who is the Jesus who proclaims the good news of God? What is the situation into which the community is placed in hearing the good news? We have already disagreed with Marxsen's view of the matter. Jesus who addresses us in the Second Gospel is not the risen Lord proclaiming his speedy return. It is false to attribute to Mark a shrunken Pauline concept of the gospel. It is even more mistaken to ascribe to him the disinterest in the past which is proper to an exaggerated existentialist hermeneutic.

The awareness of Mark's vital interest in the past has never been lost on some exegetes, and has been newly discovered by others. C. H. Dodd, for instance, is well aware that Mark is giving us a report of Jesus' life and death.[126] E. Käsemann notes the historicizing tendency of the synoptic gospels and their presentation of the *kērygma* in the form of a report.[127] Now

---

force: by means of the proclamation of the "gospel" we arrive at conversion and faith. This suggestion may be correct, but we feel that ours is more in keeping with the general thought of the Second Gospel.

[126] *Preaching*, 46-49. This does not mean that we accept his reconstruction of the background of the synoptic reports—even though a recent writer, without referring to Dodd, seems to defend a similar thesis (J. Roloff, "Markusevangelium," 80). For a criticism of Dodd's reconstruction, see D. E. Nineham, "The Order of Events in St. Mark's Gospel—an Examination of Dr. Dodd's Hypothesis," *Studies in the Gospels*, 223-39. For an overly conservative view of the Second Gospel, which, however, expresses some healthy doubts about a number of, until recently, dogmatically asserted and accepted exegetic presuppositions, see D. Guthrie's treatment of Mark in his *New Testament Introduction* (London: Tyndale, 1965) 1, 59-62.

[127] See his articles, "Blind Alleys" (in particular pp. 48-50, 62-63) and "The Problem of the Historical Jesus," *Essays on New Testament Themes* (*SBT* 41; London: SCM, 1965), 15-47 (in particular pp. 32-33). He suggests that the existence of the Gospel is due to the need felt in the church to save the *kērygma* from the arbitrariness of the enthusiasts' all too exclusive concentration on the present exalted Lord and his future coming. In order to prevent Christ and his Spirit from falling prey to human imaginings and expectations, the church instinctively took recourse in the traditions of the first disciples and emphasized the permanent and fundamental significance of the earthly Jesus, his death, and resurrection. Without contesting the evangelist's interest in the earthly Jesus, the reason suggested by E. Käsemann for this interest has been questioned recently by G. Strecker, "Die historische und theologische Problematik der Jesusfrage," *EvT* 29 (1969) 473, and by E. Güttgemanns, *Offene*

that form criticism and redaction criticism have cured us of the desire to begin anew the quest of a historically reconstructed life of Jesus, more and more voices demand that we consider Mark's Gospel as an attempt to give us a redactional "vita Jesu."[128] Our attention is called to the frequent redactionally composed temporal and geographical references in the Gospel. To regard these as means, completely insignificant in themselves, of theological elaboration, as a mere contrivance to effect a combination of pericopes and sections for purposes of mutual interpretation, does not do justice to Mark because it forces his work into the mold of a theological tract. There is a real history-telling interest in the Second Gospel; Jesus is quite recognizably a personality of the past, and his ministry is presented in a series of episodes which follow one another in a time and space sequence. It is the past event, or rather a series of past events, which is constitutive of the present; the past *kairos* determines the meaning of the present. The good news, by proclaiming the past, reveals to the present its own content and significance in the eyes of God who has saved us, once and for all, in Jesus Christ. We should emphasize that these authors are not exhorting us to embark on a new reconstruction of a biography of Jesus,[129] but that we recognize the Second Gospel for what it is: a kerygmatic life of Jesus through whom the kingdom of God has, in a hidden manner, come into the world.

Correct as these insights may be, they seem to impale their protagonists on the other horn of the dilemma under which W. Marxsen is laboring. The gospel preached in the present demands, according to him, a present, i.e., a risen, Jesus. Such authors, however, relegate certain features of the gospel which clearly have a present message and application to the past. Let us mention some examples. According to G. Strecker[130] the only kerygmatic function of Mark's messianic secret consists in driving home the great privilege of the present community which has been entrusted with the mystery kept hidden until the moment of the resurrection. Strecker may be right, of course; yet we wonder whether such a prominent feature of the Gospel should be assigned such a limited role. However questionable some

Fragen zur Formgeschichte des Evangeliums: Eine methodologische Skizze der Grundlagenproblematik der Form- und Redaktionsgeschichte (BEvT 54; München: Kaiser, 1970) 26-32.

[128] G. Strecker, "Die Leidens- und Auferstehungsvoraussagen im Markusevangelium," *ZTK* 64 (1967) 38-39; S. Schulz, "Die Bedeutung des Markus für die Theologiegeschichte des Urchristentums," *SE* 2 (= *TU* 87) 143; *Die Stunde der Botschaft: Einführung in die Theologie der vier Evangelisten* (Hamburg: Furche, 1967) 36; F. Mussner, "Gottesherrschaft," 96; J. Roloff, "Markusevangelium," 78-84.

[129] With the possible exception of J. Roloff, "Markusevangelium," 73-84; possibly we misunderstand his necessarily summary remarks on the matter.

[130] "Zur Messiasgeheimnistheorie im Markusevangelium," *SE* 3 (= *TU* 88), 103-4.

aspects of the view that Mark is struggling against portraying Jesus as a *theios anēr* may be,[131] it must be admitted that it has at least one advantage over that of Strecker: it attempts to give the messianic secret a kerygmatic role which its place in the gospel seems to call for. If the messianic secret has lost all meaning with the resurrection of Jesus, why does Mark stress it so much?

The disciples in the Second Gospel are, according to S. Schulz and J. Roloff,[132] the antithesis of the post-Easter community: what the disciples did not understand during the earthly life of Jesus the community does understand in the light of the resurrection. We will discuss the Marcan concept of understanding later; here we may point out that the verses which are generally presumed to indicate a post-resurrection understanding on the part of the disciples and the community can hardly be proven to contain such an indication. Mk 4:11 does not contain a reference to understanding; it is all too frequently read in the light of its Matthean and Lucan parallels. Neither does Mk 9:9 say anything about the disciples' understanding; it speaks of the proclamation of Jesus' divine sonship. The two verses which promise a meeting with the risen Lord in Galilee (14:28; 16:7) say nothing of the disciples' understanding. Does public proclamation of Jesus' sonship automatically involve understanding on the part of the disciples and the community? We doubt it. S. Schulz is right when he states that Mark's disciples are no model for Christians to imitate, but are they not an image of the community? It is rather difficult not to see the community through the prism of the disciples in Mk 8:27—10:52. Their confession of Jesus' messiahship is the confession of the church,[133] and Jesus' words on crossbearing (8:34-37) are clearly meant for all his followers.[134] The disciples' ambition, their refusal to serve, their contempt for little ones, and the difficulty which they experience in accepting Jesus' teaching on marriage and riches—surely all that reflects failings of the community which Mark is

---

[131] Some of the recent proponents: P. Vielhauer, "Christologie," 165-66; E. Schweizer, *Mark*, 382-83; "Zur Frage des Messiasgeheimnisses bei Markus," *ZNW* 56 (1965) 8, n. 34; U. Luz, "Das Geheimnismotiv und die markinische Christologie," *ZNW* 56 (1965) 29-30; L. E. Keck, "Christology," 347-51; T. J. Weeden, "The Heresy That Necessitated Mark's Gospel," *ZNW* 59 (1968) 145-59.

[132] S. Schulz, *Stunde*, 143-46; J. Roloff, "Markusevangelium," 91-92.

[133] See E. Haenchen, "Leidensnachfolge," *Die Bibel und wir: Gesammelte Aufsätze* (Tübingen: Mohr, 1968) 2, 109.

[134] See R. Schnackenburg, "Die Vollkommenheit des Christen nach den Evangelien," *Geist und Leben* 32 (1959) 431; G. Bornkamm, "Das Wort Jesu vom Bekennen," *Geschichte und Glaube I: Gesammelte Aufsätze III* (München: Kaiser, 1968) 29; G. Strecker, "Voraussagen," 35.

addressing.[135] Are we really allegorizing[136] when we suspect that Mark is portraying the disciples in the colors of his community? In this sort of procedure Mark is after all no trailblazer. Through the desert community in Deuteronomy we perceive the features of the seventh century Israel; Ezekiel retrojects the infidelities of later Israel into the desert period; in 4 Ezra the ancient scribe is made to pose questions which trouble a later generation. Mark's redactional structuring of the material[137] shows that his censure of the community's defects is intentional. We have no right to attribute the failure to understand the significance of Jesus' death to the disciples alone. In this, as in other defects mentioned in Mk 8:27—10:52, they are representative of Mark's community. They confess Jesus to be the Messiah, they have heard of his death and resurrection, his divine sonship has been revealed to them, they know what following in his footsteps entails, but they are afraid and fail to understand—in all of this they mirror the community which Mark is addressing.

Mark is no doubt fully aware of the difference between the earthly life of Jesus and the period after Easter. The knowledge of Jesus' divine sonship was then reserved to a few; now it is publicly proclaimed to anyone who wishes to hear. Then, they had to be silent about Jesus' messiahship; now, they must confess it and proclaim it. Then Jesus attempted to prevent the publication of his miracles; now Mark freely writes about them. The difference between the two periods should, in our opinion, be sought primarily in the fact of proclamation in the present and secrecy of the past. The failure to perceive the significance of Jesus' death, however, is characteristic not only of Mark's disciples but also of his community.

Thus Mark's "gospel" not only spans the time-gap between the earthly Jesus and the post-resurrection community, but also produces a certain contemporaneity of the two. Through the medium of the disciples the community is somehow moved backwards into Jesus' life on earth; it is being addressed by him, is witnessing his great acts, is accompanying him on his way to the cross, and is tearing itself free of all that may hinder it in following him. While Mark's disciples bear the traits of the community, his

---

[135] See K. Weiss, "Ekklesiologie, Tradition und Redaktion in der Jüngerunterweisung Mark. 8, 27-10, 52," H. Ristow und K. Matthiae (eds.), *Der historische Jesus und der kerygmatische Christus: Beiträge zum Christusverständnis in Forschung und Verkündigung* (Berlin: Evangelische Verlagsanstalt, 1960) 416-26; see also F. Hahn, *Mission*, 111-20.

[136] As J. Roloff ("Markusevangelium," 91) seems to suggest.

[137] See W. Grundmann, *Markus*, 166; D. E. Nineham, *Mark*, 278; E. Schweizer, *Mark*, 190-91.

community is "about Jesus with the twelve" (4:10, cf. 8:34).[138] It is incorrect to say, with A. Suhl, that the disciples' failure to understand remains unrelieved throughout the Gospel, that there is no progress and gradual illumination.[139] The confession at Caesarea Philippi, if nothing else, stands in the way of such a view.[140] A view like this is, of course, tempting to those who wish to see in the Second Gospel a flat, basically atemporal, theological statement and who understand the contemporaneity brought about by the "gospel" as absolute and consistent. Mark, however, will not permit us to pigeonhole him so neatly. He is writing a "vita Jesu," a "vita" which belongs to the past. Yet his report of the past events is not history but *kērygma*, for he knows that the past determines the meaning of the present. For that very reason he paints the past with the colors of the present. Thus we meet in his Gospel a clear consciousness of the difference between the two periods, that of the earthly life of Jesus and that of the community. And yet it is false to say, with Schulz,[141] that the community does not yet exist during the lifetime of Jesus. The community, confessing and ignorant, is already with the earthly Jesus, represented by the disciples. Thus we have not only a succession of periods, but also a certain type of contemporaneity within the framework of this succession.

At the close of this discussion, we may suggest an application of the view presented above. In Mk 9:9 Jesus charges the disciples "not to tell anyone what they had seen, before the Son of Man had risen from the dead." This verse, seen in the light of Mark's awareness of the difference between past and present, is a simple statement of the temporal limit of the messianic secret. Read, however, with contemporaneity in mind, the "before" may be transcategorized into a "without." Mark is thus telling us that we must not proclaim the gospel of Jesus Christ without proclaiming his death on the cross as the supreme manifestation of his sonship. Mark's gospel spans the gap between the earthly Jesus and the community. The community can understand its present only in the light of the past which the gospel proclaims. Through this medium it is also somehow drawn into the past which determines and constitutes its present.

---

[138] See W. Marxsen, "Redaktionsgeschichtliche Erklärung der sogenannten Parabeltheorie des Markus," *Der Exeget als Theologe: Vorträge zum Neuen Testament* (Gütersloh: Mohn, 1968) 23; G. Minette de Tillesse, *Le secret messianique dans l'évangile de Marc* (LD 47; Paris: Cerf, 1968) 177-78.

[139] *Funktion*, 167-68.

[140] See W. Wilkens, "Die Redaktion des Gleichniskapitels Mark 4 durch Matth.," *TZ* 20 (1964) 315.

[141] *Stunde*, 143-44.

(2) Mark 11:10

Blessed is the kingdom of our father David to come! Hosanna in the highest!

To discover the meaning of the curious expression "the kingdom of our father David" we must consider it within its immediate context, i.e., Jesus' triumphal entry into Jerusalem. We shall first determine the extent of Mark's redactional activity in the pericope; then we shall turn our attention to the message of the pre-Marcan tradition. Mark's additions to the story will aid us in seeing what elements of the story he wishes to emphasize. These preliminaries will allow us to study the meaning of vs. 10, and to perceive the degree of emphasis which the evangelist places upon it.

Critical editions agree on the text of the verse; their only disagreements concern the spelling of the name David and aspiration of the word *hōsanna*.

(A) Redactional or Redactionally Retouched Verses in Mark 11:1-10

The first three verses of the pericope contain an unusual amount of familiar Marcan vocabulary. In vs. 1 we find such expressions as *apostellō* and *hoi mathētai autou*;[142] there are, further, two historic presents (*engizousin, apostellei*),[143] and an illustration of Mark's habit of having a singular follow an impersonal plural.[144] We should also note the highly circumstantial geographical description which calls to mind 7:31, a verse which is redactional or at least redactionally retouched.[145] In 11:1, as well as in 7:31, the evangelist betrays a lack of familiarity with Palestinian geography.[146] In vs. 2 there occur *hypagō*,[147] *kōmē*,[148] *euthys*,[149] *eisporeuomai*,[150] *deō*,[151] *kathizō*,[152] *oupō*,[153] *pherō* (in the sense of "to bring");[154]

---

[142] See V. Taylor, *Mark*, 453, 205.
[143] See J. C. Hawkins, *Horae Synopticae: Contributions to the Study of the Synoptic Problem* (Oxford: Clarendon, 1909) 143-49; V. Taylor, *Mark*, 46-47.
[144] See C. H. Turner, "Usage," *JTS* 26 (1924-25) 230.
[145] See V. Taylor, *Mark*, 352; E. Schweizer, *Mark*, 154; W. Marxsen, *Mark*, 69-70.
[146] See K. Niederwimmer, "Johannes Markus und die Frage nach dem Verfasser des zweiten Evangeliums," *ZNW* 58 (1967) 181.
[147] See V. Taylor, *Mark*, 189.
[148] The word occurs seven times in Mark (against four in Mt, and twelve in Lk), in verses which are probably redactional (6:6,56; 8:26; 6:36(?); 8:23,27; 11:2).
[149] See J. C. Hawkins, *Horae*, 12.
[150] Ibid.
[151] See V. Taylor, *Mark*, 453.
[152] Ibid.
[153] See J. C. Hawkins, *Horae*, 12.
[154] See C. H. Turner, "Usage," *JTS* 26 (1924-25) 13.

added to this is the historic present *legei*. In vs. 3 we meet the verb *apostellō* again, and such typically Marcan words as *euthys* and *palin*;[155] the *ean* clause is also a rather common feature of the Second Gospel.[156] One clearly non-Marcan feature of vs. 3 seems to lie in Jesus' reference to himself as *ho kyrios*,[157] an apparently singular occurrence in the entire Gospel, though there is one other passage where Jesus may be referring to himself as "the Lord," viz., 5:19. But since Mk 5:18-20 is, in all likelihood, a redactionally constructed unit,[158] and if we consider the parallel structure of vss. 19 and 20 (*apangeilon—ērxato kēryssein*; *hosa—hosa, pepoiēken—epoiēsen*; *soi—autō*; *ho kyrios—ho Iēsous*), we are inclined to disagree with the common opinion which holds the term "the Lord" to be a reference to God.

The correspondence, however, between vss. 1b-3 and vss. 4-7a in ch. 11 shows that the units were tailored to fit each other. Whoever was responsible for the formulation of the first unit must also have had a hand in the formulation of the second. There are, indeed, some features in vss. 4-7 which betray Mark's editorial activity. In vs. 4 there are parataxis[159] and a historic present. In vs. 5 there occurs the phrase *tines tōn ekei hestēkotōn* which may also be Mark's own. In the NT "the partitive genitive . . . , while not yet extinct, is being driven out by the use of the prepositions *ek* (*apo, en*) . . . ."[160] "The genitive (alone) predominates with *tis*."[161] It is nonetheless interesting to note that the indefinite *tis* followed by a partitive genitive appears only four times in Matthew (three of these are in parallels to Mark), nine times in Luke, and nine times in Mark.[162] Luke has parallels to six Marcan passages containing this construction, but repeats the construction itself only once. A further reason for the likelihood that the phrase is redactional may lie in Mark's use of *paristēmi*, a verb charac-

---

[155] Ibid.

[156] See V. Taylor, *Mark*, 187.

[157] A common opinion holds *ho kyrios* to be a reference to Jesus. But V. Taylor (*Mark*, 455) and C. E. B. Cranfield (*Mark*, 349-50) dissent and suggest that the word refers to the owner of the animal. E. Haenchen's remark (*Weg*, 374) is pertinent in this regard: If Mark thought of the event as natural he would hardly waste five verses on it.

[158] See R. Bultmann, *History*, 419; V. Taylor, *Mark*, 284; G. Minette de Tillesse, *Secret*, 86-87.

[159] See J. C. Hawkins, *Horae*, 151; V. Taylor, *Mark*, 48-49.

[160] F. Blass and A. Debrunner, *A Greek Grammar of the New Testament and Other Early Christian Literature* (tr. R. W. Funk; Chicago: University Press, 1961) § 164,1.

[161] Ibid.

[162] 2:6; 7:1,2; 8:3; 9:1; 11:5; 12:13; 14:47(?); 15:35; most of these verses can be at least suspected of being redactionally formulated or retouched.

teristic of the Second Gospel.[163] With one exception (4:29), the form in which this verb appears is the perfect participle, and it occurs twice (14:47; 15:35) in the partitive genitive construction following an indefinite *tis*. In vs. 7a we have parataxis and a historic present of the verb *pherō* which here also has the sense "to bring."

What conclusions may be drawn from these observations? Mark's editorial activity is too evident to be disregarded. To claim, however, that he had nothing but his imagination to depend upon in composing the scene would unduly strain the evidence of the passage itself as well as the evidence of his other redactional composition in which his dependence on tradition is manifest. We would suggest that the tradition used by him already contained a description of Jesus' triumphal approach to, or entry into, Jerusalem, a reference to the disciples' securing the animal, and their assurance that the animal would be brought back. To Mark we would attribute the arrangement of the material in vss. 1-7, Jesus' command to secure the animal, and the prediction accompanying this command. The statement, "the Lord has need of it," could also be his. The question of editorial activity in the passage is complicated by evident verbal and structural similarities between Mk 11:1-7,11 and 14:12-18a.[164] Was the latter passage composed in imitation of the former?[165] Since Mark strongly edited the first of the two, we would agree with V. Taylor that "the evidence suggests that each narrative is composed by Mark on the basis of tradition." The reason why we attribute Jesus' command and prediction in 11:1-7 to Mark's redactional activity lies in the strongly redactional character of vss. 1-2 and his editorial tendencies which we shall discuss below.

Vs. 11 is, in its entirety, an editorial composition. It serves the purpose of chronological arrangement which the evangelist imposes on the material gathered in 11:1—13:37.[166] Such words as *periblepsamenos*[167] and *meta tōn dōdeka*,[168] as well as the rather awkward structure built around *eisēlthen—exēlthen* are also evident signs of editorial work.

---

[163] See J. C. Hawkins, *Horae*, 13, 35.

[164] See V. Taylor, *Mark*, 535-36; H. Schürmann, *Der Paschamahlbericht: Lk 22, (7-14.) 15-18: I. Teil einer quellenkritischen Untersuchung des lukanischen Abendmahlsberichtes Lk 22, 7-38* (NTAbh XIX.5; Münster: Aschendorff, 1953) 120-22.

[165] H. Schürmann is of this opinion.

[166] See J. Wellhausen, *Marcus*, 88; E. Klostermann, *Markus*, 110; J. Sundwall, *Zusammensetzung*, 70-71; V. Taylor, *Mark*, 450.

[167] See J. C. Hawkins, *Horae*, 13.

[168] See V. Taylor, *Mark*, 230.

(B) Marcan Source

(1) From the foregoing discussion it would appear that the traditional story of Jesus' entry into Jerusalem consisted of the following features: the disciples' securing the animal and bringing it to Jesus, Jesus riding it into the vicinity of, or into, the city amid the acclamation of the people. Should we consider this tradition to be a more or less exact reminiscence of a given event during the ministry of Jesus? Some authors[169] think that this reminiscence is responsible for the restraint of the acclamation: the crowd fails to realize what Jesus intends to proclaim by the manner of his approach to the city; its attention centers on the coming kingdom rather than on the person of Jesus. A number of reasons, however, argue against the view that the historical event can easily be reconstructed from the tradition. The phrase "the kingdom of our father David to come" is peculiar. E. Lohmeyer[170] remarks that the Jews expected God's kingdom to "come," and David's kingdom to "come again." Moreover, David is seldom referred to as "our father" by the Jews; the only examples we have of this appellation, apart from this passage and Acts 4:25, occur in late rabbinic texts;[171] the title is generally reserved for the patriarchs.[172] Since it is difficult to imagine the cry of vs. 10a on the lips of Jesus' contemporaries in Jerusalem, it may be considered a formulation of the Christian community which knows Jesus to be the Son of David.[173] "Hosanna in the highest" of vs. 10b likewise betrays a community which no longer understands the word *hōsanna* in its Hebrew sense,[174] but takes it simply to be an expression of praise. It is commonly assumed that the story has some basis in history;[175] yet there

---

[169] Ibid., 452; C. E. B. Cranfield, *Mark*, 353-54; W. G. Kümmel, *Promise*, 116-17.

[170] *Markus*, 231.

[171] See *Str-B* 2, 26.

[172] See *Str-B* 1, 918-19.

[173] See W. G. Kümmel, *Promise*, 116; E. Schweizer, *Mark*, 229; J. Schmid, *Mark*, 205; F. Hahn, *The Titles of Jesus in Christology: Their History in Early Christianity* (London: Lutterworth, 1969) 255-57.

[174] See G. Dalman, *The Words of Jesus Considered in the Light of Post-Biblical Jewish Writings and the Aramaic Language* (Edinburgh: Clark, 1902) 221; E. Lohmeyer, *Markus*, 232; C. G. Montefiore, *The Synoptic Gospels* (New York: Ktav, 1968) 1, 260; E. Lohse, "Hosanna," *NovT* 6 (1963) 117. B. M. F. van Iersel ("Fils de David et Fils de Dieu," *La venue du Messie: Messianisme et eschatologie* [*RechBib* 6; Louvain: Desclée de Brouwer, 1962] 120) suggests that *na'* at the end of *hōsanna* may have been understood as "us," thus "save us." But this seems to be negated by the dative which follows the word in Mt 21:9, as well as by Mark's *en*.

[175] See R. Bultmann, *History*, 262; M. Dibelius, *From Tradition to Gospel* (New York: Scribner, n.d.) 122.

does not remain a great deal which can be credited to a genuine reminiscence. W. Grundmann,[176] for example, is of the opinion that the memory of Jesus' practice of riding, as a Rabbi, ahead of his disciples who followed him on foot drew to itself Zech 9:9, and out of this combination the tradition arose. F. C. Burkitt[177] shows much more confidence in the historical value of the tradition when he compares the reception given to Jesus by the crowd to that customary at the feast of Dedication. Grundmann's suggestion seems to be somewhat too reserved, and Burkitt's a little too trustful—although this "feeling" is admittedly as conjectural as the opinions of Grundmann and Burkitt and many others.

(2) What role should be ascribed to Zech 9:9 in the origin and growth of the tradition? Some authors attribute a decisive function to it. M. Dibelius and E. Haenchen,[178] for instance, consider the tradition to be a cultic legend based on Zech 9:9. R. Bultmann thinks that a "report of Jesus' entry into Jerusalem with a crowd of pilgrims full of joy and expectation (at the Kingdom of God that was now coming) could provide the historical basis which became a Messianic legend under the influence of Zech 9,9."[179] We have already mentioned Grundmann's and Burkitt's views; C. G. Montefiore[180] seems to share Burkitt's opinion. H.-W. Kuhn[181] feels that the tradition which we find in Mark has been influenced not only by Zech 9:9 but by Gen 49:11 as well; he points out that the latter text was understood messianically by Qumran and by the Tannaites. There are others, however, who either deny or doubt the influence of the OT prophecy on our story. Criticizing the theory of Dibelius, W. L. Knox[182] remarks that Mark "has carelessly forgotten to mention the prophecy which appears only in Matthew ... and the Fourth Gospel . . . ." C. H. Dodd[183] says that "in Mark there is no trace of the wording of the prophecy, unless it be in the mere use of

---

[176] *Markus*, 225.

[177] "W and θ: Studies in the Western Text of St. Mark. Hosanna," *JTS* 17 (1916) 140-46; see also B. A. Mastin, "The Date of the Triumphal Entry," *NTS* 16 (1969-70) 76.

[178] M. Dibelius, *Tradition*, 122; E. Haenchen, *Weg*, 375-76.

[179] *History*, 262. See W. G. Kümmel, *Promise*, 117, n. 47, for a criticism of Bultmann's view (*History*, 261-62) according to which it is absurd to imagine that Jesus intended to fulfil the prophecy of Zech 9:9.

[180] *Gospels* 1, 259.

[181] "Das Reittier Jesu in der Einzugsgeschichte des Markusevangeliums," *ZNW* 50 (1959) 86-91; F. Hahn (*Titles*, 121, n. 121) agrees with Kuhn.

[182] *The Sources of the Synoptic Gospels* (Cambridge: University Press, 1953) 1, 77.

[183] *According to the Scriptures: The Sub-structure of New Testament Theology* (London: Collins, 1965) 49.

the terms *pōlos* and *basileia*." A. Suhl[184] likewise fails to find any trace of the Zechariah text in the story.

These doubts seem to be well founded. Mark's text does not support the view which holds Zechariah's prophecy to be at the origin of the messianic character of the story. In such a supposition one would expect it to bear a greater similarity to it. The description of the animal as one "on which no one has ever sat" need not be traced to the *pōlon neon* of the prophecy, but may simply characterize it "as in some sense sacred."[185] Suhl[186] may be right when he remarks that vs. 3 allows us to perceive a more original and more modest role of the animal: the disciples obtain it only after the assurance that it will be brought back. On the other hand, considering the attention which Mark, and to some degree the tradition at his disposal, pays to the animal, it is difficult to accept the view which refuses to attribute any influence whatever to the OT oracle. Kuhn[187] has shown that the tradition of a messianic animal was by no means unknown at the time of Jesus and the earliest community; this tradition found its origin and support in such OT passages as Gen 49:11 and Zech 9:9. The reason for the lack of direct influence of Zech 9:9 on the wording of the Marcan story may well lie in this tradition: it was the tradition as such which influenced the story; the working of OT texts was only indirect.

(3) The acclamation of Jesus by the crowd forms the climax of the story. The acquisition of the animal and homage of the disciples and others seem to be designed to lead up to and enhance the cry of greeting in vss. 9-10. Yet it is precisely on account of this prelude that the words addressed to Jesus appear to be anticlimatic; they do not give the impression of being a "full-throated Messianic homage"[188] which the preceding events would lead us to expect. Nothing seems to indicate that the people greeting Jesus would look on him as the Messiah. The words of the psalm quoted in vs. 9b could be addressed to any pilgrim,[189] and vs. 10 speaks not of the messiah but of the kingdom. If we cannot accept the view that this reserve is due to the historical reminiscence, how are we to explain it? Matthew and Luke

---

[184] *Funktion*, 57-58.
[185] V. Taylor, *Mark*, 454; he refers to Num 19:2; Dt 21:3.
[186] *Funktion*, 58, n. 153.
[187] See note 181; also E. Lövestam, *Son and Saviour: A Study of Acts 13, 32-37. With an Appendix: 'Son of God' in the Synoptic Gospels* (*ConNt* 18; Lund: Gleerup, 1961) 65.
[188] V. Taylor, *Mark*, 452.
[189] See G. S. Duncan, *Jesus, Son of Man: Studies Contributory to a Modern Portrait* (London: Nisbet, 1948) 128; E. Schweizer, *Mark*, 228.

must have felt a similar uneasiness, for they omit the reference to the kingdom of David and make the crowd greet, not merely the one "who comes in the name of the Lord," but the Son of David and the King respectively. B. M. F. van Iersel finds a note of irony in the scene: the Jews grossly misunderstand the hidden intention of Jesus' manner of coming to the holy city.[190] Yet Mark's story can hardly be understood this way, for irony, with the possible exception of 7:2-3 and 10:42, is not one of his characteristics. And we doubt whether pre-Marcan tradition would indulge in it. Should we see in the story a portrayal of the messianic secret in action? Many authors are of the opinion that this theme is at work in the story,[191] but they do not tell us how this developed. Vs. 10a is scarcely a Marcan construction since Mark shows little interest in the title Son of David.[192] Should we, then, attribute the theme of the secret to the pre-Marcan tradition? Or are we to imagine that the tradition used by Mark was more explicit on the subject of Jesus' messiahship? Since the tradition scarcely anticipated Mark's messianic secret, and since we should be hard put to prove that vs. 10 in the *Vorlage* was more explicitly messianic, we agree with F. Hahn[193] that the story as such has nothing to do with the secret.

Can we, on the other hand, imagine the Christian community to be so reticent about its Lord? To find an answer to these questions, we must examine vss. 9-10 more closely. Vs. 9 is a truncated version of Ps 118:25-26; the psalm is part of the Hallel sung at the feasts of Passover, Pentecost, Tabernacles and Dedication.[194] The words quoted in vs. 9 were originally addressed to pilgrims coming to the Temple.[195] In this story they are addressed to Jesus.[196] The question whether they have a messianic meaning in themselves has not been answered unanimously, and the answer depends on the meaning of the term *ho erchomenos*. Some authors consider the term to be a messianic title;[197] others are more hesitant since there is little solid

---

[190] "David," 120, n. 3.

[191] See E. Sjöberg, *Der verborgene Menschensohn in den Evangelien* (Lund: Gleerup, 1955) 102, 130-31; G. Minette de Tillesse, *Secret*, 284-87: T. A. Burkill, *Revelation*, 192-95; J. Schniewind, *Markus*, 144-46; C. E. B. Cranfield, *Mark*, 347, 353-54; J. Schmid, *Mark*, 204.

[192] See E. Lohse, *TWNT* 8, 489; W. Michaelis, "Die Davidssohnschaft Jesu als historisches und kerygmatisches Problem," *Der historische Jesus* (eds. H. Ristow and K. Matthiae) 320.

[193] *Titles*, 255-57.

[194] *Str-B* I, 845-49.

[195] See A. Weiser, *The Psalms* (OT Library; Philadelphia: Westminster, 1962) 729.

[196] E. Lohmeyer, *Markus*, 231.

[197] J. Schneider, *TDNT* 2, 669-71.

evidence in favor of this view.[198] The most likely opinion seems to be the one according to which vs. 9 alone cannot give a clear reply to the question; its meaning must be determined on the basis of vs. 10.[199] This verse, being the comment of the Christian community on the words of the psalm,[200] should give us a clue to the meaning of the entire acclamation. In view of the Christian provenance of vs. 10, and because "Son of David" was a firmly established messianic title in the community,[201] the phrase, "the kingdom of our father David," probably found its origin, not in "the kingdom of God," but in the messianic title just mentioned.[202]

But if "the kingdom of our father David" is derived from the title "Son of David," we must consider the latter. This title, besides enunciating Jesus' Davidic descent, designates the earthly Jesus as the Messiah—this is commonly understood to be the meaning of the pre-Pauline christological formula in Rom 1:3-4.[203] It is not the dead and risen Lord who is the Son of David, rather the earthly Master. The title does not encompass all, or even the most important, saving acts performed by him, i.e., his death and resurrection; yet it is made to cover the messianic activity which took place during his life on earth. By means of this title Jesus is designated as the one who brings about eschatological fulfilment even before his death and resurrection. Davidic sonship is a provisional dignity, a dignity superseded by that of the divine sonship at the moment of resurrection.

It is thus reasonable to conclude that "the kingdom of our father David" serves to describe the messianic dignity enjoyed by Jesus during his earthly life. The participle *erchomenē* should in all likelihood be given a present meaning, both in view of the significance of the messianic title and in view

---

[198] See J. Schniewind, *Markus*, 145. V. Taylor (*Mark*, 457) thinks that, for a time, it may have had a certain currency as such in Christian and Baptist circles.

[199] E. Lohmeyer, *Markus*, 231; E. Klostermann, *Markus*, 114; E. Schweizer, *Mark*, 228.

[200] J. Schmid, *Mark*, 205; F. Hahn, *Titles*, 255-56.

[201] See E. Lohmeyer, *Gottesknecht und Davidssohn* (FRLANT 61: Göttingen: Vandenhoeck & Ruprecht, 1953) 66-83; W. Michaelis, "Davidssohnschaft," *passim*; F. Hahn, *Titles*, 240-58. For the OT and Jewish background, see J. A. Fitzmyer, "The Son of David Traditions and Matthew 22, 41-46 and Parallels," *Concilium* 20 (1966) 77-83; and E. Lövestam, *Son*, 54-69.

[202] See F. Hahn, *Titles*, 254-56; E. Lohmeyer, *Gottesknecht*, 76; E. Schweizer, *Mark*, 228-29. P. Vielhauer ("Ein Weg zur neutestamentlichen Christologie? Prüfung der Thesen Ferdinand Hahn's," *Aufsätze zum Neuen Testament*, 141-98) subjects Hahn's work to trenchant criticism but does not object to this partcular opinion of Hahn.

[203] See E. Lohmeyer, *Gottesknecht*, 77-78; E. Schweizer, *TWNT* 7, 126; E. Lohse, *TWNT* 8, 488.

of the parallel *ho erchomenos* in vs. 9.[204] Jesus is thus being acclaimed as the Messiah.[205] The story as such nowhere indicates that the crowd misunderstands Jesus.

Mark's source, then, in 11:1-11 was a creation of the Christian community based on what was likely a significant event in Jesus' public ministry. Its primary purpose was to present Jesus as the Son of David. Jesus approached Jerusalem as the one in whom the hopes and expectations of Israel were being fulfilled; the manner of his approach and the shouts of the crowd designated him as the Messiah.

(C) Mark's Understanding of the Pericope

(1) The redactional and the redactionally retouched verses of the pericope serve as a guide in discovering how Mark saw the tradition of Jesus' approach to Jerusalem. We say approach, not entry, since Mark portrays the entry itself very prosaically, contrasting it intentionally with the triumphal scene on the way towards the city. The redactional work is found in vss. 1-7,11.

In vs. 1 Jerusalem is intended to catch the reader's attention: Jesus arrives at the gates of the city which has been his stated goal for some time. Long before the evangelist brings Jesus to the city, he gives a good idea of what Jerusalem represents for him. Significant for the understanding of 11:1 is undoubtedly 10:32-34, a passage showing evident traces of redactional elaboration,[206] and containing the last and most explicit prediction of the passion. It portrays Jesus as "walking in the lead" "on the road, going up to Jerusalem," and telling his disciples clearly that he is going to die in the city. The passage is one of the three predictions of the passion around which the entire section 8:27—10:52 is constructed; it thus takes us back to the first and second predictions in 8:31-32 and 9:30-32. From the moment of Peter's confession Mark directs our gaze toward the death which the Father's will has imposed upon Jesus,[207] a destiny freely accepted by him.[208] Mk 11:1 is the beginning of a new, clearly recognizable section of the Gospel, but it is also the climax of the preceding section, for Jesus is no longer "on the way."[209] He has arrived at the gates of the city where his

---

[204] See F. Hahn, *Titles*, 257, particularly n. 108.
[205] See J. Schmid, *Mark*, 205.
[206] See V. Taylor, *Mark*, 83; D. E. Nineham, *Mark*, 278; G. Strecker, "Voraussagen," 31.
[207] See the "had to" of 8:31 and the "what was going to happen" of 10:32.
[208] Note the "Jesus walking in the lead" of 10:32.
[209] Apart from 8:27—10:52 the phrase occurs only in 8:3.

destiny is to find its completion. Mk 8:27—10:52 is thus a prelude to, and a preparation for, 11:1. In the first half of the Gospel Jerusalem is mentioned three times (3:8,22; 7:1).[210] In two of these passages (3:22; 7:1) it is described as the home of Jesus' enemies. 11:11 confirms this impression: the triumphal procession ends outside the city, Jerusalem itself remains unmoved. "One has only to compare the development which the narrative has undergone in the Matthaean picture of all the city stirred and the children crying in the Temple ... (in the Marcan story) presumably the crowds have melted away and Jesus is left alone with His disciples."[211] A strange coldness persists between Jesus and the city.[212]

The approach to the city also illustrates and confirms Jesus' foreknowledge of what is to befall him. He is not a blind instrument of inexorable fate, but a conscious agent of the salvation decreed by God to be brought about through him.[213] Vss. 1-7 of this pericope also show his purposeful initiative in the unrolling drama of salvation. He knows about the animal which is to serve the purpose of his messianic manifestation;[214] he commands that it be brought to him. The entire scene of the triumphal procession takes place because he wills it to take place. This brings us to another theme even more evident in these verses, as well as in vs. 11, viz., that of Jesus' power.[215] He takes the animal when he needs it; the objections are stilled as soon as it is known who needs it. In vs. 11 he inspects the Temple, not as a tourist, but as its Lord in preparation for its cleansing, another act of power,[216] on the following day.

The redactional verses thus emphasize that Jesus, obedient to the will of the Father, has arrived at the city where he is to die, and where he is to be, for the first time, publicly proclaimed as the Son of God (15:39). He sets his supernatural knowledge and power in action in order to carry out the task which God has given him to fulfil.

Some questions remain disputed. Should we regard the Mount of Olives in vs.1 as a simple geographical datum, like Bethphage and Bethany which seem to serve no other purpose than that of illustrating the approach to Jerusalem, or should we look upon it as an indication of Jesus' messianic mission? It seems that at the time of Jesus and of the primitive community there was a hope abroad which expected the Messiah to appear on the

---

[210] The three verses are redactional.
[211] V. Taylor, *Mark*, 458.
[212] W. Grundmann, *Markus*, 228.
[213] See T. A. Burkill, *Revelation*, 152-53; E. Lohmeyer, *Markus*, 165-66.
[214] See J. Schmid, *Mark*, 205; E. Haenchen, *Weg*, 374.
[215] See E. Lohmeyer, *Markus*, 232; P. Vielhauer, "Weg," 154.
[216] Cf. Mk 11:27-33.

Mount of Olives.[217] This hope apparently found its origin in OT texts which mention the mountain as a place of prayer (2 Sam 15:32), as the place where the glory of God is seen to be standing (Ezek 11:23), and where God is to hold judgment over the enemies of Israel (Zech 14:4).[218] Some authors think that Mark had the messianic significance of the mountain in mind;[219] others feel that there is not sufficient evidence for such a view.[220] It is difficult to decide between the two opinions. The evangelist mentions the mountain three times (11:1; 13:3; 14:26); 11:1 and 13:3 are redactional.[221] In either case the mountain could have been introduced for the sake of its messianic significance; but it could just as well have been inserted for the sake of Jerusalem in one case and to serve the purpose of secret instruction in the other.

Some authors feel that Mark wished to stress the fact that the animal had not yet been ridden.[222] A. Suhl,[223] for instance, thinks that the clause "on which no one has ridden" was inserted by Mark because he does not repeat the clause in vs. 4 despite the parallelism which prevails between vss. 1-3 and 4-7a. This argument is not cogent since parallelism calls for a certain variation in the wording of its members. Mark would undoubtedly consider the animal thus described as consecrated for a holy purpose,[224] but it is quite impossible to determine whether he wished to emphasize this feature.

(2) A redactional reworking of 11:9-10 is unlikely. Since the messianic title "Son of David" hardly plays a role in the Second Gospel, and since Mark betrays no interest in David himself, vss. 9b-10 are given in their traditional form. To discover Mark's understanding of these verses we must consider them in the light of the pericope as a whole and of the rest of the Gospel.

Does Mark, by leaving the traditional acclamation unchanged, wish to attribute to the crowd an inadequate understanding of the significance of Jesus' coming to Jerusalem? Some authors answer affirmatively.[225] At

---

[217] See Josephus, *Ant.* 20.8,6 § 169; *JW* 2.13,5 § 262.

[218] See *Str-B* I, 840; and E. Lohmeyer, *Markus*, 229, n. 3 (for rabbinic and targumic texts).

[219] See R. Bultmann, *History*, 261, n. 1.

[220] See C. E. B. Cranfield, *Mark*, 348; A. Suhl, *Funktion*, 58.

[221] For 13:3, see R. Pesch, *Naherwartungen*, 96-100.

[222] See B. Lindars, *New Testament Apologetic: The Doctrinal Significance of the Old Testament Quotations* (London: SCM, 1961) 111.

[223] *Funktion*, 57-58.

[224] E. Klostermann, *Markus*, 113.

[225] See R. Schnackenburg, *God's Rule*, 95-96; G. Minette de Tillesse, *Secret*, 391.

first sight such an answer does not seem to have much in its favor, for in the Second Gospel non-understanding and misunderstanding affect the person of Jesus and not the kingdom which is openly proclaimed from the very beginning. However, vs. 10 is a comment on vs. 9b, and the opening words of vs. 9b are also the opening words of vs. 10; the kingdom thus defines the Messiah. Moreover, Mark extends the term gospel to the traditions of the earthly Jesus; no longer does the gospel proclaim the risen Lord only, as it did in the pre-Pauline and Pauline hellenistic communities. Parallel to this extension of the content of the gospel is the fact that Mark's Jesus, in contrast to the christology of Rom 1:3-4, does not become Son of God in virtue of his resurrection, but is already Son of God at his baptism in the Jordan. The messianic activity which Rom 1:3-4 attributes to the Son of David Mark attributes to the Son of God. The title Son of David is, as far as Mark is concerned, inadequate to describe the earthly Jesus. In this we should look for the reason why the title plays such a limited role in his Gospel. This may also explain the absence of a command of silence in 10:46-52: there is no need to forbid the proclamation of a messianic title which, unlike the titles Messiah and Son of God, only inadequately expresses the being and function of Jesus.[226] The secret is preserved by the very insufficiency of the title. The evangelist's reservations with regard to the title Son of David are voiced in 12:35-37. R. Bultmann thinks that "it is probable that Mark . . . had no specific interpretation of the saying, but included it among the controversy sayings as an example of how Jesus refuted the Scribes."[227] But this opinion stems from a false reading of the redactional context of 12:35-37. Mark has closed off the controversy section with the redactional 12:43b, "And no one had the courage to ask him any more questions." The redactional *didaskōn en tō hierō* of vs. 35 introduces Jesus' teaching to the crowd (note vss. 37b-38a). Below we will show that 12:35-37 formed a traditional unit with the preceding pericopes; if Mark redactionally sundered the unity of these pericopes we may suspect that he had some interest in the teaching given in vss. 35-37.

Hence Mark reads 11:10a in the following manner: Jesus is the Son of God bringing the kingdom of God, not the Son of David introducing the kingdom of David. People acclaiming him fail to perceive the full reality of what he is and what he is bringing about. However, we are not suggesting that Mark denies the Davidic descent of Jesus; rather, he is concerned with

[226] For explanations of this fact which differ from ours, see E. Sjöberg, *Menschensohn*, 131; T. A. Burkill, *Revelation*, 192-95; G. Minette de Tillesse, *Secret*, 283-84; an explanation similar to ours is given by C. Burger, *Davidssohn*, 61.

[227] *History*, 137, n. 4. B. Lindars (*Apologetic*, 47) and E. Best (*Temptation*, 87) express a similar view.

the theological significance of various messianic titles.[228] We suggested above that the tradition used by Mark in 11:1-11 did not contain the theme of the secret. The evangelist, however, reads the secret into this tradition, not by changing the acclamation in vss. 9-10, but precisely by leaving it as it was. The acclamation lends itself to this procedure: Mark and his readers know who *ho erchomenos* is, whereas the people who do the acclaiming do not. They realize that Jesus' coming to Jerusalem is a significant event, and they know that it is somehow connected with eschatological fulfilment; yet they conceive this fulfilment in an inadequate manner, and their notions about Jesus' role in the work of fulfilment are nebulous at best. To speculate about the content of these notions would probably be a futile undertaking. Mark would most likely set them alongside of those which he gives in 6:14-16 and 8:28.

In 11:1-11 Mark undoubtedly wishes to present a picture of Jesus' messianic approach to Jerusalem. The reference to the city itself, the presence of the messianic animal, the manifestation of Jesus' supernatural foreknowledge and power show him approaching the place where he is, in obedience to the Father's will, to suffer and to die, and then to be publicly proclaimed as the Son of God. His triumphal approach to the city is thus the first act of the Passion Narrative.[229] Yet there is a hiddenness in this arrival; only the community can in retrospect perceive that the One who "comes in the name of the Lord" is the Son of God in whom the kingdom is already present and at work. It hears in the acclamation of vss. 9-10 what it hears from Jesus' lips in 1:15: the kingdom of God is being brought about even now, in a hidden manner, by the Son of God.

(3) Concluding Remarks on Chapter I

The first word of Mark tells us that he has taken pen in hand in order to proclaim the good news of Jesus Christ. The first word of Jesus is the good news about the coming of the kingdom of God: the time of waiting has come to an end, the hopes and prophecies of the OT are being fulfilled, the kingdom of God is arriving. Mk 1:15 gives us the content of the good

---

[228] Some authors see a denial of Jesus' Davidic descent in Mk 12:35-37: B. Lindars, *Apologetic*, 46-47; A. Suhl, *Funktion*, 57-60. Their opinion is not commonly shared; see D. Daube, *The New Testament and Rabbinic Judaism* (London: Athlone, 1956) 158-68; E. Lohse, *TWNT* 8, 488-89; G. Bornkamm, *Jesus of Nazareth* (New York: Harper, 1960) 227-28; F. Hahn, *Titles*, 251-53. C. Burger, *Davidssohn*, 56-58, 65, is of the opinion that the pericope in its pre-Marcan stage denied the Davidic descent of Jesus, but that Mark affirms it.

[229] See W. L. Knox, *Sources* 1, 77; E. Lohmeyer, *Markus*, 233; E. Haenchen, *Weg*, 373.

news; it summarizes everything that is to follow, and serves as the foundation of all that Jesus is portrayed as saying and doing in the rest of the Gospel. Jesus Christ and the kingdom of God are thus inextricably linked: he proclaims its coming and, by this very proclamation, brings it. His teaching, his call to men to follow him, his exorcisms must be seen in the light and as an unfolding of his first proclamation. The kingdom is already present in his word and work, its eschatological powers are being manifested; neither demons nor natural forces nor men can resist its energy.

The manner of this manifestation, however, is paradoxical. The very fact that men must be exhorted to believe in the good news and overcome their fears in the face of sobering realities which seem to belie its happy message shows that the kingdom is not yet present in all its overwhelming glory. Men fail to understand the Son of God and to perceive the ultimate significance of his teaching and miracles. The present kingdom manifests itself in words and actions of One who will be recognized for what he is on the cross. His destiny is foreshadowed in his Precursor who dies at the hands of a tyrant, and must reecho in the life of those who will be his followers.

The present kingdom is thus a hidden kingdom, a reality which is already with us and yet is still coming, a fulfilment straining for its completion, a glory visible only to those to whom its mystery has been entrusted.

Chapter II

THE HIDDEN KINGDOM

Chapter I has shown that, in Mark's eyes, the kingdom of God is a reality already present in the word and work of Jesus, a reality which is straining for its full manifestation in the future. During the time of Jesus and the community it remains hidden. The hiddenness of the kingdom is investigated in this chapter. Three passages are studied in detail: 4:10-12, 26-29, 30-32. After a discussion of Mark's redactional treatment of the logion on the mystery of the kingdom the following phrases of the logion will be given close attention: "those outside," "in parables," "the mystery of the kingdom." The parables of the Seed Growing Secretly and that of the Mustard Seed are examined in order to discover their original message and Mark's understanding of them.

(1) Mark 4:10-12

Now when he was away from the crowd, those present with the Twelve questioned him about the parables. ¹¹He told them: "To you the mystery of the kingdom of God has been confided. To the others outside it is all presented in parables, ¹²so that they will look intently and not see, listen carefully and not understand, lest perhaps they repent and be forgiven."

The amount of literature which has been written on this rather disturbing logion of Jesus is, without exaggeration, staggering. To discuss all, or even a part of, the opinions which have been voiced up to now would demand a separate chapter. For that reason, we prefer to present our own opinion and, in the course of presentation, to indicate our indebtedness to the research done by others and our agreement or disagreement. We shall first discuss the character of the logion and Mark's redactional treatment of it; and that will be followed by a detailed study of its content. Particular attention will be given to the terms *hoi exō, en parabolais, to mystērion tēs basileias.*

The text of the logion need not detain us long. Critical editions are, for all practical purposes, unanimous. The only variant of any consequence is found in von Soden's reading of 4:10: instead of the plural *tas parabolas* he has the singular *tēn parabolēn.* In this he disagrees with the United Bible Societies edition, Westcott-Hort, Vogels, Tischendorf, Nestle and Tasker. His reading, however, can safely be dismissed as an unwarranted attempt, based on extremely weak MS evidence, to harmonize vs. 10 with the fact that the parable of the Sower alone precedes the logion. Tischendorf omits the article before the word *panta* in vs. 11b. In this he follows the Codices Sinaiticus, Bezae, Freer, Koridethi, and some others. With the majority of MSS and all other critical editions the article should be retained.

(A) THE ORIGIN OF THE LOGION AND MARK'S REDACTIONAL TREATMENT OF IT

The vast majority of exegetes are agreed on two points concerning the logion in Mk 4:11-12: first, that it was not composed by Mark but taken from the pre-Marcan tradition; secondly, that it was Mark who inserted it into an already existing traditional unit, viz., that of the parable of the Sower and its interpretation.[1]

T. W. Manson and, following him, J. Jeremias and J. Gnilka[2] have shown convincingly that the origin of the logion must be looked for in Aramaic-speaking Palestine. To quote Jeremias, "the antithetic parallelism (vs. 11), the redundant demonstrative *ekeinos* (vs. 11b), and the circumlocution thrice used to indicate the divine activity (n. 15: *dedotai, ginetai*; vs. 12 *aphethē*) are typically Palestinian. But above all we must observe that the free quotation from Isa 6:9-10 in Mk 4:12 varies widely from the Hebrew text and from the LXX, while it agrees with the Targum." The aspects in which Mark and the targum agree against the MT and the LXX are the following: (a) While in the MT and the LXX Isa 6:9b is found in the second person, Mk 4:12a and the targum have it in the third person; the Marcan participles *blepontes* and *akouontes*, furthermore, have their equivalents in the targum, but not in the MT, and only partly in the LXX.[3] (b) The phrase "and be forgiven" of Mk 4:12b diverges sharply from the MT and the LXX, which have the verb "to heal" (*rp', iasomai*), but it

---

[1] See A. Jülicher, *Die Gleichnisreden Jesu* (Darmstadt: Wissenschaftliche Buchgesellschaft, 1963) 1, 147; R. Bultmann, *History*, 325, n. 1, 444; C. Masson, *Les paraboles de Marc IV* (Cahiers théologiques de l'actualité protestante 11; Neuchâtel-Paris: Delachaux & Niestlé, 1945) 30; T. A. Burkill, "The Cryptology of Parables in St. Mark's Gospel," *NovT* 1 (1956) 252; V. Taylor, *Mark*, 255; M.-J. Lagrange, *Marc*, 98; F. D. Gealy, "The Composition of Mark IV," *ExpT* 48 (1936-37) 40; J. Dupont, "Le chapitre des paraboles," *NRT* 89 (1967) 803-4; W. Marxsen, "Erklärung," 17; E. Sjöberg, *Menschensohn*, 165, 220, 224; G. H. Boobyer, "The Redaction of Mark IV. 1-34," *NTS* 8 (1961-62) 65; D. E. Nineham, *Mark*, 137; D. O. Via, "Matthew on the Understandability of the Parables," *JBL* 84 (1965) 432; and others.

[2] T. W. Manson, *The Teaching of Jesus* (Cambridge: University Press, 1963) 75-80; J. Jeremias, *The Parables of Jesus* (The NT Library; London: SCM, 1963) 15; J. Gnilka, *Die Verstockung Israels: Isaias 6, 9-10 in der Theologie der Synoptiker* (StANT 3; München: Kösel, 1961) 13-17.

[3] The MT has infinitives absolute; the LXX has *akoē* and *blepontes*.

agrees with the targum which apparently read *rph* instead of *rp'* in its own Hebrew text. Further agreements between Mark and the targum are found in the use of the passive to avoid a direct reference to the divine name and in the plural of the last word of the OT quotation (the MT has *lô*; the LXX *autous*).[4] The vocabulary of the logion indicates that it was not composed by Mark. "*Mystērion, dedotai* as a circumlocution of the divine action, *hoi exō, ta panta, epistrephein* used in the sense of conversion, all that appears in Mark only at 4,11f."[5]

A. Suhl[6] disagrees with this view. While admitting Mark's dependence on the targum for the OT quotation, he feels that Mark himself formulated vs. 11. The fact that certain terms occur only in Mk 4:11 argues, according to him, in favor of Marcan composition rather than against it. He substantiates the validity of his argumentation by referring to Mark's disinclination to interfere with the traditional material. We fail to grasp Suhl's thought in this respect. We are learning to distinguish Marcan redactional passages from the traditional material precisely by studying the words, phrases and procedures which occur frequently and with some regularity in the Gospel. Are we to think that, on the contrary, the *hapax legomena* are the most Marcan element in the Second Gospel?

Is the logion to be traced to Jesus himself? T. W. Manson and J. Jeremias[7] think so. Their main argument in favor of its authenticity consists in its Palestinian origin. Yet the only conclusion which that fact allows us to make is that the logion was current in an Aramaic-speaking Christian community.[8] Jeremias' reconstruction of its *Sitz im Leben Jesu* may, of course, be correct, but we need weightier proofs to pass beyond the mere judgment of possibility.[9] Our present state of knowledge and the prevailing presuppositions which guide the work of exegesis allow no firm option either in favor or against the authenticity of the logion. Since our interest is directed primarily to the thought of Mark, it is not necessary to embrace either opinion. The above discussion, however, has shown the high likelihood of the view which regards Mk 4:11-12 as belonging to a pre-Marcan tradition and stemming from an Aramaic-speaking environment.

In our discussion of the second part of the generally accepted thesis, that

---

[4] J. Jeremias, *Parables*, 15. A convenient synopsis of the MT, the LXX, the targum of Isa 6:9-10 and NT occurrences of the same text is offered by J. Gnilka, *Verstockung*, 14-15.

[5] J. Jeremias, *Parables*, 15, n. 12.

[6] *Funktion*, 146-47, 150.

[7] *Teaching*, 77; *Parables*, 15, 18.

[8] See A. Suhl, *Funktion*, 147.

[9] See W. Marxsen, "Erklärung," 14, n. 9.

Mark is responsible for the place now occupied by the logion, we shall proceed in stages. There is a great deal of interdependence and repetition among authors on this matter, and we shall present the arguments which seem to be most convincing and clearest.

No one has ever claimed that Mark composed the parable of the Sower. It is also generally agreed that the interpretation of the parable in vss. 14-20 belongs, with the possible exception of some phrases,[10] to pre-Marcan tradition.[11] E. Trocmé's opinion[12] that 4:14-20 is Mark's own composition is quite untenable in view of the early Christian terminology which abounds in it and does not occur elsewhere in the synoptic gospels.[13] It is likewise highly probable that these two pre-Marcan pieces of tradition formed a unit before Mark decided to incorporate them into the Gospel; it is difficult to imagine that the interpretation would exist apart from the parable which it interprets.[14] Mk 4:11, taken in isolation from the present context, seems to have had a wider application than it is now given; this has been shown most clearly by J. Gnilka.[15] He points out, first, that the neutral *ta panta* has a summary significance, since it is placed in antithetic parallelism to the "secret of the kingdom," i.e., to a reality manifested by the entire activity of Jesus, and not by the parables alone. Secondly, the OT quotation refers to the entire activity of the prophet; it does not mention parables. Thirdly, the Marcan order of "seeing" and "hearing" in the OT quotation is the reverse of that found in MT, LXX and the targum; this reversal is likely due to the Jewish eschatological hopes in which the time of salvation was primarily an object of seeing.

It is thus difficult to escape the conclusion that 4:11-12 constitutes a foreign body in the context of 4:3-20. To this should be added the fact that 4:10-12 bursts the frame set for the parables by 4:1-2,33-34.[16] What remains to be shown, however, is that Mark himself is responsible for bursting the frame. There are authors who feel that this operation had been performed in a later stage of the tradition which preceded Mark. We shall refer to them at the close of the discussion. First, we shall present the

---

[10] J. Dupont ("La parabole de la semence qui pousse toute seule (Mc 4,26-29)," *RSR* 55 [1967] 388) considers "persecution . . . because of the word" of 4:17 to be editorial.

[11] See V. Taylor, *Mark*, 258-62; J. Schmid, *Mark*, 99-100; J. Delorme, "Aspects," 87; C. Masson, *Paraboles*, 36-38; F. W. Beare, *Records*, 112; R. Bultmann, *History*, 187; particularly J. Jeremias, *Parables*, 77-79.

[12] *La formation de l'évangile selon Marc* (Études de l'histoire et de philosophie religieuses 57; Paris: Presses universitaires, 1963) 127, n. 71, 149.

[13] See J. Jeremias, *Parables*, 78.

[14] W. Marxsen, "Erklärung," 17.

[15] *Verstockung*, 26; cf. J. Jeremias, *Parables*, 16-18.

[16] See R. Bultmann, *History*, 444; C. E. B. Cranfield, *Mark*, 147.

evidence for the view which holds that Mk 4:11-12 is Mark's redactional addition.

Arguments in favor of this view have been presented by a variety of authors. The following, however, have contributed more than others to the study of Mark's redactional procedures with regard to 4:11-12: M. Zerwick, J. Jeremias, W. Marxsen, J. Gnilka and G. Minette de Tillesse.[17] Since Minette de Tillesse's exposition is the latest, and undoubtedly the most complete and clearest, contribution to the discussion, we shall give a summary of it. He points out the need of distinguishing three levels in the material of Mark's parable chapter: the primitive tradition, later additions, and Mark's redaction. To determine the extent of these, we must, first of all, examine the formulae which introduce various units. Thus we find the phrase *kai elegen* in 4:9,26,30, and nowhere else in the Gospel; the phrase *kai elegen autois* occurs in 4:2,11,21,24; *kai legei autois* occurs in 4:13. Most likely the units introduced by *kai elegen* form the earliest layer of the source used by Mark: the parables of the Seed Growing Secretly and of the Mustard Seed, along with that of the Sower, were the first traditional pieces collected into one unit. The formula *kai elegen autois*, on the other hand, is characteristic of Mark; it is thus very likely that vss. 11-12, 21-23,24-25 were introduced into the preexisting source by Mark himself. This is confirmed by the fact that the logia in vss. 21-25 do not exhibit the customary form of the parable, and by the fact that vss. 11-12 and 21-25 speak of the same subject, viz., of revealing that which has been kept hidden. The formula *kai legei (ephē) autois* introduces Jesus' answers to disciples' questions at 7:18; 9:12; 10:11; 13:5; besides 4:13. This formula, possibly pre-Marcan,[18] introduces a later, but still pre-Marcan, layer of tradition into the source.

In Mk 4:10-13 we find two singular exceptions to Mark's customary treatment of his speech material. M. Zerwick has observed[19] that the phrase *kai elegen autois* is constantly used to introduce a new piece of traditional material into a discourse which is already in progress and thus conceived

---

[17] M. Zerwick, *Markusstil*, 38, 60-61, 67-70; J. Jeremias, *Parables*, 14; W. Marxsen, "Erklärung," 16-20; J. Gnilka, *Verstockung*, 23-24, 57-62; G. Minette de Tillesse, *Secret*, 165-73; see also V. Taylor, *Mark*, 218. H. St. J. Thackeray (*The Septuagint and Jewish Worship: A Study in Origins* [London: Oxford, 1923] 20-21) offers some interesting insights into similar redactional procedures of the Books of Kingdoms in the LXX.

[18] In this we depart slightly from G. Minette de Tillesse's opinion expressed on pp. 167, 170 of his book; W. Marxsen ("Erklärung," 17) may be right: since Mark found 4:14-20 in his source, it is possible that he also found the introductory formula.

[19] *Markusstil*, 60-61, 67-70.

as a unit, while *kai legei* (*ephē*) *autois* serves to introduce a new discourse. In Mk 4:10-13, however, there occur two exceptions to this rule: *kai elegen autois* in vs. 11 introduces the discourse, and *kai legei autois* in vs. 13 continues the discourse already begun. It is thus reasonable to conclude that Mark inserted vss. 11-12 into a preexisting source, prefacing them with his characteristic introductory formula.

Thus far Minette de Tillesse; we find his arguments convincing. There are, however, authors who think they must attribute the insertion of the logion into the present context to a stage of tradition earlier than Mark. Their main reason for doing so is the contradiction which they detect between vss. 11-12 and 33-34,[20] or the contradiction between vss. 11 and 13.[21] In vs. 33 it is implied that the crowds were able to understand something, whereas vs. 11 implies, by means of its contrast between "giving the secret" and "happening in parables," that they understood nothing. If vss. 33-34 are Marcan, would Mark not be contradicting himself by inserting vss. 11-12? If they are not Marcan, why does he not edit them away, if vss. 11-12 express his own thought? Turning to vs. 13, we notice that it contains the characteristic Marcan reproach of Jesus addressed to the disciples. This reproach is beyond doubt a redactional feature of Mark: again and again the disciples are censured by Jesus for their lack of understanding. This, in such proximity to the statement that the mystery has been given to them, argues against the supposition that both passages, vss. 11-12 as well as vs. 13, are Mark's redactional additions. These objections to the commonly held opinion, particularly that of E. Schweizer based on vs. 13, have some weight. For it is commonly admitted that Jesus' censures of his disciples are to be attributed to Mark himself; they are surely more than mere questions put to the disciples whether they, too, belong to the ranks of unbelievers;[22] a cursory look at the passages in question and at such statements as Mk 6:52; 9:32 should suffice to prove it. They also seem to be more than negatively expressed exhortations to strive for greater insight.[23] We hope to solve this difficulty later; at the moment we merely take note of it.

---

[20] See E. Trocmé, *Marc*, 127, n. 71; D. W. Riddle, "Mark 4, 1-34: The Evolution of a Gospel Source," *JBL* 56 (1937) 81, 83; H. J. Ebeling, *Das Messiasgeheimnis und die Botschaft des Markus-Evangelisten* (Berlin: A. Töpelmann, 1939) 180-83, 189; E. Schweizer, *Mark*, 92-94, 106.

[21] E. Schweizer, "Frage," 5-7; in *Mark,* 92-94, he seems to be more hesitant in his contention that vss. 11 and 13 cannot be attributed to the same editor, but he still has grave doubts about the majority opinion.

[22] As asserted by A. Suhl, *Funktion*, 146.

[23] As asserted by H. J. Ebeling, *Messiasgeheimnis*, 165, and L. Cerfaux, "'L'aveuglement d'esprit' dans l'évangile de Saint Marc," *Le Muséon* 59 (1946) 276-78.

Now we turn our attention to vss. 10 and 13. Were they already, in one form or another, in the source used by Mark? The more common opinion holds that vs. 10 was already in the source; its present form, however, is considered to be the result of Mark's redactional intervention.[24] Mark, according to this view, has changed the original singular *tēn parabolēn* into a plural. As to the awkward "those present with the Twelve," opinions are divided. Some feel that "those about him" had stood in the source and that "with the Twelve" was added by Mark.[25] Marxsen, however, defends the opposite view on account of the similarity between 4:10 and 7:17.[26] In order to support the thesis that vs. 10 is not entirely a Marcan composition, J. Gnilka[27] points out that *kata monas* and *hoi peri auton* are unusual phrases for Mark.

R. Bultmann,[28] however, observes that special instruction being given to the disciples is peculiar to Mark. He examines various cases of this type of instruction and concludes that we are dealing with the editorial work of Mark or his predecessors. G. H. Boobyer[29] is of the opinion that Mk 4:10 is to be regarded as part of the unit 4:10-12; the view that the plural "parables" replaces an earlier singular is based on the notion that the Sower was the first parable. That, however, is not the case; Jesus has already been portrayed as speaking in parables at 3:23. The plural serves to tie up linguistically and conceptually the parables in ch. 4 with what precedes. Minette de Tillesse[30] is the most eloquent defender of the view that vs. 10 should be ascribed to Mark's redaction. With R. Bultmann, he notes the fact that the theme of esoteric teaching given to the disciples is peculiar to Mark. While it is true that *kata monas* occurs only here against seven cases of the characteristically Marcan *kat' idian* elsewhere in the Gospel, we should consider the fact, according to him, that the phrase is just as rare in the rest of the tradition, occurring only at Lk 9:18, and that it is very similar to *kat' idian*. He feels that "those about him" should be ascribed to Mark because the same phrase occurs twice in the pericope just preceding the parable of the Sower. *Hoi dōdeka*, further, is a typical Marcan term; the expression *syn tois dōdeka* is not as rare as it seems to be at first

---

[24] V. Taylor, *Mark*, 255; D. E. Nineham, *Mark*, 139; C. E. B. Cranfield, *Mark*, 147; W. Marxsen, "Erklärung," 17; J. Jeremias, *Parables*, 14, n. 11; J. Dupont, "Chapitre," 804; C. Masson, *Paraboles*, 29, n. 1.

[25] J. Gnilka, *Verstockung*, 59; V. Taylor, *Mark*, 254. R. Bultmann, *History*, 325, n. 1.

[26] "Erklärung," 17-18.

[27] *Verstockung*, 58.

[28] *History*, 330-32.

[29] "Redaction," 64-66.

[30] *Secret*, 173-79.

sight if we consider 2:26 and particularly the redactional 8:34a. We tend to agree with Minette de Tillesse's conclusion, especially in view of the parallel to 4:10 at 8:34.

That 4:13 was also strongly retouched by Mark is widely admitted.[31] It contains a reproach addressed by Jesus to his disciples which is characteristic of the Second Gospel.

Hence Mk 4:11-12 is an old logion whose roots go back to the Palestinian environment; it was, before Mark came to write his Gospel, independent of its present context. Mark very likely created the introductory vs. 10 and strongly retouched vs. 13. Our way is now clear for an examination of the message which Mark wishes to convey by means of this logion.

(B) Those Outside

(1) Mark 4:11 is the only passage in the synoptic gospels in which the term *hoi exō* occurs. We encounter it frequently enough in Pauline writings where it is used to describe those who do not belong to the community. The Jewish background of the term is found in the expression *hiṣônîm* which was applied to non-canonical books, but also to heretics.[32] But we must ask ourselves: Whom does Mark designate as "those outside"? A practically universal opinion identifies them with all those who do not belong to the narrow circle of Jesus' disciples. This seems to be self-evident if we compare Mk 4:1, where a very large crowd is said to have gathered around Jesus, with 4:10, where we are told that Jesus withdrew from the crowd in order to instruct "those present with the Twelve." In the literature which we have consulted we have uncovered only three authors who call this universally held opinion in doubt: J. Coutts, L. Cerfaux and A. Farrer.[33] Coutts studies the parallels between Mk 3:20-35 and 4:10-12 and notes that, apart from 4:11, the word *exō* is used by Mark in a strictly spacial sense, and that *ekeinos* always refers to something immediately identifiable, usually from the preceding context. He concludes: "Thus Mk 4:10-12 may be seen, not as the comment on the three classes of seed which did not fall on good ground, or on the crowds who had to be addressed in parables, as opposed to the disciples, but as the ultimate pronouncement of Jesus on the enmity and opposition which finds expression in the stories of 2:1-3:6, and 3:21 and

---

[31] See T. A. Burkill, *Revelation*, 104-5; J. M. Robinson, *Problem*, 52; W. Marxsen, "Erklärung," 19; G. Minette de Tillesse, *Secret*, 179-80.

[32] *Str-B* 2, 7; J. Gnilka, *Verstockung*, 84.

[33] J. Coutts, "'Those Outside' (Mark 4,10-12)," *SE* 2 (= *TU* 87), 155-57; L. Cerfaux, "La connaissance des secrets du Royaume d'après Mt., XIII, 11 et parallèles," *NTS* 2 (1955-56) 239; A. Farrer, *A Study in St. Mark* (Westminster: Dacre, 1951) 240.

22." Cerfaux suggests the same thought: Mark is contrasting the scribes and the family of Jesus to those who are with Jesus in the house, i.e., the disciples and the crowd "sitting about him" (3:34).

Similarities between Mk 3:20-35 and 4:10-12 are indeed striking. No one doubts that Mark looked upon 3:20-35 as a unit; the section begins with a reference to the suspicions and misguided plans of Jesus' family and ends with Jesus' repudiation of his natural family ties in favor of a relationship founded on obedience to God's will. Sandwiched between these two passages we find the account of the scribes' misinterpretation of Jesus' power, and Jesus' refutation and condemnation of them. In 3:31-35 there is an unmistakable contrast between the family of Jesus, who are referred to as *hoi par' autou* in the redactional vs. 21[34] and as *exō stēkontes* and *exō* in vss. 31 and 32, on one hand, and, on the other, the crowd (*ochlos*) which is described as *peri auton* in vss. 32 and 34. The same expressions, in this case substantivized, are used in 4:10-11 to express the contrast between "those about him" and "those outside."

There are further similarities between the two passages. In 4:11 the kingdom of God is spoken of; 3:24, read, as it should be, in the context of the following two verses, speaks of Satan's kingdom. 4:11 tells us that to those outside everything happens "in parables," and in 3:23a, a redactional verse,[35] Jesus addresses the scribes "in parables."[36] Another telling parallel exists between 4:12c "lest they . . . be forgiven" and the last verse of Jesus' reply to the scribes, "whoever blasphemes . . . will never be forgiven." This reference to non-forgiveness comes at the end of both passages; it thus seems to be uppermost in the evangelist's mind; even if he was not composing freely,[37] the arrangement of the material shows that for him the main message, in one case, and the main purpose, in the other, of the address "in parables" is non-forgiveness. It may be significant that the synoptic parallels to Mk 4:12, Mt 13:13 and Lk 8:10, simply omit vs. 12c and that Mt 13:15 gives the OT quotation according to the LXX ("and turn back to me, and I should heal them").

---

[34] For indications of the redactional character of the verse, see V. Taylor, *Mark*, 235.

[35] Cf. Mk 3:13; 6:7; 7:14; 8:1,34; 10:42; 12:43; see also R. Butmann, *History*, 332.

[36] E. Schweizer (*Mark*, 93) seems to be the only one who thinks that all are being addressed. In our opinion, it is much more natural to suppose that the scribes alone are being spoken to. See W. Grundmann, *Markus*, 83; E. Klostermann, *Markus*, 37; M.-J. Lagrange, *Marc*, 65; R. Bultmann, *History*, 332.

[37] E. Schweizer (*Mark*, 83-84) is of the opinion that 3:20-35 shows strong traces of editorial composition; J. Coutts ("Outside," 156) feels that in 4:12 we have an allusion to the OT rather than a direct quotation.

Attractive as the suggestion based on these parallels may at first sight appear, it breaks down on two considerations. The first is 4:1-2, a largely redactional passage, which clearly separates ch. 3 from ch. 4, and places 4:10-12 within a framework quite distinct from that of 3:20-35. The other consideration is the sharp distinction which Mark draws between the disciples and the multitude throughout the Gospel, a distinction which is very evident in ch. 4 itself.[38]

It must be admitted, however, that the general picture which Mark presents of the multitude can hardly be harmonized with the judgment which 4:12c seems to inflict upon it. Many authors have noticed this discrepancy. T. A. Burkill[39] remarks that in the rest of the Gospel the disciples are not so sharply separated from the multitude as in 4:11. E. Trocmé notes the omnipresence of the crowds in redactional passages.[40] M.-J. Lagrange[41] points out that the multitudes do not change their attitude toward Jesus, nor does Jesus change his attitude toward them throughout the Gospel. M. Hermaniuk[42] makes the same observation, and calls attention to 10:1, "crowds gathered around him, and *as usual* he began to teach them." He notes, however, that the attitude of the crowd changes during the last part of the Passion Narrative. It seems to us that this is to be expected in a narrative in which the disciples desert Jesus and betray him. Hermaniuk notes further that Jesus' severity is directed more toward the scribes and Pharisees than toward the multitudes.[43] Mention should also be made of E. Schweizer's remark[44] that it frequently becomes evident that the real enemies of Jesus are Jewish authorities, scribes and Pharisees in particular, and not the Jewish people. It seems to us that Gnilka's portrayal of the crowd[45] does not do justice to the evidence of the Gospel, particularly in view of its presence in redactional passages.[46] We wonder whether Mark's reference to "all Jews" in 7:3 should be understood of the

---

[38] See E. J. Mally, "The Gospel according to Mark," *JBC* 2, 30.

[39] "Cryptology," 255, n. 1.

[40] "Pour un Jésus public: les évangélistes Marc et Jean aux prises avec l'intimisme de la tradition," *Oikonomia: Heilsgeschichte als Thema der Theologie: Festschrift für Oscar Cullmann* (ed. F. Christ; Hamburg-Bergstedt: H. Reich, 1967) 45-46.

[41] "Le but des paraboles d'après l'évangile selon saint Marc," *RB* 7 (1910) 10-12.

[42] *La parabole évangélique: Enquête exégétique et critique* (Bruges: Desclée de Brouwer, 1947) 305-7.

[43] *Parabole*, 323.

[44] "Frage," 3, n. 17.

[45] *Verstockung*, 84-86.

[46] See note 40; E. Trocmé refers to 1:33,45; 2:2,13,15; 3:20; 4:1; 5:21,24; 8:34; 9:14; 10:1.

multitude as easily as J. Gnilka seems to assume it.[47] It seems also that in his interpretation of Mk 12:1-12 Gnilka makes an unwarranted jump from the rejection of Judaism as a religious institution to that of the multitudes, for in the redactional vs. 12 Mark distinguishes clearly between the Jewish leaders and the crowd.

In order to ascertain whom Mark has in mind when he speaks of "those outside" we must study, first, the groups which play a distinctive role in the Gospel and, secondly, the phrase "those present with the Twelve" in 4:10.

(2) There are four clearly distinguishable groups in the Gospel: demons, enemies of Jesus, multitudes, and disciples. The differences between them stem from their varied relationship to Jesus.

The demons are characterized by their irreconcilable opposition to Jesus, and of Jesus to them,[48] as well as by the fact that they know who Jesus is. This knowledge of theirs is not a creation of Mark; he found it in the tradition where it is presented as their defense measure against the power which has come to destroy them.[49] That Mark makes this traditional datum his own and puts it in service of the messianic secret is shown by such redactional verses as 1:34 and 3:11-12.[50] These verses indicate that the demons' knowledge was, dogmatically, correct; for, whether we take the *hoti* in 1:34 as causal or as recitative, the suggestion that demons knew something which men should not yet discover is obvious. The wording of 3:12 points in the same direction: the order not to make him known implies true knowledge on their part.

Jesus' enemies are primarily, though not exclusively, scribes and Pharisees. Occasionally Herodians are associated with them, and in the Passion Narrative the chief priests play an important role. They seem to be the human counterpart of the demons, for their hostility to Jesus is as irreconcilable as that of the impure spirits. J. M. Robinson[51] has shown the similarity between exorcism stories and controversies in Mark: "The demon advances upon Jesus with hostile challenge, only to be silenced by an authoritative word of Jesus. . . . The debates with the Jewish authorities likewise begin normally with a hostile question or accusation of the opponent,

---

[47] Mark links Judeans with the Pharisees, not with the crowds; see R. Pesch, *Naherwartungen*, 146.

[48] For a good description of this mutual opposition in Mark, see J. M. Robinson, *Problem*, 35-42.

[49] Mk 1:24 is such a traditional datum; see R. Bultmann, *History*, 209, n. 1; G. Strecker, "Messiasgeheimnis," 89-90.

[50] See R. Bultmann, *History*, 341.

[51] *Problem*, 43-46.

which is frustrated by Jesus' definitive reply."[52] "Sometimes his reply consists in taking the initiative, which calls forth a further statement of the opposition before Jesus' final word."[53] Although there is no doubt that the controversies, as well as the exorcisms, in the Second Gospel owe their structure to the tradition, we can hardly question the fact that Mark accepts and elaborates the traditional portrait of the protagonists which these stories reveal. There is no need to prove Mark's acceptance of the demons' traditional image. As for the enemies, Mk 3:6, most likely redactional,[54] shows them plotting Jesus' death very early in his public ministry; other redactional verses voice the same thought.[55] Jesus' opposition to the enemies is abundantly illustrated; we have already referred to the exorcism-like features of controversies. Further confirmation is offered by passages in which there occurs, or fails to occur, a characteristic theme of the Second Gospel, namely the teaching of Jesus.[56] Jesus is portrayed as teaching (*didaskein*) the disciples (8:31; 9:31) and the multitudes (1:21, 22; 2:13; 4:1,2; 6:2,6,34; 10:1; 11:17; 12:35; 14:49); he gives his teaching (*didachē*) to the multitudes (1:22,27; 4:2; 11:18; 12:38). But he is never said to teach the representatives of official Judaism; to these he simply speaks (2:8,17,19,25,27; 3:4,23; 7:6,9; 8:12; 10:3,5; 11:29,33; 12:1,15,16,17,24,29; 14:62).[57] In 12:14 the Pharisees and the Herodians seem to acknowledge Jesus as the one who "teaches God's way of life sincerely." But their protestations are nullified by the context (vs. 13 "to catch him"; vs. 15 "knowing their hypocrisy"). On the other hand, Jesus is seldom portrayed as merely speaking to the disciples and the crowds: 1:38 occurs in the context of proclamation; in 2:2 he speaks "the word" to them; 3:33-34 is elevated by the redactional *periblepsamenos*;[58] 4:11,21, 24,26,30 are subsumed in the "teaching" of Jesus (4:2); at 4:33-34 he

---

[52] *Problem*, 44; he refers to the following texts; for exorcisms, 1:23-26; 3:11-12; for controversies, 2:6-11,16-17, 18-22, 24-28; 3:2-5; 7:5-15; 8:11-12; 12:18-27.

[53] *Problem*, 44; he refers to 5:6-13 for exorcisms, and to 10:2-9; 11:27-33; 12:13-17 for controversies.

[54] See R. Bultmann, *History*, 52, 63; E. Schweizer, *Mark*, 73-74.

[55] Mk 11:18 (see V. Taylor, *Mark*, 464); 12:12 (see V. Taylor, *Mark*, 477).

[56] That *didaskein* and *didachē* characterize the activity of Mark's Jesus has been shown by E. Schweizer, "Anmerkungen," 95-96.

[57] In the context Jesus is portrayed as angry or grieved (3:5; 8:12), as knowing their evil thoughts (2:8), branding them as hypocrites (7:6), referring to their hardness of heart (10:5), as refusing to answer their questions (8:12; 11:33), declaring them to be wrong (12:24).

[58] R. Bultmann (*History*, 332) thinks that 3:34 "may well be an editorial formulation."

again speaks "the word"; 7:14; 8:34; 9:1 are illumined by the redactional *proskalesamenos*.[59]

It could, of course, be objected that this feature of the Second Gospel should be attributed, not to Mark, but to the fact that we are dealing with controversies in which references to teaching can hardly be expected, as well as to the traditional formulation of the controversy-stories. It is not Mark's habit to change the nucleus of the traditional units; in the main he produces introductory notes to pericopes, and at the end of them he adds his own comments, summaries, or further traditional material. In controversies, however, the words of Jesus appear in the middle of the story; it could thus be argued that Mark, in view of his redactional methods, had no opportunity to change the formulae introducing the words of Jesus. In answer to this objection it should be pointed out that Mark is fully consistent in this respect; he remains faithful to his refusal to grant Jesus' teaching to the enemies even in cases in which the words addressed to them would clearly be looked upon as such by the early Christians and by Mark himself (2:17,19,25,27; 3:23; 10:5; 12:1,17,24,29). There are, moreover, at least two formulae which are redactional and which introduce pericopes resembling instruction a great deal more than controversy, viz., 3:23 and 12:1. The difference between these verses and 4:1-2 is striking and instructive. It would thus seem reasonable to conclude that Mark purposely refused to portray the enemies as recipients of "teaching," Jesus' primary salvific activity in the Second Gospel.[60] They seem to be capable only of opposing Jesus and of plotting his destruction. This is confirmed when we turn attention to one of the reactions of various audiences to Jesus' activity in the Second Gospel, viz., to the reactions of amazement and fear.

References to amazement and fear abound in the Second Gospel. Some verbs expressing these emotions (*thambeomai, ekthambeomai, ekthaumazō*) occur only in it; others are used more frequently in Mark than in the rest of NT (*thaumazō, phobeomai, ekphobos, ekplēssomai, existēmi*). The disciples wonder and fear (4:41; 5:42; 6:50,51; 9:6,32; 10:24,26,32), likewise the crowd and various individuals (1:22,27; 2:12; 5:15,20,33,36; 6:2; 7:37; 9:15; 10:32; 11:18), Jesus (6:6; 14:33), Herod (6:20), Pilate (15:5,44), women at the grave (16:5,6,8), and Jesus' enemies (11:18,32; 12:12,17). Very telling are the causes of wonderment and fear.

---

[59] See R. Bultmann, *History*, 332; it is usually the disciples who are called (3:13; 6:7; 8:1; 10:42; 12:43); in 7:14 it is the crowds, in 8:34 the crowd with the disciples, in 3:23 the scribes.

[60] See E. Schweizer, "Contribution," 422-23.

The disciples, the crowd and various individuals wonder or fear on account of Jesus' teaching (1:22,27; 6:2; 9:32; 11:18), his miracles (2:12; 4:41; 5:15,20,33,42; 6:50,51; 7:37), the presence of the transfigured Jesus (9:6,15), the imminence of his suffering (10:32), his demands (10:24,26), and in the face of death (5:36). The women at the grave wonder and fear at the news of Jesus' resurrection; Herod is afraid of John the Baptist; Pilate wonders at Jesus' silence and his unexpected death. The enemies of Jesus, on the contrary, fear the crowd (11:18,32; 12:12; cf. 14:2). Only once are they said to be amazed at Jesus (12:17 *exethaumazon ep' autō*). This amazement, however, seems to be as human as their fear of the crowd; Bauer-Arndt-Gingrich translate the verb as "wonder greatly (in the sense of grudging admiration)." It may also be significant that the verb is a *hapax legomenon,* not only in Mark, but in the NT. The preposition *epi,* which indicates the cause of amazement occurs in 1:22 and 11:18 with Jesus' "teaching," and in 10:24 with his "word"; only in 12:17 is Jesus himself the cause of amazement. It is safe to conclude, then, that the enemies' amazement in 12:17 is looked upon by Mark to be as non-religious as is their fear of the crowd.[61]

This examination of evidence is strongly indicative of Mark's attitude toward Jesus' enemies. We saw how carefully he prevents Jesus from teaching his enemies, and how liberal his Jesus is with the instruction to the crowds and disciples. The response to Jesus' mighty words and works on the part of the crowds and disciples is religious fear, wonderment, amazement; the response of the enemies is murderous hatred. There are two sets of "dramatis personae" in the Gospel who do not respond to Jesus' words and works with religious fear and wonder, viz., the demons and the enemies. Even Herod's fear of John the Baptist seems to possess a religious strain, and Pilate's amazement might be construed as religious. No matter how reprehensible Mark may consider fear and amazement to be, he obviously looks upon it as a glimmer of conversion, faith, and understanding.[62] The enemies of Jesus, however, are never taught and never respond with religious amazement.

There is a further resemblance between the demons and the enemies in the Second Gospel: their knowledge of Jesus. There seems to be a strange inconsistency in the Gospel. On the one hand, Jesus is portrayed as giving private instruction to the disciples, apart from the crowd, in the house, not

---

[61] For a discussion of wonderment in the Second Gospel, see G. Minette de Tillesse, *Secret,* 264-76. J. M. Robinson's excellent examination of amazement and fear (*Problem,* 68-73) is marred by his failure to distinguish the fear of the enemies from that of the others.

[62] See G. Minette de Tillesse, *Secret,* 264-76.

wishing his presence to become known, forbidding people to publicize his miracles, and silencing the demons. On the other hand, he flings miracles in the face of his enemies (2:3-12; 3:1-5). While he is never said to perform a miracle in order that the crowd and the disciples may learn and understand, he does so for the benefit of his enemies in 2:10. While the enemies are said to understand in 12:12, Mark goes out of his way to describe the disciples' failure to understand. It could almost be said that *ouk oida* is his technical term for the disciples' spiritual condition. An examination of the usage of the verbs employed to describe knowledge and understanding, with the disciples as the subject and a religious truth or event as the object, finds them, with one exception (13:29), accompanied by a negative particle. *Ginōskō* with the disciples as subject is found four times (4:13; 6:38; 13:28,29). It has a religious reality as its object twice: in 4:13 it is found in a question, the context of which clearly implies that the disciples do not know; in 13:29 it is found in an affirmative clause—but it should be noted that the verse speaks of the eschatological future. Wherever *oida*[63] occurs with the disciples as subject (4:13; 9:6; 10:38; 13:33, 35; 14:40,68,71), it is accompanied by a negative particle, with one interesting exception: 10:42, where the object of knowledge is a wordly and, ironically,[64] unreal matter. The texts in which the verb is negated are highly significant: the disciples fail to understand the parables, Peter is all confused on the Mount of Transfiguration, the sons of Zebedee attempt to gain the first places in the kingdom, the disciples are ignorant of the moment of the Lord's arrival, they are nonplussed in Gethsemani, and Peter denies Jesus. *Noeō* is found twice in negative question clauses (7:18; 8:17); *mnēmoneuō* occurs once (8:18) in a negative question clause; *syniēmi* occurs three times (6:52; 8:17,21), once in a negative statement and twice in negative questions; *asynetos* appears in 7:18. Admittedly Mark softens the harshness of Jesus' rebukes by expressing them in question form, but such editorial statements as 6:52 and 9:32 should suffice to prove that more is involved than mere exhortation to strive for greater insight.

Thus while the disciples' ignorance is continually stressed, and Jesus attempts to hide his miracles and sometimes his presence from the crowd, he performs miracles in the full view of his enemies who are said to under-

---

[63] C. H. Turner ("Usage," *JTS* 28 [1926-27] 360-62) states that in Mark there is no perceivable difference between *eidenai* and *ginōskein*.

[64] This is the opinion of some commentators: C. E. B. Cranfield, *Mark*, 340; V. Taylor, *Mark*, 443; E. Lohmeyer, *Markus*, 223; M.-J. Lagrange, *Marc*, 263.

stand.⁶⁵ That they do understand, not only in 12:12, but also in 2:10,⁶⁶ is shown by the redactional 3:6. There we see the same decision resulting from their knowledge as in 12:12 and 14:64: they are determined to put Jesus to death. The similarity between the demons and the enemies is striking: as they never give a sign of religious awe when confronted with Jesus' words and works, so are they the ones who are portrayed as possessing knowledge. At 1:23,34; 3:11-12 it is clearly stated or implied that the demons know who Jesus is. The enemies likewise understand Jesus' claim that he is carrying out what is reserved to God alone (2:7,10); they understand his claim to be God's Son whose death at their hands will be the cause of their own destruction (12:12).⁶⁷ They understand, while the disciples are called upon to understand. It is evident that their knowledge is entirely different from that which Jesus demands of disciples; theirs is a demonic knowledge, whereas that of the disciples is demanded as a response to Jesus' powerful teaching.

It seems to us, consequently, that we should not be attempting to find the theme of the messianic secret in the controversies of the Second Gospel, as is done by G. Minette de Tillesse in his recent book.⁶⁸ His remark on 2:3-12 that "personne n'a saisi clairement la revendication de Jésus"⁶⁹ is belied by 3:6 which shows that they understood as clearly as the high priest's court did in 14:62-64. About 2:15-17 Minette de Tillesse himself says, "Sans doute, Jésus ne dit-il pas explicitement: 'Pour que vous sachiez

---

⁶⁵ See C. Maurer, "Das Messiasgeheimnis des Markusevangeliums," *NTS* 14 (1967-68) 518.

⁶⁶ G. Minette de Tillesse (*Secret*, 117-18) is of the opinion that Mk 2:5b-10 is a Marcan redactional construction; his only argument seems to be a reference to Turner. Turner's remarks on the passage do not assert everything that Minette de Tillesse reads into them ("Usage," *JTS* 26 [1924-25] 145-46): Minette de Tillesse claims that "C. H. Turner a montré que le style de ces versets était très marcien," whereas Turner asserts only that Mark has the habit of inserting parenthetical clauses into his material; his remarks, further, are limited to 2:10-11. In this connection the view of Minette de Tillesse should be mentioned (*Secret*, 287-90) according to which the enemies in Mk 12:12 are said to have understood the parable in 12:1-9 because it had been allegorized by Mark himself. The difficulty with this suggestion lies in the fact that neither Minette de Tillesse nor Trocmé (*Marc*, 163, n. 162), to whom Minette de Tillesse refers, offers any proof of Marcan redaction within the parable. A comparison with the *Gospel of Thomas* is hardly sufficient to prove this redaction.

⁶⁷ See G. Minette de Tillesse, *Secret*, 287.

⁶⁸ *Secret*, 113-63, 287-93. R. H. Fuller, in his review of Minette de Tillesse's book (*CBQ* 31 [1969] 110), remarks that controversies are "a category frequently overlooked in discussion of the secret." We feel there is good reason for this "omission."

⁶⁹ *Secret*, 121.

que le Fils de l'Homme a sur terre le pouvoir de guérir les pécheurs . . . (c'est pour cela que je mange avec eux)'. Cela, il l'a dit une fois pour toutes en 2,10; les autres scènes doivent s'interpréter de la même façon. Ici, il est symptomatique que la réponse de Jésus ne vaut que s'il est lui-même le 'Médecin.' "[70] A discussion of redactional intervention in Mk 2:1-3:6 is not necessary to perceive the significance of Jesus' claim to be the Son of Man who has the power to forgive sins and to disregard Jewish Sabbath observances (2:10,28). But the disciples will hear Jesus referring to himself as the Son of Man only after Peter's confession. The enemies learn of Jesus' death at 2:20, the disciples must wait until 8:31. In 10:10 it is the disciples who ask Jesus for an explanation, not the enemies. It may be significant that while at 10:11-12 Jesus must supply the explanation to the disciples, it is the scribe who draws the proper conclusion from Jesus' words in 12:33. Such passages as 11:33 and 8:11-13 have nothing to do with the messianic secret; in 11:33 Jesus does not speak "avec quelque réserve"[71] of his heavenly authority; rather, he refuses to reveal it to those who have given ample proof of their unwillingness to accept it in vss. 31-33a; the same is true of 8:11-13. The enemies have heard and understood Jesus' claims; it is precisely this understanding which drives them to plot and finally to bring about his death (3:6; 11:18; 12:12; 14:61-64). If Jesus refuses to answer them it is not in the interest of the messianic secret, for there is no point in keeping away from them something which, like the demons, they already know.

The enemies of Jesus in the Second Gospel could briefly be described as the synagogue of Satan. They oppose Jesus continually; their questions are put to him not to learn but to challenge: they either state or imply that what he is doing or allowing his disciples to do is wrong,[72] or they approach him with evident malice.[73] His answers to them are designed, not to instruct, but to silence or condemn them.[74] Only once does he have a good word to say about a scribe.[75] They never receive his teaching and, like the demons, fail to be amazed at or to fear his great words and works. The demons know who Jesus is, and the enemies know who he claims to be. This knowledge makes them instruments of his death.

The next group to be discussed is the disciples. Their function is

[70] *Secret*, 123.
[71] As Minette de Tillesse asserts in *Secret*, 291.
[72] See 2:7,16,18,24; 3:4.
[73] See 10:2; 11:28; 12:14,19-23.
[74] See 12:34; 3:28-29; 12:9.
[75] 12:34.

described in 3:13-15. Mark's redactional composition[76] enumerates the characteristics of the disciples: their personal call, their life in permanent company of Jesus, and their mission to preach and exorcise. The feature most pronounced in the Gospel is their being permanently with Jesus. Another feature, not mentioned explicitly in the verses just referred to, but stressed throughout, is their reception of special instruction. This instruction would seem to serve the task of proclamation which they are to exercise after Jesus' death and resurrection, for instruction is related to the function of proclamation. Moreover, various private instructions in the Gospel seem to have a more direct bearing on the problems of the community than the rest of the material. This is evident in 4:14-20 which, with its un-Hebraic character of style and a vocabulary including several words found only in the epistles, clearly reflects community preoccupations. It is equally evident in 7:18-23 which is the result of discussions concerning ritual purity of food.[77] In 9:33-50 we sense the difficulties experienced by the community in living up to the destiny of humility and service to which its following of Jesus had committed it;[78] 10:10-12 is an adaptation of Jesus' teaching on divorce to a non-Palestinian environment.[79] Problems of Christians regarding wealth are discussed in 10:23-31, and regarding their eschatological expectations in ch. 13. Somewhat less evident, but undoubtedly present, are Christian preoccupations in such passages as 9:9-13 and 9:28-29.[80] The fundamental importance of Jesus' passion and of his sonship for the life of the community need hardly be pointed out. Private instruction given to the disciples emphasizes their position as the official interpreters of Jesus' teaching, as having the authority to decide matters, in short, as continuing the work of Jesus. Those who are portrayed as questioners in the Gospel have, in the life of the community, the function of linking it to the earthly Jesus and his message.[81]

Yet the disciples are never said to understand. As J. Gnilka has pointed out,[82] Jesus' rebukes are always expressed in question form.[83] To say,

---

[76] The redactional character of these verses is shown by V. Taylor, *Mark*, 83, 229, and particularly by R. P. Meye, *Jesus and the Twelve: Discipleship and Revelation in Mark's Gospel* (Grand Rapids: Eerdmans, 1968) 97-136, 173-91.

[77] See V. Taylor, *Mark*, 342-47.

[78] See D. E. Nineham, *Mark*, 250-59.

[79] See G. Minette de Tillesse, *Secret*, 230-34.

[80] For 9:9-13, see E. Lohmeyer, *Markus*, 181-84; for 9:28-29, see W. Grundmann, *Markus*, 191.

[81] See W. Marxsen, "Erklärung," 23; L. E. Keck, "Introduction," 364.

[82] *Verstockung*, 33.

[83] E. Sjöberg, *Menschensohn*, 163-66, argues that all references to the disciples'

with Gnilka,[84] that after Peter's confession the reproaches cease is, however, scarcely correct and narrows the field of vision. Rebukes aimed directly at their non-understanding, such as occur in 4:13; 7:18 and 8:17-21, no longer appear, indeed, and this fact should be given its full weight. But we must not fail to notice other signs of their failure to understand. The fiercest rebuke hurled at them takes place after Peter's confession in 8:33.[85] Gnilka attempts to weaken its role by consigning it to the oldest tradition.[86] But he makes this assertion without offering proof; further, he neglects to consider the structure of Mk 8:27—10:52 where Mark indicates the disciples' failure to understand after each prediction of the passion. Moreover, such verses as 10:14; 14:6,37 undoubtedly express or imply a rebuke. Mark is, in fact, just as careful to emphasize the blindness of the disciples after Peter's confession as before it; besides the passages already referred to there are to be considered 9:6,10,18,28,32,34; 10:35-37,41; 14:40,50, 66-72. These observations should not, however, lead us to false conclusions. They must not be interpreted as an indication that the disciples learn nothing throughout the Gospel. Peter's confession of Jesus' messiahship must be looked upon as an important step in their growth of knowledge. The same must be said of the revelation which they receive on the Mount of Transfiguration. The commands of silence which follow the confession and the transfiguration show that the disciples have discovered what they did not know previously and were granted an insight into the mystery of Jesus' identity before the time appointed for the public proclamation of this mystery. Probably on account of this insight Jesus' direct reproaches of the disciples' ignorance cease. There is thus an interplay of revelation and of hardness of heart, of growth in knowledge and of failure to understand. The disciples' growing awareness of Jesus' identity is continually offset by their all too human attitudes.

The image which the Second Gospel paints of the disciples is that of a

---

ignorance belong to the tradition. Even if they should—which we find impossible to admit—Sjöberg must still explain why Mark so consistently preserves these traditions; he must also explain the arrangement of material in 8:27—10:52 where each prediction of the passion is followed by a pericope showing the disciples' failure to understand. It would be rather difficult to argue that Mark inherited the section ready-made from the tradition. Apart from that, it is quite impossible to relegate such a consistent feature of a gospel to the traditional stage. See G. Minette de Tillesse, *Secret*, 227-37. That 6:52 is redactional has been abundantly proven by Q. Quesnell, *The Mind of Mark: Interpretation and Method through the Exegesis of Mark 6,52* (AnBib 38; Rome: Biblical Institute, 1969).

[84] *Verstockung*, 39.
[85] See G. Minette de Tillesse, *Secret*, 273.
[86] *Verstockung*, 39.

privileged group, called personally by Jesus to accompany him permanently, to receive special revelations and instructions, and to be entrusted with proclamation and the power to exorcise. They grow in the knowledge of Jesus despite their hardness of heart. They are the antipodes to the enemies of Jesus: as they carry out the works of Jesus, so is the enemy opposition a reflection of the attitude of Satan and the demons.

The last group is the crowd. In the portrait of it given by the Gospel there are positive and negative features. Positively Jesus is presented as teaching the crowds, and they respond with amazement to his words and works. They respond also by coming to him (2:13; 3:20), following him (2:15; 3:7), gathering about him (5:21; 10:1), proclaiming his mighty deeds (7:36). There are too many clear contrasts between the crowds and the representatives of official Judaism to be able to ascribe them all to mere chance or to Mark's inability to control the traditions which he incorporates into his Gospel. Contrast such passages as 2:15-17; 3:6-8; 7:1-5 with 7:14; 10:1-2; 11:18; or 11:27 with 11:32; or 12:12,13,18 with 12:37; 14:1-2; 15:11. While the crowds are evidently well disposed toward Jesus, his enemies are plotting his death. The redactional verses 11:18 and 12:12[87] tell us that it was Jesus' popularity with the crowd which prevented the enemies from carrying out their designs on his life. There is only one unfavorable reference to the crowd in the entire Gospel, viz., 15:11;[88] but even there we are told that it was stirred up by the chief priests to demand the release of Barabbas.

The negative nuances of "the crowd" are also obvious. First of all, Jesus does not entrust it with the instruction reserved to the Twelve. In a number of cases he purposely avoids it; the crowd disobeys his commands of silence. These traits in the Second Gospel have been frequently noted. The amazement and fear of the crowd is as ambivalent as that of the disciples. While it indicates that the crowd senses a divine power in Jesus, it is also a sign of defective faith and understanding. As the Gospel narrative progresses, Jesus pays more and more attention to his disciples. Though the crowd is not entirely absent in the latter sections, it is not as ubiquitous as it was up to 8:26.

The crowds thus seem to form a middle-ground between the disciples and the enemies of Jesus. They are not personally called, nor are they

---

[87] See V. Taylor (*Mark*, 464, 477) for indications of their redactional character.

[88] The crowd in 14:43 is clearly a group of armed men; 15:29,35 may refer to the crowd, but this is by no means certain. When J. Gnilka speaks of unmistakably negative references (plural) to the crowd (*Verstockung*, 84) he is misrepresenting the evidence. His remark that Mark seems to have little interest in the crowd is belied by the many redactional verses in which it plays a part.

committed to Jesus as the disciples are. They are nonetheless taught and healed by him and respond to him with a real, though imperfect, enthusiasm.

(3) Which one of these groups can be designated as "those outside" (Mk 4:11)? The most obvious candidates are the enemies of Jesus whose determined opposition to him makes them incapable of the understanding and faith demanded by him and of forgiveness linked with this faith. This impression is strengthened by the passages in which Jesus addresses his enemies "in parables," viz., 3:23-29 and 12:1-9. True, we cannot consider 3:23-29 and 4:11-12 as parts of one contextual unit; but it is instructive to consider 3:23-29 as well as 12:1-9. These three passages have in common, besides the phrase "in parables," a similar content; in each case hardness of heart is involved. That the sin of blasphemy against the Holy Spirit carries with it the trait of obdurate impenitence has been shown in a recent work of E. Lövestam.[89] The repeated refusals of the tenants to give of the produce of the vineyard to its owner which culminate in the murders of his messengers and of his son show the same persistent unwillingness to submit to the will of God. All three passages end with a condemnation: 3:28-29 announces that those who radically misinterpret the great works of Jesus can never be forgiven; 12:9 predicts the murderers' destruction and the substitution of them by others. The resemblance to 4:11-12 is thus striking in form as well as in content. The difference lies in the point of view; while in the two parables the prime concern is human perversity bringing upon itself divine condemnation, the chief interest in the logion lies in the divine judgment which results in the hardness of heart. Two causes of the enemies' rejection are juxtaposed: their guilt is stressed in 3:23-30 and 12:1-9, but the sovereign will of God is emphasized in 4:11-12. Mark does not attempt to work out an intellectually satisfying harmonization of these aspects of the enemies' spiritual ruin; but he is clearly aware of them.

The evident contrast between 4:1-2, which depicts the crowds gathered about Jesus and listening to him, and 4:10, where we find him alone in the narrow circle of disciples, seems to suggest, however, that the crowds must also be numbered among "those outside." The disciples alone are given the explanation of the parable, while the crowd is condemned to ignorance. The difficulty with this conclusion lies in vs. 12c which condemns those outside to impenitence and non-forgiveness. The image which the rest of the Gospel paints of the crowd does not justify such a judgment.

---

[89] *Spiritus blasphemia: Eine Studie zu Mk 3,28f par Mt 12,31f, Lk 12,10* (Scripta minora regiae societatis humaniorum litterarum lundensis, 1966-1967:1; Lund: Gleerup, 1968), particularly 51-57, 62-68; see also W. Beyer, *TDNT* 1, 622; C. K. Barrett, *The Holy Spirit and the Gospel Tradition* (London: S.P.C.K., 1966) 103-5.

They do not know who Jesus is, but they do not radically misconstrue the source of his power as do the enemies in 3:22,30.[90] Nowhere else in the Gospel are they similarly condemned.

This leads us to a consideration of vs. 12 itself. Is it possible to understand it in a way which removes its sting? The answer to this question is given by the particles *hina* and *mēpote*. Let us begin with *hina*. According to one opinion,[91] the conjunction should be taken as representing *hina plērōthē*, thus introducing an OT prophecy being fulfilled in the work of Jesus. The hardening of hearts is accordingly not to be seen as the purpose of Jesus when he speaks in parables, but as the realization of the divine plan laid down in the Scriptures, which is set in operation by means of the parables. This solution, however, exposes itself to the danger of eisegesis. Further, it has been rendered dubious by the work of A. Suhl[92] on the OT quotations and innuendoes in the Second Gospel. He questions whether Mark thinks and writes within the framework of promise and fulfilment. In Mark OT quotations have a qualifying or interpretive function, and not, as in Matthew, that of pointing to Jesus as the fulfilment of OT prophecies. Another opinion holds that the particle introduces a result clause.[93] M.-J. Lagrange[94] long ago objected to this solution and pointed out that the conjunction has such a meaning with verbs that express request; with other verbs it can indicate only purpose. J. Gnilka[95] observes that purpose always originates with an agent; only where a thing or condition is presented as bringing about a given state can the particle be understood as introducing a result clause. In Mk 4:11, however, it is clearly implied that God is carrying out his judgment. Many exegetes take the conjunction as expressing purpose, at least in Mark's understanding of it.[96] This is, we feel, the most natural meaning to be given to it. The fact

---

[90] Mk 3:30 is generally regarded as a redactional comment of the evangelist restating the principal reason for Jesus' condemnation of his accusers. It is one of the "context supplements" which are so characteristic of the earlier chapters of the Second Gospel; see J. C. Hawkins, *Horae*, 125-26.

[91] Held by M.-J. Lagrange, *Marc*, 99; "Le but des paraboles," 29; J. Jeremias, *Parables*, 17; W. Marxsen, "Erklärung," 25; E. F. Siegman, "Teaching in Parables," *CBQ* 23 (1961) 176; J. Gnilka, *Verstockung*, 48; G. Minette de Tillesse, *Secret*, 192-93; also the translators of the *NEB*.

[92] *Funktion, passim*, particularly pp. 66, 94; see also S. Schulz, "Markus AT," 188; J. Dupont, "Chapitre," 806.

[93] C. H. Peisker, "Konsekutives *hina* in Markus 4,12," *ZNW* 59 (1968) 126-27; A. Suhl, *Funktion*, 149-50.

[94] "Le but," 28.

[95] *Verstockung*, 46-47.

[96] E. Schweizer, *Mark*, 93; E. Sjöberg, *Menschensohn*, 124; V. Taylor, *Mark*, 256-57; C. E. B. Cranfield, *Mark*, 155; W. Manson, "The Purpose of the Parables: A Re-

that Matthew (13:13) changed it into *hoti* seems to indicate that he found it as disturbing as modern exegetes.

What meaning should be assigned to *mēpote*? Jeremias[97] states that the word has two meanings in Greek: "in order that not" and "lest perhaps." It may be a translation of the Aramaic *dlm'*, which, in certain contexts, has the sense of "unless." Jeremias accepts this last meaning for Mark in view of rabbinic exegesis. Yet it would seem that the presumption should be in favor of one or the other Greek meaning of the term. Mark, after all, wrote in Greek, and one would imagine that he would use a different conjunction if his understanding of it were that attributed to him by Jeremias.[98] Many authors are of the opinion that the meaning assigned to *mēpote* should follow that of *hina*.[99] The only NT passages in which the particle does not have the final meaning are Lk 3:15; Jn 7:26 and 2 Tim 2:25; in all three cases it introduces a question clause.[100] We would conclude that, no matter what sense it carried in the original form of the logion, it should be taken as introducing a final clause in the Second Gospel.

It is thus most likely that, according to Mark, the purpose of the parables is to keep those outside in ignorance and impenitence. Obviously the knowledge spoken of in vs. 12 is not that of the demons and the enemies of Jesus, for that knowledge, instead of saving, condemns; rather, the knowledge of vs. 12 is that to which the disciples are being exhorted; it is an understanding which is practically identical with faith.

But on which element in vs. 12 does Mark place the accent: on the failure to understand in 12ab or the condemnation in 12c? Since he is not formulating freely but is bound by a traditional logion, such a distinction may be

---

Examination of St. Mark IV. 10-12," *ExpT* 68 (1956-57) 132; J. Dupont, "Chapitre," 806; T. A. Burkill, *Revelation*, 99, 110-11; M. Hermaniuk, *Parabole*, 304, 314; J. Delorme, "Aspects," 89, 91; M.-J. Lagrange, "Le but," 28; J. Gnilka, *Verstockung*, 48-50. H. Windisch ("Die Verstockungsidee in Mc 4,12 und das kausale *hina* der späteren Koine," *ZNW* 26 [1927] 203-9) decisively refutes the suggestion that it has a causal sense.

[97] *Parables*, 17; see also T. W. Manson, *Teaching*, 78-79. W. Marxsen ("Erklärung," 25-26) and A. Suhl (*Funktion*, 150) think it possible that Mark understood it in the sense suggested by Jeremias.

[98] Jeremias admits that he does not know in which sense the word was understood by the targumist.

[99] For references, see note 96. A. Jülicher (*Gleichnisreden* I, 131) observes that the sentence becomes the wildest yes-no construction if we should admit that *mēpote* has the sense of "si quando." W. Marxsen ("Erklärung," 26) thinks that the meaning "unless" is consonant with the Marcan context, according to which the withholding of the mystery is not permanent and definitive (4:21-25).

[100] See Blass-Debrunner-Funk, *Grammar*, § 370(3); in Heb 9:17 *mēpote* occurs in conjunction with *epei* and has a consecutive sense.

permissible. When the narrower context of the logion, i.e., ch. 4, is considered, it would seem that the stress lies on the first two members of the verse. Vs. 13, in particular, suggests that while those outside do not understand, the disciples do perceive the meaning of the parables because Jesus supplies the explanation. Matthean and Lucan parallels which omit vs. 12c may serve as a confirmation. Yet there are strong reasons to hesitate about this attempt at solving the problem. What stands in its way is, above all, Mark's concept of understanding which does not permit a separation of 12ab from 12c; the verse is of one piece as far as Mark is concerned. But even apart from that, there are reasons which speak strongly against the supposition that Mark wishes to accentuate the failure to understand at the expense of impenitence and non-forgiveness. There is, first, the absence of *gnōnai* in the Marcan version of the logion, a fact which must be considered and given proper weight; secondly, the *mystērion* cannot simply be equated with the explanation of parables. There is, thirdly, on the assumption that understanding is given the chief emphasis in vs. 12, a contradiction between vss. 10-12 and 13: immediately after having been assured that the understanding of the mystery has been granted them the disciples are reproached for their failure to understand the parable of the Sower. Finally, Mk 8:14-21 is to be considered. One of the reasons why Jeremias[101] thinks that the logion in 4:11-12 is pre-Marcan is its interpretation of Isa 6:9-10 "to mean solely *hoi exō* whereas Mark himself extends it to the disciples," as is shown by 8:14-21. That passage, whose formation is to be attributed to Mark's redactional work,[102] applies to the disciples (8:18) the words which 4:12 predicates of "those outside." It contains the severest and most prolonged of all the rebukes which Jesus administers to the disciples in the Second Gospel. Yet despite 8:18 we feel that Jeremias' argument is superficial. Missing from 8:18 is the last line of 4:12 which contains the damning characteristic of those outside, viz., the fact that they will not "repent and be forgiven." The disciples may not understand, but the entire Gospel breathes with the confidence that they will eventually understand, that they will repent and be forgiven. The difference between those outside and the disciples is to be sought primarily in this: the disciples will be converted and forgiven, those outside will not. There is, moreover, a certain similarity observable in the structure of 4:1-13 and 8:1-21. The Feeding of the Four Thousand in the latter passage corresponds to the parable of the Sower in the former. There follows Jesus' refusal of a sign to the Pharisees (8:11-13), corresponding to 4:11b,12. Finally there

---

[101] *Parables*, 15, n. 12.
[102] See V. Taylor, *Mark*, 83; E. Schweizer, *Mark*, 160-61.

comes the rebuke of the disciples. The suggestion in both sections seems to be that, while there is hope that the disciples' hardness of heart will be overcome, there is no hope for those outside because they have closed themselves, or have been closed by God, to the power emanating from the mighty words and works of Jesus. Thus, while some elements of the narrower context seem to imply that Mark's interest in 4:12 lies in non-understanding of the parables, other considerations, taken from the narrower as well as the broader context, suggest that he lays the stress on the divine judgment of impenitence and non-forgiveness.

The problem remains: the image which Mark paints of the crowd in the rest of the Gospel does not harmonize with the one he paints in 4:10-12. One could explain this lack of harmony by suggesting that Mark is a somewhat awkward compiler who has failed to master various strands of tradition which he incorporates in his work. This type of solution is, in these days of redaction criticism, an admission of defeat or, at best, a last resort in which we may take refuge only after all other possibilities have been explored and found unacceptable. Another solution could take the function of the crowds in the Gospel to be purely "christological," i.e., the crowd is merely a chorus of no value of its own, serving as a sounding board to echo and amplify the words and works of Jesus. Yet it appears too frequently, and is endowed by the evangelist with characteristics which are too clearly delineated to permit this hypothesis.

The solution to the problem lies in the phrase "those present with the Twelve" of 4:10. R. P. Meye has subjected this phrase to a close scrutiny.[103] His concern is to disprove the thesis that in 4:10 we meet another group of disciples besides that of the Twelve. For Mark, according to Meye, "the Twelve are *the* disciples—those who constitute *the school* of Jesus."[104] This interpretation of Meye is, in our view, quite correct. Most welcome too is his well-reasoned rejection of the thesis that the Second Gospel is a polemic against the disciples.[105] However, we cannot agree with his claim that the phrase "those present with the Twelve" refers to a smaller group within the number of the Twelve. His argumentation based on 13:3 fails to prove his thesis; he also fails to do justice to the many-layered character of the Gospel message. In Chapter I we attempted to indicate the continual

---

[103] "Messianic Secret and Messianic Didache in Mark's Gospel," *Oikonomia*, 61-66; "Those about Him with the Twelve," *SE* 2 (= *TU* 87), 212-17; *Twelve*, 152-56.

[104] "Secret," 63 (his italics).

[105] This view has been voiced recently by J. B. Tyson, "The Blindness of the Disciples in Mark," *JBL* 80 (1961) 266-68; T. J. Weeden, "Heresy," *passim*; S. G. F. Brandon, "The Apologetical Factor in the Markan Gospel," *SE* 2 (= *TU* 87), 42-46; S. Sandmel, "Prolegomena to a Commentary on Mark," *JBR* 31 (1963) 298.

interplay of the past and present in the Gospel: while the past determines the evangelist's present, his present colors the past. How various levels of Mark's message mesh is particularly evident in 13:37; 8:34 and, we think, in 4:10. Mk 13:37, "What I say to you, I say to all: Be on guard!", transfers the reader abruptly from the plane of instruction given to four chosen disciples in the past to his own time and to the needs of the hour in which he lives.[106] Mk 8:34a, whose resemblance to 4:10 is striking, likewise stresses the universal applicability of the words which follow.[107] The difference between this verse and 13:37 lies in the method of bringing out the general validity of the message: in one case "all" are being addressed, in the other the "crowd" is spoken to. The phrase "those present with the Twelve" of 4:10 serves the same purpose: Mark has the Christian community in mind.[108] Neither another group of Jesus' disciples nor a smaller group within the number of the Twelve, but Mark's Christian readers are being assured that they are the chosen recipients of the mystery of the kingdom.

What significance does this suggestion have for the meaning of *hoi exō?* The term is clearly contrasted with *hymin*. The redactional vs. 10 tells us who these *hymeis* are: not only the Twelve, but the Christian community with them. The contrast thus suggests that "those outside" should not be looked upon merely as a group or groups depicted as playing a role during the earthly life and ministry of Jesus. Those outside, like those about him with the Twelve, are not a reality of the past alone but of the present also; primarily of the present, in fact, since vs. 10 shows that the words of the logion are addressed chiefly to the Christian community. The community now shares what was, before the death and resurrection of Jesus, the exclusive privilege of the Twelve; the community is "with the Twelve" because it has accepted the revelation which Jesus communicated to them during his life on earth. Since the term *hoi exō* is not Mark's own, but has been taken by him from the tradition, we may surmise that it describes, as it does in Pauline writings, those who do not belong to the Christian community. The context within which Mark places the logion suggests, however, that he adds further precision to the term: the OT quotation in vs. 12,

---

[106] It is normally admitted that in this verse Mark is directly addressing his community; see R. Pesch, *Naherwartungen,* 202; W. Grundmann, *Markus,* 272; E. Schweizer, *Mark,* 280.

[107] See R. Schnackenburg, "Vollkommenheit," 430-31; W. Grundmann, *Markus,* 174. The redactional *proskalesamenos* and *hoi mathētai,* as well as the sudden and unexpected presence of the crowd in the company of the Twelve, betray Mark's editorial hand and intention.

[108] See W. Marxsen, "Erklärung," 23; G. Minette de Tillesse, *Secret,* 177, 274.

as well as the interpretation of the Sower parable, suggest that "those outside" are men who have heard the Christian message but have refused to believe. They are like the enemies of Jesus during his life on earth who indeed knew what he claimed to be, but refused to understand his words and works in a manner which leads to salvation; their understanding became their condemnation and destruction. The crowd belongs neither to "those outside" nor to the disciples. It has not been called by Jesus, it does not yet know who Jesus claims to be; it therefore cannot accept or reject him definitively.

(C) IN PARABLES

(1) E. Schweizer is of the opinion that *parabolē* should be considered to be a Marcan redactional term.[109] There is little doubt, of course, that the term was already present in the pre-Marcan tradition; yet it is just as evident that it occurs in a number of passages which owe their origin, or at least their present form, to Mark's editorial work. To gain an insight into Mark's concept of the parable, we must examine the verses in which the term occurs.

Mk 3:23 is, without doubt, a redactional passage; *proskalesamenos* and *elegen autois* are sure signs that Mark is personally at work. A very similar verse is 12:1a; *ērxato* is typically Marcan, and the fact itself that it forms the introduction to a unit gives reason to think that the verse is redactional.[110] 3:23a and 12:1a have a number of features in common. They introduce addresses given to the enemies; Jesus is portrayed as speaking only, not as teaching or preaching the word; surprisingly 12:1a has the plural of the term whereas only one parable follows; something similar may be detected in 3:23 also, for the redactional 3:30 reaches back to 3:22, seeming to imply that Mark looked upon 3:23b-29 as a single parable. The most likely explanation of this feature is that we have here a generalizing plural: Jesus addresses them in a parabolic manner.[111] Only in these two verses is Jesus portrayed as speaking "in parables" to his enemies; both passages thus introduced close with an unmistakable condemnation of those addressed. This automatically recalls the phrase "in parables" of 4:11.

The passage which should be considered next is 4:33-34. Is it redactional?

---

[109] "Anmerkungen," 97, and "Contribution," 423.

[110] See K. Grobel, "Idiosyncracies of the Synoptists in Their Pericope Introductions," *JBL* 59 (1940) 405-10.

[111] See W. Grundmann, *Markus*, 238.

R. Bultmann[112] represents the common view, suggesting that vs. 33 formed the conclusion of the pre-Marcan collection of parables which has found its way into ch. 4 of the Second Gospel. Equally commonly assumed is that vs. 34 is, in whole or in part, a Marcan correction of the previous verse. Some authors, however, disagree with the commonly held view. E. Schweizer[113] considers both verses redactional, and finds in them Mark's theology of parables which is, according to him, one of the most important aspects of Mark's message. To support his thesis of the redactional nature of these verses, he points to the phrase *elalei autois ton logon* which occurs also in 2:2 and 8:32; the latter passage is of particular weight, since it forms the counterpart to 4:33-34: Jesus no longer speaks in parables, but "plainly." Schweizer's suggestion is based primarily on the fact that the term "parable" frequently occurs in redactional passages. Yet the mere presence of the term is not sufficient to conclude that the verse containing it is redactional; of this Schweizer himself is fully aware.[114] In order to establish the editorial character of a given verse, its entire vocabulary must be examined. For 4:33-34 this examination has been made by J. Gnilka,[115] who observes that vs. 34 contains three *hapax legomena* in Mark: *chōris, idioi mathētai* and *epilyein*. Despite the attempts of G. Minette de Tillesse[116] to show that these *hapax legomena* need not be taken as certain evidence of the pre-Marcan origin of the verse, we feel that it is methodically more correct to conclude, with Gnilka, that vs. 34 is not a product of Mark's redaction.[117] Unless we pay attention to vocabulary in such sensitive cases as is vs. 34, we may fall prey to our own preconceptions about Marcan thought. The argument based on similarity of thought in 4:11 and 34[118] does not prove the redactional origin of the latter verse. We would agree that vs. 34 is an intentional correction of vs. 33. It is difficult to attribute both verses to the same hand; for vs. 33 clearly suggests that the crowds

---

[112] *History*, 332, 444; see also E. Sjöberg, *Menschensohn*, 168; C. Masson, *Paraboles*, 47-48; J. Jeremias, *Parables*, 14, n. 11; W. Marxsen, "Erklärung," 19-20; G. Minette de Tillesse, *Secret*, 181-85. We would disagree with G. H. Boobyer, "The Secrecy Motif in St. Mark's Gospel," *NTS* 6 (1959-60) 59, who presents as common the opinion that vs. 33 belongs to the editorial material.

[113] *Mark*, 105-7. E. Trocmé (*Marc*, 127, n. 71) also thinks that vss. 33-34 come from Mark's pen; so also H. J. Ebeling, *Messiasgeheimnis*, 189, and J. Dupont, "Chapitre," 804.

[114] See "Anmerkungen," 97.

[115] *Verstockung*, 59-60.

[116] *Secret*, 183-84.

[117] *kat' idian*, however, may well be owing to Marcan redaction.

[118] G. Minette de Tillesse, *Secret*, 182-83.

were able to understand, whereas vs. 34 presupposes that parables had to be explained in order to be understood.[119] It is most natural to assume that vs. 34 was added by the pre-Marcan redactor who had inserted the interpretation of the Sower into the preexisting collection of parables of growth.[120] Thus neither verse is Marcan—for if vs. 34 is not redactional, vs. 33 is less so; it is hardly likely that Mark would be suggesting that crowds were able to understand, even to a limited degree. Yet the evangelist gladly subscribes to vs. 34—he probably added the phrase *kat' idian* —since it expresses a procedure which he frequently adopts or constructs. Our conclusion, then, would be that "parable" in 4:33-34 is not redactional.

This brings us to a consideration of 4:2a. It is generally agreed that in 4:1-2 we have to do with a fair amount of Marcan redactional work.[121] Vs. 1 is to be ascribed to the evangelist;[122] such typically Marcan words as *palin, ērxato, didaskein, synagetai* (historic present), *thalassa*,[123] and *epi tēs gēs*[124] stamp it as redactional. Vs. 2 is likewise considered by many to be redactional;[125] its second half containing, as it does, the characteristically Marcan *kai elegen autois* and *didachē* is surely the evangelist's formulation. Vs. 2a, however, could well have stood in the source. Since vs. 33 very likely formed the conclusion of the first stage of the pre-Marcan collection of parables, it is reasonable to suppose that the collection also had an introduction; vs. 2a probably served that purpose.[126] This is confirmed by the strange formulation of the verse. M.-J. Lagrange remarked[127] that vs. 2 would contain "un pléonasme intolérable," should 2b be considered to be a simple repetition of 2a. Mark, moreover, does not indulge in fanciful parallelisms when introducing his speech material. It is thus likely that vs. 2a was in the source already, and that Mark added

---

[119] For a discussion of *epilysis*, see J. Gnilka, *Verstockung*, 62-64.

[120] See W. Marxsen, "Erklärung," 20; his opinion that vs. 34a is Marcan is weakened by J. Gnilka's observation that *chōris* is a *hapax legomenon* in Mark.

[121] Here we feel that G. H. Boobyer ("Secrecy," 59) limits the common opinion too severely when he claims it finds editorial material in 4:1a only.

[122] See R. Bultmann, *History*, 444; J. Dupont, "Chapitre," 804; F. D. Gealy, "Composition," 40; H. J. Ebeling, *Messiasgeheimnis*, 189; E. Trocmé, *Marc*, 127, n. 71; W. Marxsen, "Erklärung," 19; J. Jeremias, *Parables*, 14, n. 11; D. E. Nineham, *Mark*, 134.

[123] See W. Marxsen, *Mark*, 57-66.

[124] See below the treatment of the parable of the Mustard Seed, pp. 124-25.

[125] See the authors mentioned in note 122, except for Bultmann who thinks that vs. 2 belonged to the source. See particularly T. A. Burkill, *Revelation*, 98, and W. Marxsen, "Erklärung," 19, whose view we share and present in greater detail.

[126] See W. Marxsen, "Erklärung," 20.

[127] *Marc*, 90.

vs. 2b in order to emphasize that parables were a manifestation of Jesus' teaching to the disciples and the crowds.

Mk 4:30 stems from the source which Mark employed; this is suggested by the double question in the Q form of the Mustard Seed parable at Lk 13:18. The term parable in this verse retains the fundamental meaning of the Hebrew *māšāl*, namely that of comparison.[128] In 13:28a, however, the term could well be redactional;[129] the verse forms the transition from one section of the apocalyptic discourse to another. 12:12 is very likely redactional.[130] We have already discussed 4:10-13, and uncovered many Marcan interventions. Should 7:17 be looked upon as redactional?[131] We have opted for the opinion according to which 4:10,13 are largely redactional. W. Marxsen argues, however, that at 7:14-23 and 4:10-20 (minus 11-12) we find a traditional formulation, strongly edited by Mark in 4:10-13 and somewhat less so in 7:17-18. Since interpretations of parables in 4:14-20 and 7:19-23 are pre-Marcan[132] it is likely that some formula joined the parables with their interpretations in the source. Yet the sequence, disciples' question—rebuke of Jesus, is a very regular feature of the Second Gospel. This seems to indicate that the passages under consideration are Mark's own; in case they are not, they are no less indicative of his thought since he has obviously adopted them as one of the means of shaping his Gospel.

Consequently, in our opinion these cases of the term "parable" derive from Mark: 3:23; 4:10,13; 7:17; 12:1,12; 13:28. The following probably stem from various sources at his disposal: 4:2,11,30,33,34. We would thus not ascribe to the term the same predominantly redactional character as does E. Schweizer.

(2) It is not necessary to enter into the discussion about the parables of Jesus as such. A number of studies in the last few decades have examined the question from various points of view. The direction taken by E. Lohmeyer, E. Fuchs, and E. Jüngel[133] seems to be the most fruitful, leading

---

[128] See A. R. Johnson, "*mšl*," *Wisdom in Israel and the Ancient Near East, presented to H. H. Rowley* (*VTSup* 3; eds. M. Noth and D. W. Thomas; Leiden: Brill, 1960) 162-63; F. Hauck, *TDNT* 5, 747-48.

[129] R. Pesch, *Naherwartungen*, 176; V. Taylor, *Mark*, 520, 642.

[130] V. Taylor, *Mark*, 477; G. Minette de Tillesse, *Secret*, 219.

[131] For the view that the formulation is pre-Marcan, see W. Marxsen, "Erklärung," 16-17; for the opposite view, see G. Minette de Tillesse, *Secret*, 174-75.

[132] For 4:14-20, see above; for 7:19-23, see V. Taylor, *Mark*, 342-43, 346-47.

[133] E. Lohmeyer, "Vom Sinn der Gleichnisse Jesu," *Urchristliche Mystik: Neutestamentliche Studien* (Darmstadt: Wissenschaftliche Buchgesellschaft, 1958) 123-57; E. Fuchs, *Hermeneutik* (Bad Cannstatt: Müllerschön, 1963) 211-30; "Bemerkungen zur Gleichnisauslegung," *Zur Frage nach dem historischen Jesus: Gesammelte Aufsätze*

us out of the blind alley of the *Sitz im Leben Jesu* into which C. H. Dodd and J. Jeremias conducted us. (This remark is not intended to deny their great contributions to our understanding of the parables.) Lohmeyer, Fuchs, and Jüngel have also begun to tread the path away from the arbitrariness to which the search for the *tertium comparationis* has frequently condemned us.

But what is Mark's concept of the parable? The redactional passages in which the term occurs enable us to proceed with some confidence. A common opinion considers Mark's parables to be riddles: he saw them as means designed to bring about blindness in "those outside," or, at best, thought that the result of their being heard by those outside was blindness. This understanding of Marcan parable is derived from the contrast, in 4:11, between "to you the mystery of the kingdom of God has been confided" and "to the others outside it is all presented in parables." This contrast alone is not sufficient to explain the common understanding of the Marcan parable. Decisive is another presupposition which is taken as self-evident and considers the phrase "to you the mystery of the kingdom of God has been confided" as a reference to communication of hidden knowledge. Attitudes taken toward this parable theory go all the way from F. Grant's remark that "Mark's theory can only be described as perverse,"[134] to E. F. Siegman's efforts to show that "riddle," as a translation of *ḥîdāh*, misses the point, in some OT instances at least.[135]

It would be difficult to dispute this common opinion. It is well-known that in the Greek OT *parabolē* generally translates the Hebrew term *māšāl*. And it has long been recognized that in the NT *parabolē* can have the same variety of meanings as the OT *māšāl*.[136] One of the meanings of this OT literary form is that of enigmatic statement. This is shown by such OT texts as Prov 1:6 and Ps 49:5 where *māšāl* and *ḥîdāh* are parallel, and by the easy conflation of *ainigma* and *parabolē* in Sir 39:3; 47:15; indeed, in

(Tübingen: Mohr, 1965) 2, 136-42; E. Jüngel, *Paulus und Jesus: Eine Untersuchung zur Präzisierung der Frage nach dem Ursprung der Christologie* (Hermeneutische Untersuchungen zur Theologie 2; Tübingen: Mohr, 1967) 87-215. A fine summary of these views is given by J. Blank, "Marginalien zur Gleichnisauslegung," *Bibel und Leben* 6 (1965) 50-60.

[134] "The Gospel according to St. Mark," *The Interpreter's Bible* (New York: Abingdon, 1951) 7, 700.

[135] "Teaching," 175-76.

[136] See M.-J. Lagrange, "Le but," 5, 13; R. A. Stewart, "The Parable Form in the Old Testament and the Rabbinic Literature," *EvQ* 36 (1964) 134, 140; E. F. Siegman, "Teaching," 175-76; R. E. Brown, "Parable and Allegory Reconsidered," *NovT* 5 (1962) 37-38; C. E. B. Cranfield, *Mark*, 148, 159; G. von Rad, *Theology* 2, 302-3; and particularly E. Lohmeyer, "Gleichnisse."

Sir 47:17 *parabolē* translates the Hebrew *ḥîdāh*. That Mark himself is no stranger to the breadth of meaning the term can have is shown by the editorial insertion of vss. 11-12 and 21-25 in his parable chapter; the typical proverbs in vss. 21-25 should be categorized along with the parables proper in ch. 4, viz., vss. 3-8; 26-29 and 30-32. Mark has no difficulty in referring to the logion in 7:15 as parable (7:17). For that reason it is unnecessary to suppose with W. Marxsen[137] that the NT parable had to undergo a drastic change in order to become a *māšāl* by the time that Mark undertook to write his Gospel. Rather, it was a *māšāl* all along. Marxsen does not appreciate the elasticity of this OT literary form. To mention but one example, the term is used, according to A. R. Johnson[138] "in the book of Ezekiel to denote a composition which offers in . . . colourful and . . . elaborate allegorical language a forecast of some impending event which is . . . envisaged by the speaker in terms of Yahweh's purposeful action—in each case, as it happens, a warning of imminent doom for Ezekiels's contemporaries in the southern kingdom." He refers to Ezek 17:1ff.; 21:1-5; 24:3ff. In the intertestamental literature the term has undergone further development.[139]

In the NT *parabolē* has, of course, travelled its own road and been subject to the stresses peculiar to this tradition. The pre-Marcan explanations at 4:14-20 and 7:18-23 are sufficient evidence of the fact that the community had to interpret parables to itself long before Mark decided to write. Even when spoken by Jesus, parables were not as easily and automatically understood as is sometimes assumed. Since they spoke of God's kingdom and its coming, realities of their nature ineffable, they revealed and concealed at the same time.[140] To quote E. Jüngel, "the parable can conceal only because it is designed to disclose."[141] For it is no mere illustration of a general truth or a moral maxim, but a call for decision and commitment.[142] The loss of its original context, the fact that it afforded various possibilities of application once the original context was lost, and the allegorizing and parenetic tendencies of the early Christian proclamation conspired to create the impression that a parable needed to be interpreted in order to be understood.[143]

---

[137] "Erklärung," 21-23.
[138] "Mashal," 168; see also G. Minette de Tillesse, *Secret*, 206-7.
[139] See G. von Rad, *Theology* 2, 302-3; F. Hauck, *TDNT* 5, 747-51.
[140] See E. Lohmeyer, "Gleichnisse," 142-45; J. Schniewind, *Markus*, 61-65; E. Jüngel, *Paulus*, 120-39.
[141] *Paulus*, 137 (my translation).
[142] See E. Linnemann, *Gleichnisse Jesu: Einführung und Auslegung* (Göttingen: Vandenhoeck & Ruprecht, 1966) 38-41.
[143] See W. Marxsen, "Erklärung," 22; D. E. Nineham, *Mark*, 128-31.

It was natural for Mark who was writing a book of hidden epiphanies[144] to seize on this, by his time already traditional, datum. In 4:13,34 and 7:18a he makes it abundantly clear that he views the parable, without explanation, as unintelligible even to the inner circle of disciples. In this, as in other respects, Mark is not creating something entirely new; rather, he is editorially emphasizing a traditional feature. It is against this background that we should read the phrase *en parabolais . . . ginetai* of Mk 4:11b.

The meaning of this phrase has been carefully investigated by J. Jeremias, E. Sjöberg and J. Gnilka. According to Jeremias[145] the verb *ginetai* "renders an Aramaic $h^a w\bar{a}\ l^e$ 'to belong to somebody, to be assigned to somebody'. It is followed by b^e (*en* Mk 4,11) for instance in Gen 15,1. . . . Hence Mk 4,11b must be translated: 'But to those who are without all things are imparted in riddles,' i.e., they remain obscure for them." Sjöberg[146] agrees with Jeremias' opinion in general but is not convinced by his rendering of *ginesthai en parabolais* as "to remain obscure." The parallels to which Jeremias refers are of a different nature: in all of them *en* retains the force of a local preposition. He prefers to understand the verb *ginesthai* in the sense of "to become"; Jesus' words become riddles to those outside because they do not understand the mystery of the kingdom. Gnilka[147] comes, by another route, to the same conclusion as Sjöberg. By referring to a number of parallel cases he contends that *ginesthai* should be rendered "to become, to turn into" with an inchoative force: "to those outside everything turns into riddles." Jeremias' or Sjöberg's opinion may well apply to the meaning of the logion immediately after it was translated from the Aramaic. We wonder, however, whether we should be allowed to assume that Mark understood it that way. Was this Aramaic construction so commonly used by Greek-speaking persons whose language was still under the influence of their Semitic background that we can infer, without adequate proof, that it was understood in its Aramaic sense? We feel that Jeremias and Sjöberg owe us such proof if we are to accept, for Mark, the meaning which they propose. Gnilka's method of arriving at his conclusion, however, is not convincing. The parallels which he adduces[148] do not seem to be real parallels to Mk 4:11b. From the examples which he offers we can conclude that *ginesthai en* with a noun can be understood in the sense "to become" only when the noun itself already contains some reference to a state or activity,

---

[144] M. Dibelius, *Tradition*, 230.
[145] *Parables*, 16-17.
[146] *Menschensohn*, 223-25.
[147] *Verstockung*, 26-28.
[148] *Verstockung*, 27, n. 23; cf. Bauer-Arndt-Gingrich, *Lexicon: ginomai* II, 4a.

natural or supernatural, of the subject.[149] To prove the point which Gnilka is trying to make, Phil 2:7 should read *en anthrōpois genomenos*, and not, as it does, *en homoiōmati anthrōpōn genomenos*.

We would propose a simpler solution. The preposition *en* should be taken in its instrumental or sociative function.[150] There are at least two very close parallels to Mk 4:11b in the NT: 1 Thes 1:5 *to euangelion hēmōn ouk egenēthē eis hymas en logō monon*; and 2:5 *oute gar pote en logō kolakeias egenēthēmen*. It seems more likely that Mark understood the preposition *en* in the sense which it has in 1 Thessalonians than in that proposed by Jeremias and Sjöberg. Where Jesus is portrayed as speaking "in his teaching" or "in parables" the two terms seem to have an instrumental function. We would thus translate vs. 11b: "to those outside everything comes in parables." The context itself in the logion, in ch. 4, as well as the rest of the Gospel, indicates that the parable in this case is to be understood as a dark, unintelligible statement. Against Jeremias,[151] who thinks that "the contrasting parallelism of the two clauses vs. 11a and vs. 11b requires that *mystērion* and *parabolē* should correspond," we agree with Sjöberg[152] who sees the contrast between the gift of the mystery and the utterance in parables. For mystery too remains unintelligible until it is given.

(3) This, however, does not exhaust the message of the logion; the unintelligibility of the parables serves a purpose, viz., that of blinding and condemning those outside. To perceive the relationship between unintelligibility and impenitence, we must pause to consider Mark's concept of understanding as well as his concept of Jesus' teaching.

While the demons and the enemies of Jesus are said to have understood, the disciples are reproached for their failure to understand and thereby are exhorted to arrive at understanding. These two types of understanding, that which the demons and enemies possess and that to which the disciples are exhorted, cannot be distinguished terminologically. Such verbs as *oida* and *ginōskō* are used indiscriminately of the disciples, the enemies and others; *oida* is used of the demons also; the only verb which may be approaching the status of a technical term to describe the understanding of the disciples is *syniēmi*.[153] Yet the two types of knowledge are undoubtedly fundamentally distinct. The knowledge of the enemies brings about Jesus' death (3:6;

---

[149] The examples he gives are the following: *en agōnia, en pneumati, en heautō, en penthei, en synnoia, en tyrannidi kai stasesi, en dialogismō, en orgē, en nosō, en ethei*.

[150] See M. Zerwick, *Biblical Greek* (Scripta Pont. Inst. Biblici 114; Rome: Biblical Institute, 1963) 116-19.

[151] *Parables*, 16.

[152] *Menschensohn*, 224.

[153] It occurs in Mk 4:12; 6:52; 8:17,21; 7:14.

12:12; 14:62-64); the understanding of the disciples, on the contrary, is the response demanded of them to Jesus' words and works (4:13; 7:14; 8:17-21). In our search for Mark's concept of the disciples' understanding, two redactional verses, 6:52 and 8:17, deserve special attention. In both of them failure to understand and hardness of heart are clearly associated.[154] We meet here the OT theme of the stiff-necked resistance to God's will;[155] if any confirmation of this is needed, it is supplied by the quotation of Isa 6:9 in 8:18, the OT text in which the power of the prophetic word is depicted as reaching the very hearts of the Israelites and hardening them against subjecting themselves to the word of God.[156] That Mark understood the prophetic word in substantially the same sense is shown by 4:11-12. The disciples' failure to understand is thus not the result of an intellectual inability to grasp the literal meaning of Jesus' words, but a moral blindness that manifests itself in their inability to tear themselves away from earthly concerns and trust the divine power and guidance revealed in Jesus' word and work. They are in grave danger of becoming like the enemies of Jesus who are quite able to perceive intellectually the meaning of Jesus' words, but whose radical refusal to admit the presence of divine saving power in him condemns them to total blindness and impenitence.[157] Parables and miracles "are signs with a potency either for revelation or for hardening the heart. Even the disciples are in grave danger of completely misapprehending their meaning."[158]

We should arrive at the same result if we consider the association of the disciples' failure to understand with their fear and amazement. Fear, for Mark, is a sign of defective faith. Fear and awe "are attributed to unclarity as to Jesus; fear is due to lack of faith (4:40), confusing Jesus with a ghost (6:49), and not knowing that 'it is I' (6:50); awe expresses itself with the question 'Who is he?' (4:41), and is explained (6:52) as follows 'for they did not understand about the loaves, but their heart was hardened.'"[159] Fear and lack of understanding are associated in the redactional verses 9:6,32.[160] How Mark conceived the disciples' non-understanding is par-

---

[154] See J. M. Robinson, *Problem*, 76; it will be noticed that our discussion owes a great deal to his treatment of the question in *Problem*, 73-78.
[155] K. L. and M. A. Schmidt, *TDNT* 5, 1027.
[156] See G. von Rad, *Theology* 2, 151-55.
[157] See W. Grundmann, *Markus*, 143-44; E. Schweizer, *Mark*, 141-42, 160-61; D. E. Nineham, *Mark*, 181-82, 213-16; V. Taylor, *Mark*, 331, 366; J. M. Robinson, *Problem*, 76.
[158] C. K. Barrett, *Holy Spirit*, 89.
[159] J. M. Robinson, *Problem*, 69-70.
[160] For the redactional character of these verses, see R. Bultmann, *History*, 261, 332; E. Schweizer, *Mark*, 180-82; W. Grundmann, *Markus*, 182.

ticularly evident in 9:30-36. In vs. 32 they are said to have failed to understand the prediction of the passion and to have been afraid to ask Jesus about it. A concrete example of their failure to understand is given in the redactionally formulated vss. 33-34.[161] Mark is not suggesting that the literal meaning of the passion prediction escaped them; rather, their non-understanding consisted in failing to perceive the bearing of Jesus' death upon their own lives and attitudes: "on the way they had been arguing about who was the most important."[162] The third prediction of the passion (10:32-34) is likewise accompanied by fear and awe, as well as by a blatant case of moral blindness on the part of the sons of Zebedee and other disciples (10:35-45); again Jesus must outline the consequences which his death has for the life of his disciples: "whoever wants to rank first among you must serve the needs of all." Peter's strong protest which follows the first prediction (8:31) is a clear indication of the same moral blindness.

Hence Mark associates the failure to understand with fear and hardness of heart, and contrasts these attitudes with those of faith and insight. Insight, in the OT and in Qumran literature, is a gift of God for which man must, nonetheless, strive.[163] The same notion is present in Mark: the demand that the disciples strive for insight is implied by Jesus' reproachful questions, by the repeated *oupō* in connection with faith and understanding (4:40; 8:17,21), and by the continual exhortation to listen to his words (4:3,9,23,24; 7:14). On the other hand, Mark knows that this insight can only be given by God. He states it in 4:11, and suggests it—in the opinion of many exegetes—by means of the miracles reported at 7:31-37 and 8:22-26. These miracles have for the evangelist a symbolic significance: the divine power residing in Jesus is about to open the eyes and ears of the disciples' minds and hearts as it gives sight to the blind and hearing to the deaf. These miracles are a prelude to Caesarea Philippi and to the divine voice on the Mount of Transfiguration.[164] The understanding demanded of the disciples in the Second Gospel thus consists in knowledge, but it is a knowledge which is primarily an acceptance of Jesus as the Messiah and Son of God, a willingness to take up one's cross and follow him, and an obedience to his word.

However, do not interpretations of the parables, which Mark either gives

---

[161] That these verses are redactional has been shown by V. Taylor, *Mark*, 403; R. Bultmann, *History*, 149. See below, ch. III.

[162] See D. E. Nineham, *Mark*, 250-51.

[163] H. Conzelmann, *TWNT* 7, 888-89.

[164] See A. Kuby, "Konzeption," 52-53, 58; D. E. Nineham, *Mark*, 218; E. Schweizer, *Mark*, 163-64; W. Grundmann, *Markus*, 164-65; J. M. Robinson, *Problem*, 77; W. Wilkens, "Gleichniskapitel," 315.

or refers to, militate against this concept of understanding? Interpretation seems to be a matter of intellectual clarification; what is dark in the parable is presumably made clear in the interpretation. In answer to this objection one should note, first, that in the book of *Enoch* the term *māšāl* applies to the visions of what is to happen at the end as well as to the interpretations of these visions.[165] Interpretation thus shares in the nature of the *māšāl*. One must, secondly, ask what precisely an interpretation is. It is not primarily the translation of the imagery of the parable into abstract or non-metaphorical terminology; it is, rather, a further appeal to the listener. Its chief purpose is to make the listener or the reader aware of the eschatological message of the parable.[166] It is a parenetic application much more than a rational exegesis of the images and symbols. Mark's interpretations are certainly of this type: 4:14-20 is primarily an admonition to the Christian community not to lose heart and not to fall away because of the failure of God's word in so many cases.[167] It closes with the solemn assurance, and without explaining the image, that there are men in whom the word will meet with success, for the word is stronger than all the obstacles it encounters.[168] The chief aim of the interpretation of the "parable" in 7:15 is the practical, and for many Christians painful, conclusion that all foods are clean. For it seems that the short statement at the end of 7:19, *katharizōn panta ta brōmata,* is Mark's redactional insertion into a traditional pericope.[169] The interpretation of the prohibition of divorce in 10:11-12 is

---

[165] For visions, see 38:1; 45:1; 58:1; for interpretations, see 43:4. See also F. Hauck, *TDNT* 5, 749-50; W. Marxsen, "Erklärung," 23; L. Cerfaux, "Secrets," 245.

[166] See E. Lohmeyer, "Gleichnisse," 154.

[167] See R. Schnackenburg, *Das Evangelium nach Markus* (Geistliche Schriftlesung 2/1; Düsseldorf: Patmos, 1966) 107-9; S. Schulz, *Botschaft,* 150.

[168] See J. Schniewind, *Markus,* 67.

[169] See V. Taylor, *Mark,* 345; C. E. B. Cranfield, *Mark,* 241; J. Schmid, *Mark,* 139; E. Lohmeyer, *Markus,* 142; J. M. Robinson, *Problem,* 76; C. H. Turner, "Usage," *JTS* 26 (1924-25) 149. Well known is the suggestion made by M. Black, *An Aramaic Approach to the Gospels and Acts* (3d ed.; Oxford: Clarendon, 1967) 217-18, according to whom the participle *katharizōn* is a mistranslation of an Aramaic passive which referred originally to food. The hypothesis is highly unlikely; if Mark was translating from the Aramaic directly—which is not probable—he, on M. Black's own admission, misunderstood or reinterpreted it. We quote G. Minette de Tillesse (*Secret,* 146): "Avouons que la signification donnée ainsi au logion est pour le moins assez platte, pour ne pas dire vulgaire, et il n'était vraiment pas nécessaire, semble-t-il, pour Jésus, d'attirer ses disciples "à l'écart de la foule . . ." pour leur expliquer cela. D'ailleurs le texte syriaque sur lequel s'appuie Black est presque seul à présenter cette leçon. . . ." There is another, more serious, objection to the opinion that Mark inserted the brief comment: in 7:3-4 there is a description of Jewish practices of purification, presumably intended for an audience which is no longer familiar with them. This objection has

simply a restatement of the same prohibition in terms of a different environment. Mk 10:24b,25 serve to stress more strongly the saying of 10:23. The parable of the fig tree at 13:28 is "explained" in vs. 29 by an exhortation to watch for the signs of the end.

Mark thus sees a close nexus between non-understanding and impenitence. In fact, for him non-understanding and impenitence are two aspects of the same reality: that of being imperviously closed to God's saving action in Jesus Christ. "Those outside" may well perceive the literal meaning of the words addressed to them, but they do not believe their message and refuse to accept the One speaking to them as sent by God to bring about eschatological salvation. Disciples too are in danger of failing to overcome their blindness; only God's gift of the mystery of the kingdom can save them from it; they can be saved by knowing Jesus as the Son of God and confessing him as the Messiah. This insight, knowledge and confession, however, are not a theoretical exercise of their mental powers, but a matter of decision which gives a totally new content to their lives.

We have thus sought to clarify the connection between the unintelligibility of the parables and impenitence. This unintelligibility is a quality which has above all ethical consequences. Mark does not reason as we are inclined to reason: those outside cannot respond to the message which they cannot understand intellectually. Rather, he reasons: the parables are an appeal for a decision, a call to man's heart; in "those outside" they create moral blindness and they darken not so much their minds as their hearts. They harden instead of converting. This will be further substantiated by considering Mark's concept of Jesus' *didachē*.

By having constructed the introductory verse to the parable chapter, and even more by having added vs. 2b, Mark emphasizes that parables are

---

been voiced recently by W. Paschen, *Rein und Unrein: Untersuchung zur biblischen Wortgeschichte* (StANT 24; München: Kösel, 1970) 172-73 (he refers to A. Schlatter, *Markus, das Evangelium für die Griechen*, 136). Mk 7:3-4 is very likely a redactional passage (see V. Taylor, *Mark*, 335). Mark's community would thus no longer be concerned with Jewish dietary prescriptions, and would not need to be especially informed that Jesus cleansed all foods. Two observations might be made about this objection. First, are we certain that the main purpose of 7:3-4 is information? Secondly, are we certain that the difficulties in Mark's community with the purity of foods stemmed only from Jewish dietary laws? Such scruples may have come from sources other than Judaism, as Col 2:20-23 seems to indicate. Mark, further, is the only evangelist who makes redactional use of the traditional term *pneuma akatharton* (1:27; 3:11; 5:8; 6:7). This would seem to indicate that for him the phenomenon of impurity went far beyond the limits set by Jewish dietary prescriptions. The very fact that he included the section into his Gospel seems to indicate that he wished to address a definite problem which bedevilled his community.

vehicles of Jesus' teaching. Teaching is the most characteristic activity of Mark's Jesus.[170] To quote E. Schweizer,[171] "Mark usually says nothing about the content of Jesus' teaching or gives merely one example or simply a very brief summary of it. This shows that the fact of his teaching is decisive, not its content. This is buttressed by the almost exclusive usage of *ekplēssesthai* and *thambeisthai* in redactional paragraphs of Mark which describe the amazing success of Jesus' teaching. It is the same with *pas ho* or *ochlos ho* describing the coming of the whole country to Jesus." What is important to Mark is "the way in which God's own power was present in Jesus' teaching, his 'authority.' " In this Mark differs from Matthew: "What Mark states is simply the fact that God himself encountered men in Jesus' teaching." Jesus' teaching is thus parallel to the miracles which are "demonstrations of the divine wrath against Satan or of the authority of the victor liberating the world from all demons." Schweizer concludes: "Thus, Mark, by referring to the mighty deeds or speeches of Jesus, merely describes the 'dimension' in which Jesus lives, acts and speaks. It is the heavenly dimension in which God himself breaks into the world. . . ." Jesus' teaching, then, is for the Second Evangelist an act of divine power; while Matthew and Luke clearly distinguish between miracles and teaching, Mark sees both activities as one and the same manifestation of this power. We could say that the word which teaches is the same as the word which expels demons.[172] "The Rabbis taught, and nothing happened. Jesus taught, and all kinds of things happened."[173] The teaching of Jesus is the sign of the inbreaking kingdom; it is a manifestation of the presence of this kingdom. It is a communication of knowledge, of course, but of a knowledge which takes possession, not only of man's mind, but of his heart and existence. By describing parables as teaching, Mark is suggesting that through them God's saving power is reaching out to men. Parables teach, i.e., instruct, but this instruction is no mere information, it aims at involving men in a new condition of life. They are a facet of the kingdom which is coming and is present in Jesus' words and deeds.[174] In them the Spirit of God is at work bringing about eschatological salvation. The phrase *en tē didachē autou* suggests also the following thought: it is significant that of the five

---

[170] See E. Schweizer, "Anmerkungen," 95-96.

[171] "Contribution," 422-23.

[172] See J. Delorme, "Aspects," 84-85; see also L. E. Keck, "Introduction," 360-61; K. H. Rengstorf, *TDNT* 2, 140-41.

[173] W. Manson, *Jesus the Messiah* (London: Hodder and Stoughton, 1961) 35; see also J. Coutts, "The Authority of Jesus and of the Twelve in St. Mark's Gospel," *JTS* ns 8 (1957) 116.

[174] See E. Lohmeyer, *Markus*, 83; E. Jüngel, *Paulus*, 125.

occurrences of the term *didachē*[175] four are qualified by *autou*; the absolute use of the term, by itself, already points to the One teaching more than to the content of the teaching. The addition of the pronoun strengthens the concentration on the person of Jesus. Thus the parables which, in the hand of Mark, have become "his teaching" are a manifestation not only of the saving presence of God's eschatological Spirit, but also of the essential role played by Jesus in setting up God's kingdom.

The coming of the kingdom has, however, also its destructive, punitive side: the demons are expelled, the enemies of Jesus are silenced and condemned. "He will come and destroy those tenants and turn his vineyard over to others" (12:9); "whoever blasphemes against the Holy Spirit will never be forgiven. He carries the guilt of his sin without end" (3:29).[176] The parables share in this work of destruction and condemnation. For those to whom the mystery of the kingdom is not given they remain dark and blinding sayings. The passive form in the logion at 4:11 indicates that they are intended by God to remain unintelligible and not to deliver their message to those outside in order that these may not understand and be forgiven.

There is a note of finality about this logion which recalls parables in the intertestamental literature describing the final judgment.

"The first Parable:
When the congregation of the righteous shall appear,
And sinners shall be judged for their sins,
And shall be driven from the face of the earth:
. . . .
Where then will be the dwelling of the sinners,
And where the resting-place of those who have denied the Lord of Spirits?
It had been good for them if they had not been born."

(*1 Enoch* 38:1-2)

"And this is the second Parable concerning those who deny the name of the dwelling of the holy ones and the Lord of Spirits.
And into the heaven they shall not ascend,
And on the earth they shall not come,
Such shall be the lot of the sinners
. . . .
Who are thus preserved for the day of suffering and tribulation."

(*1 Enoch* 45:1-2)

---

[175] 1:22,27; 4:2; 11:18; 12:38; in Mt the term occurs three times, and in one case it refers to the teaching of Pharisees and Sadducees; in Lk it occurs once.

[176] See J. Gnilka, *Verstockung*, 81.

"For just as the husbandman sows much seed upon the ground and plants a multitude of plants, and yet not all which were sown shall be saved in due season, nor shall all that were planted take root; so also they that are sown in the world shall not all be saved."

(*2 Esdras* 8:41)

"And I answered and said: . . . Show me . . . whether in the Day of Judgment the righteous shall be able to intercede for the ungodly, or to intreat the Most High in their behalf . . .
And he answered me and said: . . . the Day of Judgment shall be the end of this age and the beginning of the eternal age that is to come; wherein
corruption is passed away,
weakness is abolished,
infidelity is cut off;
. . . .
So shall no man then be able to have mercy on him who is condemned in the Judgment, . . ." (*2 Esdras* 7:102,114-15)

These apocalyptic parables and predictions differ from those of Mark in one important respect: while they still look forward to a future judgment, in the Second Gospel the division of spirits is already taking place; for the kingdom is mysteriously present in Jesus' words and works.

Before we close this discussion we should make two observations. First, Mark is not unaware of the other side of the coin in the matter of divine condemnation and the blinding of those outside. Human guilt is manifestly recognized in the interpretation of the Sower parable immediately following the logion at 4:11-12, as well as in 3:22-30 and 12:1-9. Secondly, Mark looks upon the enemies of Jesus and those outside primarily as types, and not as past or contemporary groups of individuals to whom the way to God has been definitively barred. In this he is simply following the OT tradition:[177] it is their attitude which he principally condemns.

(4) The difference between the Marcan version of the logion in 4:11, on the one hand, and the Matthean (13:11) and Lucan (8:10), on the other, must be considered in every discussion of Mark's parable theory. While Mark has *hymin to mystērion dedotai,* Luke and Matthew have *hymin dedotai gnōnai ta mystēria.* We have to do here with one of the minor agreements of Matthew and Luke against Mark. We cannot retrace the discussion which has accompanied this problem over the years.[178] In a recent

---

[177] See K. L. and M. A. Schimdt, *TDNT* 5, 1024.

[178] Two recent contributions to the discussion suggest that Luke used Matthew as one of his sources: W. Wilkens, "Zur Frage der literarischen Beziehung zwischen Matthäus und Lukas," *NovT* 8 (1966) 48-57; R. T. Simpson, "Agreements."

article, S. McLoughlin attempted to reduce the number of significant minor agreements. He feels that the minor agreement here is not significant for two reasons: Matthew and Luke could have followed the same line of reasoning on the subject of the Marcan logion; Mark's logion is of such a nature that it calls for a correction.[179] But this type of explanation is convincing only if it can be proven that Matthew and Luke had no other source for this logion apart from Mark. We do not mean another written source; but it is conceivable that another form of the logion was known to them independently of Mark and each other. Perhaps the other two Synoptics are even giving us an earlier form of the logion. Three arguments favor this suspicion: first, the fact that Mark never attributes knowledge to the disciples; secondly, the constant connection between mystery and knowledge in intertestamental and Qumran literature; thirdly, Matthew's treatment of the Marcan text.

The first point has been dealt with above. With regard to the second, we refer to L. Cerfaux[180] who argues that the Mt-Lk form of the saying preserves the tradition better than the Marcan form, and for three reasons. It sounds more like similar forms in the Dead Sea Scrolls; *rz* in Qumran literature is found in the plural; "mysteries" are generally linked with knowledge of them on the part of men. But R. E. Brown[181] replies to L. Cerfaux' second argument, pointing out that *sôd*, which also has the sense of "mystery," is found in the singular. Brown is not convincing, however, when he claims that the Marcan form could, for that reason, be original, for he does not consider sufficiently Cerfaux' third argument. When we examine the intertestamental and Qumran texts in which *rz*, *sôd*, and other terms occur to describe mystery, we see that they are most frequently accompanied by a reference to man's, or God's knowledge of them.[182] In thirty-three of the Qumran texts quoted by E. Vogt "mystery" is found linked with knowledge,[183] whereas in ten of them it occurs

---

[179] "Les accords mineurs Mt-Lc contre Mc et le problème synoptique: Vers la théorie de deux sources," *ETL* 43 (1967) 24, 26; for another, most unconvincing, attempt, see A. Suhl, *Funktion*, 146, 150.

[180] "Secrets," 241.

[181] "The Semitic Background of the NT *mystêrion*," *Bib* 39 (1958) 428-29.

[182] A complete list of texts in which "mystery" occurs in the Dead Sea Scrolls is given by E. Vogt, "Mysteria in textibus Qumran," *Bib* 37 (1956) 247-57. References to "mystery" in the intertestamental literature are found in R. E. Brown's article referred to in the previous note and its continuation (*Bib* 40 [1959] 70-87), and particularly in "The Pre-Christian Semitic Concept of Mystery," *CBQ* 20 (1958) 417-43.

[183] The usual forms are hiphil and inf. cstr. of *yd'*.

without a direct reference to knowledge.[184] A few of the texts are too fragmentary to permit any certainty. Thirty-three of the intertestamental texts cited by Brown link "mysteries" and knowledge, while in only two texts "mystery" occurs without an express reference to knowledge.[185] Thus it is more likely that Mt-Lk form of the logion is original. The impression of the greater originality of the Mt-Lk version is likewise strengthened by the Qumran usage of the verb *ntn* with God as the subject and insight as the object.[186]

But is it not precisely this background which explains the agreement between Matthew and Luke? To answer this objection we must point to Matthew's redactional treatment of Mark. Matthew's interest in stressing the understanding of the disciples is evident;[187] on this point he repeatedly corrects or completes Mark. The verb which he uses for this purpose is *syniēmi*.[188] Of the nine cases of this verb in Mt six are found in ch. 13, the parallel to Mark's parable chapter. In Mk 4 the verb occurs but once, viz., in the OT quotation in vs. 12. It is particularly interesting to observe Matthew correcting Mark, by inserting the verb in 13:19 (par Mk 4:15) and 13:23 (par Mk 4:20). At the end of Matthew's parable chapter (13:51) Jesus asks: "Have you understood (*synēkate*) all this?" They said to him, "Yes." This verse, then, seems to be a correction of Mk 4:13. Hence had Matthew changed Mark's text in 4:11 on his own, he would undoubtedly have inserted, instead of *gnōnai*, an infinitive of the verb *syniēmi*. The fact that he has *gnōnai* argues in favor of the assumption that he is reproducing an independent tradition, and not correcting Mark's text.

Should we admit the greater originality of the Mt-Lk form of the logion, there is really no way of deciding whether it was Mark or the tradition which he was using that was responsible for the omission of the reference to knowledge. In either case, the absence of this reference is in keeping with the tenor of his Gospel, for Mark never predicates understanding of the disciples. It is, therefore, significant for his concept of revelation and, indirectly, of the parable as the vehicle of revelation.

---

[184] 1Q27:5 (mystery being carried out); 1Q27:7 (all who adhere to mysteries); 1QM 3:8-9 (God's secrets to destroy godlessness); 14:9 (secrets of hate); 14:14 (God's secret plans); 16:11 (through God's secrets troops begin to fall); 17:9 (God's secrets keep them strong); 1QH 5:36 (mysteries of sin); 8:6 (hidden source of life); 8:11 (mysterious seal).

[185] *1 Enoch* 63:3; 68:5.

[186] 1QH 11:27-28; 12:11-13; 13:18-19; 14:8; 18:27; see also 1QH 16:11-13; 17:17; 10:27.

[187] G. Bornkamm, G. Barth, H. J. Held, *Tradition and Interpretation in Matthew* (*NT Library*; Philadelphia: Westminster, 1963) 109-10; H. Conzelmann, *TWNT* 7, 893; W. Wilkens, "Gleichniskapitel," 313-14.

[188] See Mt 16:12 par Mk 8:21; Mt 17:13 par Mk 9:11-13.

In what does this significance consist? We cannot help but notice the continual struggle waged by Jesus against the disciples' failure to understand. This failure is not only momentary, as it is for Matthew, but radical: only a miracle can cure the disciples of their blindness as to the true being of Jesus and of their deafness with regard to his message. Their confession of Jesus as the Messiah and their hearing the divine voice proclaiming him to be the Son of God are the result of such a miracle. It is a step forward in their growth of understanding, yet only a step.[189] Throughout the section 8:27—10:52, which is devoted primarily to their instruction, the disciples show that the scales have not yet fallen from their eyes. Revelation for Mark is thus a dynamic process: "To be with and follow Jesus through the varied situations of his ministry—teaching, deed, passion, and resurrection—is to be finally brought to sight, to understanding. In this sense the total "Christ Event" is the revelation. However, having said this it must likewise be observed that each of these factors was crucial for Mark in his description of the Revelatory Event."[190] There is growth in revelation as well as in understanding. Understanding is, correlative to the revelation, not an accomplished fact from the beginning of the Gospel; it is rather the final aim of the Gospel.[191] The absence of *gnōnai* in 4:11 is thus not accidental, for the struggle is still going on. The disciples are only at the beginning of their association with Jesus; they are receiving only the first glimmers of understanding.

The above remarks take only the historical dimension of the Gospel into consideration. Yet Mk 4:10 speaks of those about him with the Twelve, i.e., of the Christian community. The logion is addressed primarily to the present. Mark's image of the disciples is colored by the conditions and problems of the Christian community. It is not the disciples alone who are in danger of remaining stiff-necked; the Christian community too is in continual peril of judging, like Peter, not "by God's standards but by man's" (8:33). The struggle for understanding never ends; the resurrection of Jesus was a decisive step in the growth of understanding, but it was not the final step. The community is being persecuted (4:17; 10:30; 13:9-13), is afraid (13:11), is tempted to be ashamed of its Lord (8:38), is prey to ambition (9:33-37), is too impatient for the coming of the end (13:7), is in danger of falling away (4:14-19). It knows that Jesus has died and risen from the dead, but this does not remove the possibility that this knowledge becomes a meaningless riddle which, instead of fostering understanding, takes away the little light which it had for a time engendered. The community, in short,

---

[189] See J. Gnilka, *Verstockung*, 35.
[190] R. P. Meye, *Twelve*, 220.
[191] See W. Wilkens, "Gleichniskapitel," 310-16; see also J. Delorme, "Aspects," 90; H. B. Swete, *The Gospel according to St. Mark* (London: Macmillan, 1898) 72.

has been given the mystery of the kingdom, but the kingdom is yet to come with power (9:1; cf. 13:26). The greatest act of God has not yet taken place, and the revelation already given makes sense only in the light of, and as a preparation for, that final manifestation of the divine salvific purpose.

The three parables of Mark's parable chapter look forward to this final saving act of God in Jesus Christ. What was happening in the past when Jesus taught, worked, died, and rose, and what is happening now in the community has its purpose in what is yet to come and receives its significance from the end toward which it strives. Sowing is done for a purpose: its meaning lies in the harvest, and the various soils are judged, not in the light of sowing, but in the light of the harvest.[192] A glorious consummation gleams through humble beginnings and makes failures bearable. Hiddenness and apparent weakness are not permanent conditions of the kingdom; when the promised coming in power takes place the full meaning of the parables will come to light. Until then they must remain parables, they cannot but be enveloped in a shroud of darkness, they need interpretations which reemphasize their call for a decision for or against the future which they promise. Only those to whom the mystery of the kingdom is given can pierce their darkness. Yet even their understanding is incomplete and insecure; it must grow and be buttressed, for until the harvest they will not understand as they should understand.

(5) Finally we may point out some event-like features of Mark's parable. There is, first of all, the word *ginetai* in 4:11. V. Taylor[193] has noted that a number of Marcan MSS had replaced *ginetai* by *legetai*; he remarks that the phrase is a strange expression to describe instruction. Yet parables for Mark are teaching as we have noted. Teaching, however, is no mere communication of knowledge but speech full of power claiming man's total adherence. Parables, furthermore, do not merely describe the nature of the kingdom; they are also a facet of its coming, a manifestation of the divine will to save the world. The verb *ginetai* is also in harmony with *mystērion* in the singular. The plural of the term suggests that various aspects of the kingdom, known to God alone, are being revealed to the disciples, while "the singular would imply that the mystery is a single event."[194] The phrase "the secret has been given" undoubtedly contains the idea of the communication of knowledge, but it is nonetheless significant that the event-like coloring clings to it in a greater measure than to the Mt-Lk version. The same characteristic seems to recur in Mark's redactional insertion into the inter-

---

[192] See E. Fuchs, *Hermeneutik*, 224-28.
[193] *Mark*, 256.
[194] E. F. Siegman, "Teaching," 172.

pretation of the saying at 7:15. One may be tempted to look upon the logion of 7:15 as a general statement of truth. Yet Mark's redactional interpretation of it in 7:19, *katharizōn panta ta brōmata*, makes it evident that he looked upon it as a decision of Jesus which changed the character of all foods.[195] That 3:28-29 is not to be considered as an instruction on forgivable and unforgivable sins, but as a condemnation, is shown by the redactional vs. 30; this verse refers the entire speech of Jesus "in parables" to the concrete situation, viz., that of the rumors spread by the scribes concerning the source of Jesus' power. Mark thus portrays Jesus as pronouncing God's judgment on the scribes. 12:1-9 is likewise not an illustration of a truth, describing the fate of unfaithful tenants in general, but an announcement of what will befall the Jewish authorities. Vs. 9 is not a piece of information, but a verdict of condemnation and rejection by God.

Let us briefly summarize our discussion of Mark's concept of parable. There can be no doubt that he looks upon it as a dark saying. This notion he has not created, but has inherited from the tradition and utilized in his editorial work. The parable, being an instrument of Jesus' teaching and thus a manifestation of the inbreaking of the kingdom, brings about enlightenment and conversion in those to whom the mystery of the kingdom has been given. By means of its unintelligibility it also participates in the destructive aspects of the coming kingdom: in the hearts of those outside it produces utter darkness and impenitence. Its primary message is the kingdom of God, a reality which is already present in a hidden manner, but is still waiting to manifest itself with power. The parable is thus still shrouded in darkness; it is still a riddle to those outside, and only the firm faith in Jesus, the Son of God, and a firm hope in the glorious future of the hidden kingdom brought by Jesus can penetrate this darkness. What the parable promises will be fully known only when its promise is fulfilled. Thus Mark's parable is a continual invitation to understand, urging and prodding man to allow God's eschatological gift to penetrate his hardened heart. There is no point in time at which it is fully understood and fully assimilated. It is like the good news in 1:15, the verse which serves as the summary and foundation of the Second Gospel, and which proclaims the saving act of God and calls for conversion and faith: it is a permanent challenge, a permanent demand for man's response; there never comes a moment when this response is so perfect that the demand would lose its relevance. The entire Gospel of Mark is a challenge and an invitation to be converted, to believe, to understand. This understanding grew during the

---

[195] The RSV translation, "Thus he declared all foods clean," does not do justice to *katharizōn;* Jesus did not declare them clean, he cleansed them.

ministry of Jesus, it crossed an all-important threshold with his resurrection, and yet the community's hold on it can be very tenuous. Until the moment when the kingdom comes in power the community must pray with the epileptic boy's father: "I do believe! Help my lack of trust!"

(D) THE MYSTERY OF THE KINGDOM

(1) Mk 4:11 with its parallels is the only passage in the synoptic gospels in which the term *mystērion* occurs. Our present knowledge of the rich OT, intertestamental, and Qumran background of the term absolves us from the need of searching for its origin in non-Palestinian and non-synoptic sources. For it was very much at home in the place and time in which Jesus and the primitive community lived, thought, and preached. This background has been well presented by others.[196] E. Vogt defined the *mystērion* thus:[197] "Mysteria sunt imprimis arcana consilia sapientiae atque decreta voluntatis divinae." This definition applies not only to Qumran literature, but also to the notion encountered in the apocalyptic literature, canonical as well as extracanonical. The term *rz*, a Persian loanword, of which *mystērion* is the most frequent Greek equivalent, and other terms (*sôd*, *nstr*) are applied to various hidden realities: evil mysteries, cosmic mysteries, mysteries of man's action, mysteries of divine providence. But paramount importance is assigned in all apocalyptic literature to God's mysterious design for the end of time.

What is the content of "mystery" in Mark? Exegetes are fairly unanimous in their answer to this question: the mystery of the kingdom consists in its dynamic presence in the person of Jesus.[198] This conclusion is correct, although the methods employed in arriving at it occasionally leave much to be desired. It is reached, in some cases, by referring to the fact that the term *mystērion* is in the singular; the content is then assigned in view of

---

[196] See R. E. Brown, "Mysterion"; "Pre-Christian Semitic Concept"; H. Ringgren, *The Faith of Qumran: Theology of the Dead Sea Scrolls* (Philadelphia: Fortress, 1963); F. F. Bruce, *Biblical Exegesis in the Qumran Texts* (Grand Rapids: Eerdmans, 1959); E. Vogt, "Mysteria"; B. Rigaux, "Révélation des mystères et perfection à Qumrân et dans le Nouveau Testament," *NTS* 4 (1957-58) 237-62; G. Bornkamm, *TDNT* 4, 813-17; L. Cerfaux, "Secrets."

[197] "Mysteria," 256-57.

[198] See T. A. Burkill, *Revelation*, 102; "The Hidden Son of Man in St. Mark's Gospel," *ZNW* 52 (1961) 204-5; R. Schnackenburg, *God's Rule*, 189; G. Bornkamm, *TDNT* 4, 818-19; E. F. Siegman, "Teaching," 172; C. E. B. Cranfield, *Mark*, 153; G. H. Boobyer, "Redaction," 67; E. Schweizer, "Frage," 2; W. Grundmann, *Markus*, 92; J. Delorme, "Aspects," 88; J. Jeremias, *Parables*, 16; G. E. Ladd, *Jesus and the Kingdom: The Eschatology of Biblical Realism* (New York: Harper & Row, 1964) 218. H. B. Swete (*Mark*, 72) identifies the mystery with the content of the Gospel; W. Manson (*Jesus*, 55) thinks it is the totality of revelation made in Jesus.

the over-all message of the Gospel.[199] In other cases it is simply assumed without an attempt at proof. It is also commonly agreed that the mystery of the kingdom is to be identified with the messianic secret. This becomes particularly evident in the discussion of Mk 4:21-22, a passage which is understood to be a supplement to, and a correction of, vss. 11-12. Vss. 21-22 speak of the provisional reign of the messianic secret which is, for a time only, entrusted to a few, but is destined to be manifested to all very soon.[200]

J. Gnilka[201] has discussed the content of the mystery in some detail. He studies the Gospel passages in which the disciples are rebuked by Jesus for their failure to understand. Those passages give us an indication of what Mark considers to be of particular importance for the enlightenment of disciples. They are the key to the content of the mystery of the kingdom, for in 4:11 the disciples' reception of the mystery is contrasted with the outsiders' reception of unintelligible riddles. The first passage discussed is Mk 4:35-41; and Gnilka shows that the main point of the Marcan story of the calming of the storm lies at the end, in vss. 40-41. "Jesus' questions are intended to force the disciples to reflection and to move them to an understanding of his self-revelation";[202] this reflection does take place in vs. 41. The same message is conveyed by the story of Jesus' walking on the water (6:45-52): Matthew's disciples (14:33) put into words what the event should have communicated to those of Mark, viz., that Jesus is the Son of God. These two pericopes show that the mystery of the kingdom is closely linked with that of the person of Jesus. The discussion about the leaven of the Pharisees and of Herod (8:14-21) brings out that fact even more clearly. Whatever the value of Gnilka's interpretation of Mark's understanding of the leaven,[203] it seems that the question of Jesus' identity is involved.[204] After this, Gnilka studies explanations given to the disciples in 4:11-25; 7:17-23; 10:10-12; 9:28-29, and in the apocalyptic discourse. He

---

[199] See W. Wrede, *Messiasgeheimnis*, 58-59; G. Bornkamm, *TDNT* 4, 818-19; T. A. Burkill, *Revelation*, 102; "Son," 204-5.

[200] See M.-J. Lagrange, "Le but," 14-15; *Marc*, 109-13; M. Hermaniuk, *Parabole*, 336-37; V. Taylor, *Mark*, 262; E. Schweizer, *Mark*, 100; E. Trocmé, *Marc*, 149; T. A. Burkill, *Revelation*, 98, 111; J. Jeremias, *Parables*, 221, n. 66; E. F. Siegman, "Teaching," 168-69; J. Delorme, "Aspects," 87; C. E. B. Cranfield, *Mark*, 164-67; F. W. Beare, *Records*, 113; G. Minette de Tillesse, *Secret*, 172, 201, 214, 216, 280-81; S. E. Johnson, *Mark*, 93; A. E. J. Rawlinson, *St. Mark (Westminster Commentaries*; London: Methuen, 1956) 54; J. Schmid, *Mark*, 102-4; E. P. Gould, *Mark*, 78; E. Klostermann, *Markus*, 43; F. C. Grant, "Mark," 702; H. B. Swete, *Mark*, 77-78; G. Strecker, "Messiasgeheimnis," 98-100; C. H. Dodd, *Parables*, 107, 110, n. 39.

[201] *Verstockung*, 34-44.

[202] *Verstockung*, 35 (my translation).

[203] *Verstockung*, 36-37.

[204] See E. Lohmeyer, *Markus*, 158; R. Schnackenburg, *Markus*, 203.

sums them up by saying that Jesus in the Second Gospel gradually unveils the mystery of his person and the essential content of his teaching.

Another passage, however, should be added, viz., the editorial remark of the evangelist at the end of Jesus' refutation of his enemies' claim that he is possessed and that his exorcisms are due to the power of the prince of demons in 3:30. Vs. 30 gives the reason for the condemnation uttered in vss. 28-29; the enemies refuse to acknowledge the true source of Jesus' power, and declare it to be the very opposite of what it is. Before he condemns them, Jesus shows how untenable their perverse interpretation of his mighty works is, for his exorcisms are a sign that One stronger than Satan is at work in him, and that, as a result, Satan's kingdom is falling apart. To put the matter positively, the mystery of the kingdom consists in Jesus' visibly carrying out the divine counsel, God's eternal plan to set up his kingdom at the end of time, a plan which has been kept hidden up to now. We may conclude then with Gnilka: "the mystery is to be sought in one definite perception: in the knowledge of the fact that the kingdom has already set in with the coming of Jesus, the hidden Messiah."[205]

(2) Here something must be said about the theme of the messianic secret in the Second Gospel. A full discussion of the theme is, of course, impossible within the scope of this work, and we limit ourselves to those features that bear on our problem.

The traits in which the theme appears are the following: Jesus forbids demons to make known his divine sonship, commands silence about his miracles, and tries on occasion to avoid crowds. The disciples alone are given special instructions which allow them to understand the parables and the sayings of Jesus; they alone recognize him as the Messiah and hear the divine voice proclaiming him to be the Son of God. But they are bound to silence. Their failure to understand seems to be a correlative of the secret; despite their growing insight they prove again and again that they do not perceive the meaning of Jesus' words and destiny. With the death and resurrection of Jesus the duty of silence ceases; from that moment on the church proclaims him as the Messiah and Son of God, and what the disciples had heard in private the evangelist tells everyone who reads his book.

These features have given rise to a number of interpretations since W. Wrede's classic attempt, *Das Messiasgeheimnis in den Evangelien*, first published in 1901, to discover a single motif that would explain all of them. He felt that the phenomena connected with this Gospel theme must not be treated separately, as they had been up until his time, but should rather be

---

[205] *Verstockung*, 44 (my translation).

interpreted in the light of one guiding principle.[206] Wrede's postulate has prevailed ever since,[207] and only in the last few years has it been called into question.[208]

G. Minette de Tillesse has proposed the following thesis: "In Mark the messianic secret expresses Jesus' irrevocable and free decision to embrace his suffering, because such is God's will."[209] The repeated injunctions of silence serve to prevent his divinely willed way to the cross from becoming a triumphal march which could not possibly end upon the cross. The obedient Son of God manifests divine salvation in the very event which, to the Jews, disproved his messiahship. Thus, the closer the hour of his death the more he permits the veil of the secrecy to be lifted: his messianic titles appear more frequently after the confession of Peter and injunctions of silence decrease in number, until the moment of death when Jesus can be publicly proclaimed the Messiah and the Son of God. He dies because he is the Messiah, and he is confessed as the Son of God in his death.[210]

Unlike Minette de Tillesse, we fail to find the theme of the messianic secret in the Marcan controversies. Two other points of criticism have been voiced with regard to his book and his view of the secret: first, he omits any reference to the view according to which Mark is combatting the *theios anēr* conception of Jesus; and secondly he interprets Mark's emphasis on the cross too exclusively as an apologetic against Jewish messianic ideals.[211] This criticism is justified: the opinion just mentioned is held by too many exegetes to be simply disregarded, and we suspect that Mark's emphasis on the cross is designed to address Christians primarily. It represents an attempt to counter tendencies manifesting themselves in the community itself.

Many exegetes today[212] hold that the theme of messianic secret serves Mark as a weapon against an image, widespread among hellenistic Christians, which painted Jesus as a *theios anēr*.[213] Mark is, according to them,

---

206 See *Messiasgeheimnis*, 36-38.
207 See G. Minette de Tillesse, *Secret*, 33-34.
208 By U. Luz, "Geheimnis," and J. Roloff, "Markusevangelium," 84-92.
209 *Secret*, 321 (my translation).
210 *Secret*, 317-26.
211 See R. H. Fuller's review of G. Minette de Tillesse, *Secret*, CBQ 31 (1969) 110-11; the interpretation of Mark's emphasis on the cross is found in *Secret*, 323-24.
212 See H. Conzelmann, "Gegenwart und Zukunft in der synoptischen Tradition," ZTK 54 (1957) 277-96; P. Vielhauer, "Christologie"; E. Schweizer, *Mark*, 382-83; "Leistung"; L. E. Keck, "Christology," 347-51; T. J. Weeden, "Heresy"; J. Schreiber, "Christologie"; *Vertrauen*; U. Luz, "Geheimnis."
213 For a presentation and discussion of this hellenistic image, see L. Bieler, *THEIOS ANĒR: Das Bild des "göttlichen Menschen" in Spätantike und Frühchristen-*

attempting to correct this misconception by his theology of the cross. We must consider two aspects of this theory; first, the assertion that the tradition of Jesus' exorcisms and miracles had been, in the course of oral transmission, strongly colored by the hellenistic image of the "divine man"; second, the assertion that Mark is combatting this image. With regard to the first, L. E. Keck has convincingly contended that the picture is by no means so simple as is sometimes assumed. For there are two streams of miracle tradition to be detected in Mark: "one closely related to the Palestinian scene and the message of Jesus in its native setting; the other relatively unrelated to Jesus' message (except in the thought of Mark). The former stands under the rubric "the strong man" (3:27); the latter under the stamp of the hellenistic *theios anēr* whose divine power is manifested on earth."[214] As signs of the latter material he enumerates the following features:[215] the material contains no conflict with Judaism, no debate about Jesus' healing, or the Sabbath, or the authority by which he works; in it there is no stated connection with the kingdom or forgiveness of sins. But there are references to the powers resident in Jesus and to his being different from other men; there are few references to faith. Keck feels that these features are present in Mk 3:7-12; 4:35—5:43; 6:31-52,53-56. While T. A. Burkill's criticism[216] of this opinion may have weakened Keck's contention that in 3:7-12 traces of a pre-Marcan summary may be discovered, it has not been quite as effective regarding the various streams of miracle material present in the Second Gospel. Keck's suggestion finds indirect support in H. C. Kee's careful study of the terminology of Mark's exorcism stories,[217] in particular in the background of the verb *epitimaō*. The OT and Qumran equivalent of this verb (*g'r*) carries "the connotation of divine conflict with hostile powers, the outcome of which is the utterance of the powerful word by which the demonic forces are brought under control";[218] this utterance is expressed by *g'r—epitimaō*. In hellenistic miracle stories this verb never occurs; in Greek papyri it is never used in the sense of bringing hostile powers under control. Kee concludes: "While it is true that the gospel ac-

---

tum (Darmstadt: Wissenschaftliche Buchgesellschaft, 1967); A. Oepke, *TDNT* 3, 567-68; 4, 609; R. Bultmann, *Theology of the New Testament* (London: SCM, 1965), I, 130-31; J. Bieneck, *Sohn Gottes als Christusbezeichnung der Synoptiker* (*ATANT* 21; Zürich: Zwingli, 1951) 31-34; J. Schreiber, *RGG*³ 6, 119.

[214] "Christology," 350-51.
[215] "Christology," 349-50.
[216] "Mark 3, 7-12 and the Alleged Dualism in the Evangelist's Miracle Material," *JBL* 87 (1968) 409-17.
[217] "The Terminology of Mark's Exorcism Stories," *NTS* 14 (1967-68) 232-46.
[218] "Exorcism," 238.

counts—especially in the later stages of tradition—do serve this function (i.e., "to create a supernatural aura around an esteemed figure of the past"), at the beginning the exorcisms were understood on a far wider background than the purely Christological question, Who is Jesus? That background was nothing less than the cosmic plan of God by which he was regaining control over an estranged and hostile creation, which was under subjection to the powers of Satan."[219] The tacit assumption that all miracle tradition is of a "divine man" quality is thus much too unqualified and too general to be accepted as it stands.

Should we accept the theory that Mark is polemicizing against such a view of Christ, by making Jesus forbid the publication of his miracles "before the Son of Man had risen from the dead"? At first sight this seems to be an attractive suggestion. Yet here again we must guard against a pitfall, viz., of attributing our knowledge and reconstructions of the past to the evangelist. If Mark is struggling against the image of Jesus as a "divine man," why does he enjoin the proclamation of the one miracle which seems to bear the unmistakable character of a hellenistic exorcism tale? The presence of *horkizō*—a hellenistic exorcizing verb—and other features[220] should lead him to hide the deed reported in 5:1-20 more fervently than any others. Since vss. 18-20 are probably redactional,[221] it is all the more surprising, if Mark is combatting the attribution of "divine man" features to Jesus.[222]

---

[219] "Exorcism," 246.

[220] See M. Dibelius, *Tradition*, 87-89; H. C. Kee, "Exorcism," 241.

[221] See M. Dibelius, *Tradition*, 74; G. Minette de Tillesse, *Secret*, 86-87; V. Taylor, *Mark*, 284.

[222] Some attempt to explain away the seeming contradiction to the secret presented by 5:18-20. T. A. Burkill (*Revelation*, 94-95) thinks that 5:19 is traditional and has nothing to do with the theme of the secret. One would think that, even if the verse were traditional, Mark would feel compelled to excise such a flagrant contradiction to his theory; he was, after all, quite capable of adding verses which expressed the secret. W. Bousset (*Kyrios Christos: A History of the Belief in Christ from the Beginnings of Christianity to Irenaeus* [Nashville: Abingdon, 1970] 107, n. 99) suggests that we should understand Jesus as commanding the man to tell his own people only; but this suggestion has nothing in the text to support it. M. Dibelius (*Tradition*, 74) feels that the main point of the verses lies in Jesus' unwillingness to accept the man among his disciples. But surely the message about the exorcism is given predominant attention in the passage. H. Sahlin ("Die Perikope vom gerasenischen Besessenen und der Plan des Markusevangeliums," *ST* 18 [1964] 163) explains it by saying that there is no danger of misunderstanding in a pagan land. P. Vielhauer ("Christologie," 158) says that injunctions of silence are laid only upon, and among, the Jews. But it would seem that by means of the messianic secret Mark is trying to solve a problem of his contemporaries. He is not struggling against past Jewish misconceptions but against present Christian ones; the "divine man" view of Jesus and his work was a danger in hellenistic communities, not Jewish ones. G. Minette de Tillesse (*Secret*, 87) feels that

The cure reported at 5:24b-34 would also seem to call for an injunction of silence, for it clearly speaks of a quasi-magic supernatural power residing in Jesus. To say that such an injunction in the given circumstances would be absurd is no answer, since Mark is obviously not concerned with historical credibility in enjoining silence. His evident preoccupation with the question, "Who is Jesus?," along with these facts, make questionable the assumption that he is wrestling with a view of Jesus as a "divine man," as conceived by some present-day exegetes. Mark was rather opposing tendencies in the conception which the community had of its Lord and which had arisen from an unduly hellenized tradition of his mighty works. Chief among these would be the tendency to make him an independent agent seeking his own glory.

A recent contribution to the discussion of the messianic secret has been made by J. Roloff.[223] According to him, one should distinguish between Jesus' silencing of the demons, the miracle secret, and the commands of silence imposed on the disciples. Only in connection with the last does Mark develop an independent theory of the secret. Silencing the demons is a datum which he takes over from tradition and expands redactionally, without fundamentally changing it. The miracle secret likewise stems from tradition and preserves some genuine historical reminiscences. The Marcan redaction stresses that despite Jesus' desire for secrecy miracles are being publicized; what is being publicized, however, is not Jesus' divine sonship but the amazing outward events which accompany his ministry. The miracle secret is not to be taken as a kerygmatic motif but as a means of Mark's presentation of history. Injunctions of silence given to the disciples concern not the manifestation of Jesus' deeds but of his person and destiny. Within this framework the failure to understand finds its true context. Mark places the disciples in the pre-resurrection situation as Jesus' constant companions and the recipients of his teaching; but he knows that only the resurrection revealed the mystery of Jesus' person and destiny to them. Since he cannot produce an unmessianic image of the earthly Jesus, he helps himself with the idea that the disciples did indeed receive revelations, but failed to understand them until their encounter with the risen Lord. The secret granted to the disciples thus

---

there is no danger in proclaiming to a people who, by asking Jesus to go away, have shown their failure to understand him. But the cured man is said to "proclaim" to them, and we are told that "they were all amazed," i.e., the message to them is of the same nature as that given to the multitudes and they react in the manner of the multitudes. The comparison with 2:1-12 and 3:1-6, which Minette de Tillesse draws, is quite invalid, for the enemies of Jesus are not amazed; they rather plot his death.

[223] "Markusevangelium," 84-92.

serves Mark as a principle of composition which enables him to present the disciples' communion with Jesus as an event of the past without having to interfere with the traditional image of the earthly Jesus already richly embroidered with the traits of the risen Lord. The disciples of Mark are the instruments of historical continuity between Jesus' activity on earth and the church; however, in their failure to understand and their betrayal and desertion during the passion they are the antithesis of the church.

It may be unfair to judge an opinion presented on only a few pages and breaking with a strong exegetical "tradition." But some reservations may be expressed; some of them have already been voiced. Is it correct to base a theory of the secret on a datum never mentioned in the Gospel, viz., the reception by the disciples of insight into the mystery of Jesus' person and destiny at their meeting with the risen Lord? The confession of the centurion in 15:39 militates against this assumption of Roloff. The likelihood that the disciples in 8:27—10:52 are cast in the mold of Mark's community throws doubt on his contention that in the matter of understanding the Twelve are an antithesis of the church. Mark's concept of understanding does not permit us to think of his church as a community which has arrived at the full measure of insight and knowledge.

Of the three opinions presented above we would opt for the second one, with the reservations indicated. Of the many varieties of this view the one which is most convincing is that of E. Schweizer. By means of his theology of the cross Mark is struggling against tendencies arising from a tradition which, too one-sidedly, stressed the "divine man" characteristics of the earthly Jesus. Only in the light of Jesus' death in obedience to God's will and of his resurrection can his acts of power be seen aright. Jesus is not an independent agent, not a wonderworker who wishes to attract attention to himself; rather, he sees himself as an instrument in God's saving hands. True insight will come to the disciple only if he is willing to follow Jesus on his way to the cross.

(3) The content of the mystery of the kingdom and of the messianic secret is the same: the kingdom is brought into being by Jesus' obediently going to the cross. In his word, work, and destiny the eschatological powers of the kingdom have become operative in the world. Those to whom this mystery has been entrusted have accepted him as the instrument in the hand of God for the definitive salvation of the world and the destruction of all satanic forces. In order to understand a little more fully the implications of the first part of the logion in Mk 4:11 we must consider the immediate context in which it has been placed by Mark.

This context is the parable of the Sower and its interpretation, by means of which Jesus exercizes his characteristic eschatological activity of teaching.

The parable of the Sower has received such varied interpretations,[224] that some exegetes refuse to decide on its original meaning.[225] The basic problem lies in the question where one should place the accent. Some feel that the difference in soils is being stressed;[226] the parable would be an illustration of the dispositions with which the Word is received and of the responsibility connected with hearing it. Related to this is the interpretation which sees in the parable an illustration of the fate of the word.[227] Others, and these seem to be in the majority today, feel that the main point of the parable is found at the end, i.e., in the harvest.[228] Accordingly the parable speaks of the coming kingdom. This interpretation seems most likely, especially because of its association—an association undoubtedly effected long before Mark came to write his Gospel—with two parables which speak expressly of the kingdom. Another reason lies in the parable itself: the harvest is the natural climax of the story; in its light everything else is judged. It is the harvest which gives meaning and purpose to the work of the sower. The failure of the seed which falls on the path, on the rocky ground, or among thorns lies in the fact that it contributes nothing to the final success of the harvest. Having said this, however, we must admit that it is somewhat difficult to regard the seed which fails as a mere foil to the seed which bears fruit; this feature of the story is too clearly pronounced to be completely neglected. An interpretation which does justice to both elements, i.e., the paramount importance of the harvest as well as the broad description of the failure, is that of Jeremias. He sees it as a contrast parable: "In spite of every failure and opposition, from hopeless beginnings God brings forth the triumphant end which he had promised."[229] No matter how humble the beginning may seem to be, it contains the seed of future glory.

[224] For a summary presentation of these interpretations, see V. Taylor, *Mark*, 250-51; for another summary, see R. Schnackenburg, *God's Rule*, 146-50.

[225] See R. Bultmann, *History*, 200; E. Klostermann, *Markus*, 39; A. M. Hunter, "Interpreting the Parables," *Int* 14 (1960) 171-72.

[226] See C. Masson, *Paraboles*, 39; C. E. B. Cranfield, *Mark*, 148-51; G. E. Ladd, *Kingdom*, 225-26; G. Lundström, *Kingdom*, 237-38; E. Linnemann, *Gleichnisse*, 120-23; A. George, "Le sens de la parabole des semailles," *SacPag* 2, 166-67.

[227] B. Gerhardsson, "The Parable of the Sower and Its Interpretation," *NTS* 14 (1967-68) 186-87; H. Kahlefeld, *Parables and Instructions in the Gospels* (Montreal: Palm, 1966) 17-23.

[228] See J. Jeremias, *Parables*, 149-51; R. Schnackenburg, *God's Rule*, 149-52; E. Fuchs, *Hermeneutik*, 226; D. E. Nineham, *Mark*, 135; V. Taylor, *Mark*, 254; W. Grundmann, *Markus*, 90; E. Schweizer, *Mark*, 91; J. Schmid, *Mark*, 101-2; M.-J. Lagrange, *Marc*, 111; N. A. Dahl, "The Parables of Growth," *ST* 5 (1952) 152-54; E. Jüngel, *Paulus*, 151, n. 4.

[229] *Parables*, 150.

But how did Mark look upon the parable of the Sower? That he connected it with the theme of the kingdom seems indicated by his insertion of the logion on the mystery of the kingdom between it and its explanation. But did he see in it a contrast between the humble beginnings of the kingdom in the ministry of Jesus and its glorious coming in his return? The explanation in vss. 14-20 clearly applies the details of the parable to the problems and the difficulties of the primitive church. That Mark made his own the understanding of the parable contained in the explanation is shown by the redactional vs. 13. For him that explanation contains deeper insights than the parable itself, for it forms a part of the teaching given to the disciples in private.[230] The attention of the explanation no longer lies in the harvest, as it does in the parable, but in the causes which bring about the failure of the word;[231] its interest lies in various evils that hinder the acceptance of the Christian message.[232] Parenesis has the upper hand; instead of proclamation a warning is addressed to the community which is well aware that many have refused to accept or have rejected the good news, and is itself in constant danger of becoming untrue to its calling. Christians are admonished not to become like the first three groups of hearers.[233]

But has this explanation deprived the parable of its eschatological message? Has the parable been applied to the present condition of the community in such a way that it has lost all references to the future? V. Taylor thinks that those who hear and accept the word in the explanation "are a mere foil to the discreditable types."[234] Yet Mark's redactional insertion in vss. 11-12 with its sharp contrast between those who are given the mystery and those outside argues in favor of the view that he sees the parable and its explanation in a similar light.[235] The breadth of description in vs. 20 also indicates that the fourth group of hearers is no foil to the other three groups. Vs. 20 tells the community where it must stand and assures it that failure is not the last word.[236] The question, however, has not yet been answered. Does not vs. 20 voice merely confidence in the success of the Christian mission and the spread of the church? Here we must recall what has already been said about the interpretation of parables in general.

[230] See J. Gnilka, *Verstockung*, 41.
[231] See J. Jeremias, *Parables*, 150; V. Taylor, *Mark*, 258; S. E. Johnson, *Mark*, 91; J. Schmid, *Mark*, 101-2; D. E. Nineham, *Mark*, 140; G. H. Boobyer, "Redaction," 66; C. Masson, *Paraboles*, 37-38.
[232] See T. A. Burkill, *Revelation*, 111; D. O. Via, "Understandability," 430.
[233] See E. Linnemann, *Gleichnisse*, 123-25; R. Schnackenburg, *Markus*, 108-9; E. Haenchen, *Weg*, 169-70.
[234] *Mark*, 261.
[235] See J. M. Robinson, *Problem*, 77.
[236] See D. E. Nineham, *Mark*, 140.

Such interpretations are not primarily translations of images into their allegorical counterparts; rather, they are a call to the community to attend to the message of the parable which is being interpreted. They spell out the demands which the future announced by the parable imposes on the present. If sowing is given greater play in vss. 14-20 than in the parable this does not mean that the harvest has been forgotten. C. E. B. Cranfield remarks: "The harvest of v. 20 is eschatological, not psychological, and the implication of vv. 14-20 as a whole is that the seriousness of the question how the Word is received derives from the fact that it is the Word of the Kingdom of God that has come near to men in Jesus, and that their final destiny depends on their reception of it."[237]

Thus we cannot agree with S. Schulz[238] who regards the parables of Mark 4 as a thoroughly de-eschatologized parenesis speaking of the church and its mission. He bases his view on the fact that vss. 11-12 are addressed to the community first and foremost and on the interpretation of the Sower in vss. 14-20. But, we repeat, in vss. 14-20 it is the sowing that is interpreted, not the harvest. The community sees itself, its disappointments and dangers in the first member of the contrast which the parable presents; the other member remains for the future. To say, with Schulz, that Mark identifies the kingdom of the parables with the church amounts to a disregard of a very important—in fact, essential—aspect of his Gospel, an aspect which dominates ch. 13 and appears in sharp contrast to the present suffering and temptations in such passages as 8:34—9:1 and 14:62. The community has, indeed, been given the mystery of the kingdom, but its present duty consists in cross-bearing. The powers of the kingdom are active in it, the teaching of Jesus received by it has set it in an eschatological dimension, but it still awaits the coming of the kingdom with power; it is not to be identified with the kingdom. The time of the community is the time of sowing; the harvest is yet to come.

(4) Here we should discuss Mk 4:21-25. It was Mark who placed these proverbs in the parable chapter.[239] The Marcan editorial phrase, *kai elegen autois,* in vss. 21 and 24 is a clear indication of this. Moreover, their content is very similar to the other editorial insertion, vss. 11-12. The manner in which these logia probably came to be collected is described by Jeremias:

---

[237] *Mark,* 161; A. George ("Semailles," 169) expresses a similar opinion.

[238] *Botschaft,* 149-56.

[239] See, among others, W. Wrede, *Messiasgeheimnis,* 70; E. Schweizer, *Mark,* 99-100; D. W. Riddle, "Source," 80-81; J. Gnilka, *Verstockung,* 61-62.

From the analysis of these verses we find (a) that, as is shown by Mt 5,15 and Lk 11,33, the originally independently transmitted metaphor of the lamp . . . attracted to itself as an explanatory comment the similarly independently transmitted logion in Mk 4,22 (cf. Mt 10,26; Lk 12,2); (b) that a similar process was repeated in the case of the word about the measure (Mk 4,24, cf. Mt 7,2; Lk 6,38), which, as a result of the verbal association *prostethēsetai/dothēsetai* attracted to itself by way of explanatory comment Mk 4,25 (cf. Mt 25,29; Lk 19,26).[240]

Jeremias is, further, of the opinion that the two pairs of logia had been joined before Mark. To us this seems to be less likely. If Mark had found them joined in the tradition he would have introduced them with a single introductory formula. It is more probable that he joined the two sets of logia;[241] that he also added vss. 23 and 24a in imitation of 4:3,9.[242] *Blepete* is a redactional term which he employs in apocalyptical contexts.[243] These two verses indicate that he looked upon vss. 21-22 and 24-25 as parables.

We need not try to discuss the meaning of these logia on the lips of Jesus; as it is, all that has been produced so far in this regard is guesswork. About their meaning in the Second Gospel there is a surprising degree of unanimity among exegetes. Almost everyone seems to think that vss. 21-22 speak of the provisional reign of the secret which is, for a time, entrusted to a few but is destined to be manifested to all very soon. There is thus a reference to the missionary activity of the community which takes place after Jesus' death and resurrection. They are a supplement and a correction, as it were, of vss. 11-12.[244] Vss. 24-25 speak of the ever deeper perception of those who open themselves to the message of the parables; the responsibility laid upon the hearers of the word and the reward of attentive hearing are stressed.[245] Some commentators find in these verses

---

[240] *Parables*, 91. T. Soiron (*Die Logia Jesu* [NTAbh VI/4; Münster: Aschendorff, 1916] 76-77) thinks that Mark is responsible for combining the last phrase of vs. 24 and vs. 25 with vss. 21-24ab. K.-G. Reploh (*Lehrer*, 61-62, 132) suggests the possibility that Mark himself combined the four logia; he refers to R. Schnackenburg ("Mk 9, 35-50," *Synoptische Studien: Festschrift für A. Wikenhauser* [München: K. Zink, 1953] 196-97) who suggests that the particle *gar* may have been a Marcan means of redactional combination of independent logia. With the reservations mentioned in the text, Jeremias' opinion seems to be more probable.

[241] See G. Minette de Tillesse, *Secret*, 172-73.

[242] See E. Schweizer, *Mark*, 100; R. Bultmann, *History*, 326; J. Jeremias, *Parables*, 91.

[243] See E. Lohmeyer, *Markus*, 86; T. J. Weeden, "Heresy," 151, n. 15.

[244] For references, see note 200.

[245] Ibid.; add C. Masson, *Paraboles*, 40-42; B. Gerhardsson, "Sower," 173-74; H. J. Ebeling, *Messiasgeheimnis*, 192.

also a reference to the missionary activity of the community:[246] the gift will grow according to the measure in which it is being communicated to others.

The interpretation of vss. 21-22 must be closely linked with our understanding of vss. 10-12. If the phrase "those about him with the Twelve" refers to the community and, consequently, the message of vss. 11-12 is addressed primarily to the community, then vss. 21-22 speak to the community first and foremost, for there is little doubt that these verses speak of the same reality as vs. 11. We cannot shift our stance from the present in connection with the earlier passage to the past in connection with the later one: if vs. 11 impresses on the community that it has already received the gift of the mystery, it is difficult to imagine that vss. 21-22 speak of something which, for the community, lies in the past.

We should recall Mark's dynamic view of revelation. He depicts Jesus during his lifetime struggling against the disciples' failure to understand, and this struggle is by no means superfluous after the resurrection. The final *hina* is characteristic of the Marcan version of these sayings.[247] True, an all-important stage in the process of enlightenment had been reached with Jesus' resurrection, but the process is not yet complete. A witness to this is the exhortation which follows in vss. 24-25 and Mark's redactional insertions in vss. 23 and 24a. The possession of the message on the part of the community is still insecure; the word has not yet permeated their hearts and minds to such a degree that it could not again become a parable to them which, instead of being salvific teaching, would turn into a verdict of condemnation. The happy news of the coming harvest becomes an admonition to be worthy of the harvest; the proclamation of the arrival of light naturally glides over into a warning not to fail this light. The spiritual condition of the community is thus not entirely different from that of the Twelve during Jesus' life on earth. The community must also struggle against hard-heartedness; it must be continually reminded of the dangers of the surrounding darkness and urged to be true to the light.

Vss. 21-22 speak of the inner dynamism of the revelation of the kingdom brought by Jesus. It was given step by step to a narrow circle of men during the earthly ministry of Jesus; now it is being granted to the church and proclaimed to the world. These two stages, differ as they may, have a provisional character; their goal is the final and definitive manifestation of the divine salvific plan which is yet to come. Vss. 24-25 stress man's

---

[246] M.-J. Lagrange, *Marc*, 115; A. E. J. Rawlinson, *Mark*, 55; J. Coutts, "Outside," 157; K.-G. Reploh, *Lehrer*, 70-71.

[247] See Mt and Lk parallels to vs. 22; also W. Wilkens, "Gleichniskapitel," 313, n. 32.

responsibility in the face of the divine gift of light. The deeper his understanding, the greater his absorption in its dynamic expansion and its growth toward the final goal. It is likely that the missionary effort of the church is envisaged in these verses as well as in vss. 21-22, for the entire passage speaks of the manifestation of the mystery in the community and in the world.

The structure of vss. 21-25 should be noted, particularly because Mark is most likely responsible for joining the two pairs of logia. This structure seems to be the same as that of 1:15. The first half of the passage announces the coming of light, the second spells out the response required on the part of man.

(5) We conclude with a summary of the main results of our discussion.

The contrast in the logion in vss. 11-12 between those outside and those about Jesus with the Twelve leads to the conclusion that "those outside" are not primarily a historical group of people in Jesus' ministry but a group that belongs to the period after Easter. We base this suggestion on the likely thesis that the phrase "those about him" in vs. 10 refers to the Christian community. In Pauline writings the term "those outside" is applied to people who do not belong to the community. It seems to have the same significance for Mark. The content and the context of the logion would, however, point to a more precise definition: vs. 12 and the interpretation of the parable of the Sower suggest that "those outside" are men who have heard the Christian message[248] but have rejected it for various reasons. The two post-resurrection groups, i.e., the community and those who have rejected the good news, have their counterparts in the period of the earthly ministry of Jesus. The predecessors of "those outside" are the enemies of Jesus who hear and perceive intellectually the meaning of his words, but refuse to subject themselves to them and to the claims they contain; they are thus incapable of true understanding. The predecessors of the community are the disciples whose hardness of heart and fear place severe obstacles on the way of true insight, yet the divine gift of the mystery overcomes these obstacles step by step.

The parable, being a vehicle of Jesus' teaching, shares in the power of his eschatological activity. It is an instruction about the kingdom, but above all it is a call to man's heart; as teaching, it is a means of the arrival of the kingdom and a manifestation of its hidden presence. It participates not only in the saving aspect of this arrival, i.e., enlightening, calling and promising, but also in its punitive aspect, by blinding those who have not

---

[248] It is generally assumed that *ho logos* has this meaning; see J. Schniewind, *Markus*, 66; V. Taylor, *Mark*, 259.

been granted the mystery. Some of its darkness continues to cling to the parable even within the community, for what it proclaims has yet to be fully manifested.

The mystery of the kingdom, finally, consists in the fact that the divine plan of salvation is already at work in Jesus and the community. The kingdom is present, but it is still a hidden one, tending, with all its being, toward full revelation. Its inner dynamism brings about the revelation of the messianic secret to the Twelve during the public ministry of Jesus, to the church and through it to the world after Easter. Mark is convinced that the period of hiddenness will soon be over.

(2) Mark 4:26-29

He also said: "This is how it is with the kingdom of God. A man scatters seed on the ground. [27]He goes to bed and gets up day after day. Through it all the seed sprouts and grows without his knowing how it happens. [28]The soil produces of itself first the blade, then the ear, finally the ripe wheat in the ear. [29]When the crop is ready he 'wields the sickle, for the time is ripe for harvest.' "

(A) The Text and the Original Form of the Parable

(1) The differences among the critical editions in regard to the text of the parable of the Seed Growing Secretly are of little consequence for its interpretation. In vs. 26 Vogels alone adds *ean* to *hōs*. That *hōs* alone should be retained as the *lectio difficilior* is evident from the variant readings in some MSS which attempt to improve the text by replacing *hōs* by *hōsper*, by adding *ean* to *hōs*,[249] or inserting *hotan* after *anthrōpos*. In vs. 28 von Soden and Vogels prefer *eita* to the *eiten* of all other critical editions, which base their choice on the *prima manus* of the Codices Vaticanus and Sinaiticus. "The Ionic form *eiten* is not uncommon in the papyri."[250] *Plērē siton* is the reading most frequently found in MSS; it is accepted by Vogels, von Soden, Westcott-Hort, Merk, and the United Bible Societies edition; the reading *plērēs sitos*, found in the Codices Vaticanus, Bezae, and Freer (the last two, however, insert the definite article between the words), is accepted by Tischendorf, Nestle, and Tasker. V. Taylor[251] thinks that "the reading *plērēs siton* (C), in which the adj. is an indeclinable, is probably

---

[249] Bauer-Arndt-Gingrich, *Lexicon*, *hōs* II, 4c: "It is likely that *an* . . . once stood before *anthrōpos* and was lost inadvertently," since the phrase *hōs anthrōpos balē* "is gravely irregular fr. a grammatical viewpoint."
[250] V. Taylor, *Mark*, 267.
[251] Ibid.

original; it accounts best for the variants." For our purpose it is not necessary to opt for any one of these readings preferred by the critical editions or commentators. In vs. 29 Vogels and von Soden take the reading *paradō* in preference to *paradoi* of other critical editions. "The form *paradoi* is subj., a vernacular ending well illustrated in the papyri."[252]

(2) E. Lohmeyer has observed that the basis and the beginning of the parable are Aramaic:[253] only in the Orient does the day begin in the evening, so that people sleep first and then rise. He thinks, however, that the phrase *hōs ouk oiden autos* is influenced by Latin usage, and that the phrase *hotan paradoi ho karpos* is possible only in *Koine* Greek. E. Percy[254] objects to Lohmeyer's suggestion that the first phrase was formed under the influence of the Latin "dum nescit ille," since this is not a common Latinism; and secondly, the role which Lohmeyer attributes to *hōs* finds no analogies in Greek. As to the second phrase, M. Black[255] has shown that it can be traced back to the Aramaic. The paronomasia in his tentative retranslation of the parable into Aramaic is striking.

R. Bultmann[256] has questioned the originality of the introductory words of the parable, "The kingdom of God is," and thinks that it began with *hōs anthrōpos balē*. He is followed by G. Harder[257] (who suggests, however, that indicatives had stood in the place of the subjunctives), E. Jüngel, and E. Grässer.[258] J. Dupont [259] thinks that the introduction may be redactional. Bultmann gives two reasons for his opinion: "It is not easy to relate this similitude to the Kingdom of God, and . . . it gives the impression of being one of the introductory formulae that are frequently added." The first reason can be accepted only on the supposition that we know all the possible nuances of Jesus' concept of the kingdom and that the precise message of the parable has already been decided upon—neither of which is really the case. W. G. Kümmel[260] sees no reason for striking the introductory sentence, and E. Percy[261] considers it unlikely that a parable would begin with an

---

[252] V. Taylor, *Mark*, 268.

[253] *Markus*, 86.

[254] *Die Botschaft Jesu: Eine traditionskritische und exegetische Untersuchung* (Lund: Gleerup, 1953) 203, n. 2.

[255] *Approach*, 164-65.

[256] *History*, 173.

[257] "Das Gleichnis von der selbstwachsenden Saat," *Theologia viatorum* 1 (1948-49) 52.

[258] *Paulus*, 149; *Das Problem der Parousieverzögerung in den synoptischen Evangelien und in der Apostelgeschichte* (BZNW 22; Berlin: Töpelmann, 1960) 145.

[259] "Semence," 383.

[260] *Promise*, 128, n. 81.

[261] *Botschaft*, 204.

"as" without some reference to what its narrative is being compared. Bultmann's second reason is by itself hardly sufficient to throw doubt on the originality of the introductory phrase. Dupont's suggestion is hardly more convincing; he offers no instance of an insertion like that of Mk 4:26a in the rest of the Gospel.

In the past there were exegetes[262] who considered vs. 29 to be secondary; with the exception of A. Suhl[263] who accepts A. Jülicher's opinion[264] that the verse is either an addition or an elaboration of Mark, hardly anyone today still defends that view. The quotation of, or rather the allusion to, Joel 4:13 may be secondary,[265] but the growth described in the parable almost requires the harvest at the end. J. Wellhausen's remark, which Suhl quotes with approval,[266] that vs. 29 overshoots the mark is exposed to the same objection as that voiced against Bultmann's reason for omitting the introductory words of the parable. Apart from recalling Jülicher's and Wellhausen's views, Suhl hardly offers any proof that it was Mark who inserted the OT quotation. His linking of vs. 29 with the rest of the parable chapter and 13:10[267] serves only to determine Mark's understanding of the verse. If the OT quotation is secondary, it was much more likely added before Mark. It seems, in fact, impossible to discern Marcan redaction in the parable.[268]

Hence Mark reproduces the parable as he found it in the tradition, and, with the possible exception of the allusion to Joel 4:13, the parable as a whole has its home in an Aramaic-speaking environment. It does not seem reasonable to question its substantial authenticity simply because one or another exegete's understanding of its message does not tally with his understanding of Jesus' concept of the kingdom.[269]

(B) Interpretation of Jesus' Parable

(1) "The variety of interpretations is truly astounding." This statement of F. W. Beare[270] elicits only a despondent Amen. It must be admitted,

---

[262] For references, see W. G. Kümmel, *Promise*, 128, n. 82.
[263] *Funktion*, 154-155.
[264] *Gleichnisreden* 2, 545.
[265] See G. Harder, "Saat," 53; J. Dupont, "Semence," 381. W. G. Kümmel (*Promise*, 128, n. 82) and N. A. Dahl ("Parables," 149) feel there is no reason to consider the OT quotation to be secondary.
[266] *Funktion*, 154; he gives the reference to J. Wellhausen.
[267] *Funktion*, 155-57.
[268] See J. Dupont, "Semence," 388.
[269] G. Harder ("Saat," 69-70) and E. Grässer (*Parousieverzögerung*, 145) think it possible that the community produced the parable.
[270] *Records*, 113.

of course, that the form of the parable lends itself to this variety. J. Dupont outlines the possible centers of attention within it:[271]

(a) We can place the emphasis either on the fate of the seed or on the behavior of the sower or on both.

(b) If the seed is chosen, is its growth or the certainty of the harvest to be stressed?

(c) If the sower is chosen, is it his inactivity at the time of growth or his activity at the harvest which is important?

(d) What should the conduct of the sower illustrate: the conduct of God in setting up his kingdom, the conduct of Jesus in fulfilling his mission, the conduct of Zealots who are being exhorted to patience, or the conduct of discouraged disciples who are urged not to lose heart?

Dupont is aware that this outline does not exhaust all the possibilities of interpretation; yet it represents the lines of investigation in the past and present. It is superfluous to trace the history of the earlier interpretations of the parable in detail; the determining factor in this history has been the understanding which various interpreters have had of the kingdom. Summaries of opinions have been given by V. Taylor,[272] G. Harder,[273] Dupont,[274] and others.[275] The most thorough of these is Harder's, but the one most germane to the matter is Dupont's. This summary of earlier views we shall present briefly.

The "classical" interpretation finds in the parable a teaching on the progressive development of the kingdom in the world. It speaks of the time of the church during which the seed sown by Jesus grows irresistibly and independently of men's attitude toward it; its inner energy is strong enough to overcome all obstacles. Its principal message consists in an appeal for confidence in the future of the kingdom in the world. The principal names mentioned in connection with this view are those of A. Jülicher, P. Feine, and D. Buzy.

Advocates of thorough-going eschatology believed they found in the parable a confirmation and illustration of what they considered to be the specific trait of Jesus' teaching on the kingdom, viz., its imminence. The ministries of John the Baptist and of Jesus are signs of its nearness, insignificant

---

[271] "Semence," 374-75; on pp. 368-75 he gives a wide conspectus of recent and less recent literature.

[272] *Mark*, 265-66.

[273] "Saat," 53-58.

[274] "Semence," 368-73.

[275] See F. Mussner, "Gleichnisauslegung und Heilsgeschichte: Dargetan am Gleichnis von der selbstwachsenden Saat," *TTZ* 64 (1955) 257-61; J. Gnilka, *Verstockung*, 74-75.

though they may seem to be in comparison to its expected definitive manifestation. The eschatological harvest is ripening; divine intervention is not far away. The most eloquent proponent of this theory was A. Schweitzer.

C. H. Dodd saw the parable in the light of his theory of realized eschatology: "The parable in effect says, Can you not see that the long history of God's dealings with His people has reached its climax?"[276] Jesus' ministry is not a preparation for the imminent arrival of the kingdom; the time of preparation is past, and in Jesus the eschatological harvest is already taking place.

J. Jeremias interpreted the parable with a different concept of Jesus' eschatology in mind: neither the realized eschatology of Dodd nor the thorough-going eschatology of A. Schweitzer, but an eschatology in the process of realization. According to him, the parable contrasts the patient waiting of the farmer and the harvest which constitutes the reward of his waiting. "Thus with the same certainty as the harvest comes . . . , does God when his hour has come, . . . bring in the Last Judgement and the Kingdom."[277] The parable was told to dampen the impatience of the disciples who were under the influence of zealot mentality and disturbed by Jesus' apparent inactivity. N. A. Dahl is also of the opinion that "the essential contrast is not that between the passivity of the husbandman and the growth of the seed, but that between his passivity during the time of growth and his hurry to put in the sickle at the moment the grain is ripe."[278] But the parable has more to say: "between the coming of the Kingdom and the ministry of Jesus in Israel there is a relation similar to that between the harvest and the time of sowing and growth which has to precede it. . . . the forces of the Kingdom are at work."[279]

Thus far Dupont's summary of earlier interpretations. Under the influence of the nineteenth-century reaction to the theories of D. F. Strauss, the "classical" view of Jülicher gave insufficient attention to the future aspect of the kingdom.[280] This view was swept aside by a theory diametrically opposed to it, viz., that of thorough-going eschatology. That in its turn gave rise to Dodd's realized eschatology, and with Jeremias a movement set in that tried to do justice to the future as well as to the present aspects of the kingdom in the parables.

[276] *Parables*, 135; V. Taylor (*Mark*, 266) accepts this view.
[277] *Parables*, 151-52.
[278] "Parables," 149.
[279] "Parables," 150.
[280] See G. R. Beasley-Murray, *Jesus and the Future: An Examination of the Criticism of The Eschatological Discourse, Mark 13, with Special Reference to the Little Apocalypse Theory* (London: Macmillan, 1954) 1-32, especially p. 28.

(2) Some further interpretations have been offered in recent years.

According to R. Schnackenburg,[281] the ending of the parable gives the rest its meaning and significance. At the harvest, the symbol of the definitive arrival of the kingdom, the growth has its meaning. It is not the farmer, in his idleness and activity, who is the nub of the story; his inactivity and failure to notice the growth enhance the irresistibility of the "automatic" ripening of the crop. The parable stresses that God alone is at work in setting up the kingdom: no amount of human activity can hasten its arrival, and it cannot be discerned in historical development. The harvest has not yet arrived; at harvest time one does not think of the process of growth. And the calm composure of the farmer speaks against the interpretation that finds in the parable an annunciation of the speedy arrival of the kingdom. The hidden presence of the kingdom in the activity of Jesus is implied. The parable probably served to counteract the desire that he fulfil the mission of a political messiah.

J. Gnilka's interpretation[282] is similar. The salient point of the parable is the harvest; contrasted to it is the apparently unconcerned attitude of the farmer during the period of growth. Its message can only be that the coming of the kingdom is as certain as is the harvest, once the seed has been sown; the beginning contains the guarantee of the glorious climax. It thus suggests that with Jesus' ministry the kingdom is already setting in in a mysterious manner. The time between sowing and reaping is in the hands of God who is bringing about the full realization of the kingdom according to a predetermined plan. Following O. Kuss[283] and N. A. Dahl,[284] Gnilka rejects Jeremias' opinion that "the people of the Bible, passing through the plough-land, look up and see miracle upon miracle, nothing less than resurrection from the dead."[285] In Dahl's words, "the growth of seed and the regularity of the life in nature have been known to peasants as long as the earth has been cultivated."[286] Biblical man was convinced that nature and history are subject to God's government, but he did not look upon these processes as miraculous. Unlike Schnackenburg, Gnilka thinks that the farmer holds the center of the stage. He is not certain whether this figure should be allegorized or not, and if so, whether it represents God or the Messiah. Presumably in order to keep the possibility

---

[281] *God's Rule*, 153-54.
[282] *Verstockung*, 74-78.
[283] "Zum Sinngehalt des Doppelgleichnisses vom Senfkorn und Sauerteig," *Bib* 40 (1959) 648-52.
[284] "Parables," 140-47.
[285] *Parables*, 149.
[286] "Parables," 141.

of allegory open, he characterizes the unconcern of the farmer as apparent. However, we must confess that we find him somewhat confusing in this regard. If we must admit the possibility that the farmer represents God and at the same time hold to the idea of God's bringing about the kingdom according to a set plan, how are we to reconcile this with the fact that the parable contrasts the farmer's idleness with the lively fruitfulness of the earth?

E. Fuchs[287] believes that the main point of the parable lies in the difference between the work of the earth and the passivity of man: during the time which elapses between sowing and reaping the farmer need not worry about the fate of the seed, for the earth is taking care of it, and he is free for other things. The parable thus indicates the difference between the present of man's freedom and the future of God's kingdom. Man is free in the present precisely because of the certainty that the future is in the hands of God; his freedom is conditioned by his decision to accept this certainty. This freedom is a responsible freedom, for the present is not unlimited. If Jesus spoke this parable, this is the only meaning that it could have had; he was addressing it to the present, in which the word of God comes to us, making us share in his own certainty of God and his future.

E. Jüngel's interpretation[288] follows, to a degree at least, in E. Fuchs's footsteps. The eschatological *euthys* shows that the ending of the parable is decisive for its interpretation. Yet the word *automatē* also has an eschatological significance.[289] On these two words the parable rests. The two terms teach us the difference in times: the first indicates that the time of work is past, the second that the time of idleness is past. What is taken for granted by the farmer thus constitutes the specific characteristic of the kingdom: man can do nothing for it, though everything depends on it. As the certainty of the harvest is the cause of the farmer's calmness, so does the composure of the one who listens to the parable spring from his utter trust in the future which lies in God's hands. His present is thus a time of freedom from the past (sowing) and freedom for the future (harvest). The parable grants its listeners time to hear the word.

[287] "The Theology of the New Testament and the Historical Jesus," *Studies of the Historical Jesus* (SBT 42; London: SCM, 1964) 179-84; "Jesus' Understanding of Time," *Studies*, 133-34.
[288] *Paulus*, 149-51.
[289] In this he disagrees with G. Harder ("Saat," 61), for whom the word has a negatively concessive value (i.e., even though the sower leaves the seed to itself for a time, he will care again when it is ripe).

Fuchs admits that freedom is not mentioned in the parable;[290] it is, however, self-evident, according to him, once we perceive that the earth has liberated the farmer from the worry about the seed. This interpretation may be correct; but, in order to persuade oneself that the central message concerns the freedom of those who share Jesus' certainty of the future which is in God's hands one must share Fuchs's confidence that he has discovered Jesus' understanding of existence.[291] There is something repellent in his statement that the parable can be attributed to Jesus only on condition that it expresses this understanding.[292] There comes to mind the remark made by the Jewish scholar, S. Sandmel:[293] "The Jesus of the nineteenth century scholarship which Schweitzer surveyed never existed. The new quest can at best turn up a twentieth century Jesus who never existed." Regarding Jüngel's assertion that what is taken for granted by the farmer constitutes the specific characteristic of the kingdom, one wonders whether the farmer's freedom from care about his field is quite as obvious as he seems to assume. What is taken for granted is clearing, ploughing, harrowing, worry about drought and storms.[294]

H. Kahlefeld[295] has travelled a rather independent road in his interpretation: as in the parable of the Sower, so also in this one the seed is a symbol of the word of God which is being proclaimed. Its message: The proclaimer of the word is freed from the care and worry which exceed his strength. Power resides in the good news itself, for the word of God is irresistible. However, this interpretation exposes itself to at least one objection: the parable does not attribute the power of growth to the seed, but to the earth.[296] Another recent interpretation is that of H. Baltensweiler.[297] According to him the sower does not know that growth is taking place because he does not want to know. The parable should thus be entitled "The Unbelieving Farmer." It is the grotesque story of a man who sows without confidence. Its background lies in the fact that some disciples have left Jesus (Jn 6:64,66). It was told to stress the truth that the kingdom will come despite their falling away. J. Dupont's comment on this inter-

---

[290] "Understanding," 134.
[291] This understanding is presented in his *Hermeneutik*, particularly pp. 149-58, 175-76, 196-97, 239-48.
[292] "Theology," 180.
[293] "Prolegomena," 300.
[294] See E. Schweizer, *Mark*, 102.
[295] *Parables*, 23-24.
[296] See J. Dupont, "Semence," 370, n. 10.
[297] "Das Gleichnis von der selbstwachsenden Saat (Mc 4, 26-29) und die theologische Konzeption des Markusevangelisten," *Oikonomia* (ed. F. Christ) 69-75.

pretation:[298] "Ces élucubrations se fondent évidemment sur une mauvaise lecture du texte."

The most recent interpretation is that of Dupont.[299] He begins his study with an outline of its structure: Vs. 26 describes the sowing, vss. 27-28 the growth, and vs. 29 the harvest. The act of sowing is a mere preliminary. Growth is described in detail, but greater attention is paid to the farmer. The purpose of the term *automatē* is not to stress the fruitfulness of the earth, but to point out that the farmer need not worry. The center of interest is the sower, not the seed. The message of the parable is to be sought in the contrast between vss. 27-28 and vs. 29; the sudden transition from idleness to intense activity is brought out particularly by the words *de* and *euthys* of vs. 29. The farmer is to be taken as a symbol of God, and this for two reasons: the ending of the parable speaks of God's judgment; the principal figure of a kingdom parable should normally be presumed to symbolize God, for the coming of the kingdom is his work. The problem which Jesus attempted to resolve lay in the apparent absence of God during the time of his ministry; his ministry does not give the impression of being the judgment which John the Baptist had predicted (cf. Mt 3:10,12). Jesus replies: God has already begun the process which leads to the definitive establishment of the kingdom, but he is waiting until this ministry produces its fruit. The present, characterized by an apparent absence of God, is precisely the period which immediately precedes his final intervention. The parable thus enunciates the contrast between two periods, that of Jesus' ministry and that of God's judgment; but it also enunciates the continuity, for the two periods are intimately linked to each other—Jesus' ministry is defined by its connection with the final judgment. Like Fuchs and Jüngel, Dupont finds the difference of times described in the parable. But his interpretation is better founded exegetically, and not forced to serve preconceptions which may, or may not, correctly assess Jesus' understanding of existence. Doubts about his interpretation arise on other scores. Does he not allegorize unduly? Can *hōs ouk oiden autos* be made to mean no more than an apparent unconcern? Can the entire vs. 28 be taken only as an illustration of the fact that the farmer need not take anxious care? With these questions in mind we take a closer look at the structure of the parable.

(3) Dupont's outline of the structure of the parable is quite obvious; sowing, growth and harvest are three clearly discernible steps in the progress of the narrative. Yet more should be said about the arrangement of the

---

[298] "Semence," 374, n. 18; but see below, note 329.
[299] "Semence," 376-88.

clauses and the "dramatis personae," particularly in vss. 27-28. Vs. 26 presents the actors; the man who sows, and the ground which has been seeded. Vs. 27 begins by taking our eyes away from the grain in the ground and directing them to the man:

>> he goes to bed and gets up
>> day after day;

the other actor is brought to the stage:

>> the seed sprouts and grows;

enter the man again:

>> how, even he does not know.

E. Lohmeyer[300] noted that the parable is divided into three sentences, the first of which is divided into two clauses, and the second and third into three clauses. The story has an internal rhythm. But the third clause of vs. 27, though harmonizing well with the overall rhythm, sharply breaks the pace of the verse itself:

>> he goes to bed and gets up
>> day after day;
>> the seed sprouts and grows.

The last clause, however, brings a sudden end to this placid scene; by stating the farmer's ignorance it raises a question. Two words within this clause are emphasized: *autos* at the end, [301] and *hōs* at the beginning. How should *hōs* be translated? Bauer's *Lexicon*[302] takes the particle as corresponding to *houtōs* and translates the clause: "(in such a way) as he himself does not know = he himself does not know how, without his knowing (just) how." The particle could also be relative-interrogative, as it appears in such phrases as *oisth' hōs meteuxei* and *oistha . . . hōs nyn mē sphalēs*,[303] *exēgounto . . . hōs egnōsthē autois* (Lk 24:35).[304] It is most natural to take *hōs* as expressing an indirect question since the clause is preceded by the description of growth and followed by the emphatic *automatē* which clearly supplies the answer to the question implied in the clause.[305] The

---

[300] *Markus*, 86.
[301] See E. P. Gould, *Mark*, 80.
[302] Bauer-Arndt-Gingrich, *Lexicon*, *hōs* I, 1; see also H. G. Liddell and R. Scott, *A Greek English Lexicon: A New Edition Revised and Augmented throughout by H. Stuart Jones with the Assistance of R. McKenzie* (Oxford: Clarendon, 1961), *hōs* A a.
[303] Ibid., A c.
[304] Bauer-Arndt-Gingrich, *Lexicon*, *hōs* I, 2d.
[305] See A. Jülicher, *Gleichnisreden* 2, 540; E. Percy, *Botschaft*, 203; D. E. Nineham,

man's ignorance concerns the manner, not the fact, of growth. Like the emphatic *automatē* at the beginning of the following sentence, *hōs* is placed at the head of the clause for the sake of emphasis.

In the first two verses of the parable, there is the quick switching of stagelights from one actor to the other: man, seed in the ground, man, seed, man. The movement slows down in vs. 28; the entire verse is devoted to the earth with its seed, replacing the sower of vs. 27c. The word *automatē* is strongly emphasized; and this argues against the assumption that its only function is to stress the man's freedom from care. The minute detail, moreover, with which the growth and its results are described, not only in vs. 28 but also in vs. 29 two of whose three clauses are devoted to the seed, strongly suggests that this aspect of the narrative should be considered the most important. G. Harder's opinion[306] that the stages of growth serve only to increase tension can hardly be correct. What tension could have been produced in an audience thoroughly familiar with the process of growth?

Vs. 29 is set off from the rest of the narrative by means of *de* and *euthys*. These words indicate that something decisive has happened, that the climax has been reached: not only has the ground with its seed produced fruit, but the man also becomes suddenly active again. Note too that the stagelights continue switching in the same order as before; in vs. 29a they center on the seed; in 29b they focus on the farmer; and in 29c they return to the now ripe fruit.

The principal role, then, should be assigned, not to the man, but to the irresistible growth and its climax. The attention of the story alternates indeed between the farmer and the growth, yet the farmer is a foil to the growing seed. This seems to be the thrust of vs. 27c, for the question it raises with such abruptness concerns the manner of growth. Even in vs. 29 the farmer, though active, seems to be passive with regard to the ripe seed. The middle clause, which alone describes his action, is flanked by clauses which suggest that his resumption of work is determined by the fact that the harvest has arrived. The last clause in particular points out the man's dependence on the earth and its seed, restating what has already been said in the first. It is, admittedly, part of the OT quotation; but this should not be overstressed, for the freedom with which the OT text is handled, here as well as in 4:32, makes it an allusion rather than a quota-

---

Mark, 143. Objections of G. Harder ("Saat," 60) and of H. Baltensweiler ("Saat," 72) are answered by the examples of Greek usage given above. Lohmeyer's suggestion has already been discussed.

[306] "Saat," 61.

tion.³⁰⁷ Keeping that in mind, we need not, with M.-J. Lagrange,³⁰⁸ assume that the last verse makes an allegory out of the parable. Interpretations which make of the sower the salient point of the story³⁰⁹ tend to allegorize it and do not do justice to the emphatic statement of vs. 27c.

(4) There is little doubt that we should interpret the parable from the point of view of its ending, i.e., the harvest.³¹⁰ The entire narrative tends toward it, and the manner in which vs. 29 is set off from the rest of the story is a clear indication of this. But we must not sacrifice the term *automatē* to the ending; it is too evidently stressed to be given such treatment. These two elements of the parable must be taken in conjunction; not only is the growth toward maturity described, but also its manner. The coming of the kingdom is being proclaimed, but that is not the whole message; the parable has something to say about the manner of its arrival. Its coming is, first of all, irresistible and certain; once the seed has been sown, the harvest becomes a necessary consequence. This coming is, secondly, totally independent of human effort; man sleeps and rises, and has no inkling how the growth is taking place. Vs. 26 should not be urged against this, for it simply sets the stage; it lies in the nature of things that seed must be sown by someone. The term *automatos*, infrequent in the OT as well as in the NT, may still carry with it OT overtones of spontaneous growth.³¹¹ Since the arrival of the kingdom is not to be attributed to man, it must be the work of God alone.

Not only is this coming certain and irresistible, and totally independent of man's will and action; it is also inscrutable and ineffable.³¹² Vs. 27 seems to stress the farmer's ignorance of the manner of growth more than his idleness. Translations and interpretations which render vs. 27c as "without his noticing it" or "without his taking anxious thought"³¹³ fail to do justice to the emphatic affirmation of the man's failure to understand the process

---

³⁰⁷ See E. Fuchs, "Theology," 180, n. 1.

³⁰⁸ *Marc*, 117-18.

³⁰⁹ See the remarks above on Gnilka and Dupont. Harder ("Saat," 61-62) offers the suggestion that during the period of growth no one can tell the owner of the field, whereas at the harvest he becomes visible again.

³¹⁰ Besides the opinions summarized above, see G. Harder, "Saat," 60-62; F. Mussner, "Gleichnisauslegung," 265; W. G. Kümmel, *Promise*, 128; J. Schmid, *Mark*, 105; H. Baltensweiler, "Saat," 71.

³¹¹ Lev 25:5,11; 2 Kgs 19:29 (LXX); see H. B. Swete, *Mark*, 80.

³¹² See E. Lohmeyer, *Markus*, 87; A. E. J. Rawlinson, *Mark*, 55-56; J. K. Howard, "Our Lord's Teaching Concerning his Parousia: A Study in the Gospel of Mark," *EvQ* 38 (1966) 69.

³¹³ J. Jeremias, *Parables*, 151; R. Schnackenburg, *God's Rule*, 153; E. Fuchs, "Theology," 180; D. E. Nineham, *Mark*, 143.

of growth. Vs. 27c is not a mere restatement of 27a; it introduces a new thought and asks a new question.

How should we think of this growth? We have already referred to the opinions of Dahl, Kuss, and Gnilka in this matter. Dahl has shown that the idea of organic growth was not unknown to the Jews of Jesus' time, either in agriculture or in history.

History is not understood as an immanent evolution; at all stages it is governed by God, or by the struggle between God and the evil forces. But history also follows a determined order, where one period follows another according to the ordinance of God. This conception lies at the bottom of the apocalyptic conception of the periodicity in history.[314]

To prove his point, he refers to *2 Esdras* 7:74; 4:28-29,35-40; *2 Apoc. Bar.* 70:2; Jas 5:7-8—passages in which the divine governing of history is compared to the process of growth in nature. He concludes:

We have every reason to assume that also in the parables of growth the idea of growth is used in a similar way. To the growth which God in accordance with his own established order gives in the sphere of organic life, corresponds the series of events by which God in accordance with his plan of salvation leads history towards the end of the world and the beginning of the new aeon.[315]

Bultmann objects to this notion: "But is it possible to think of the apocalyptic idea, according to which many events should happen before the eschatological fulfilment, as an organic growth?"[316] Yet is this no more than a battle of words? It must be conceded to Bultmann that the term "organic" suggests to the present-day mind a growth due to a power inherent to the thing growing. When one speaks of irresistible growth toward maturity, one does not necessarily refer to the notion, dear to the "classical" interpretation, that the parable is meant to illustrate the stages of spiritual growth of the kingdom. G. E. Ladd summarizes this notion: "Just as there are laws of growth resident within nature, so there are laws of spiritual growth through which the Kingdom must pass until the tiny seed of the gospel has brought forth a great harvest."[317] Ladd himself opposes this idea,[318] along with practically all modern exegetes.[319] What the image of

---

[314] "Parables," 143.

[315] "Parables," 146.

[316] *History*, 418.

[317] *Kingdom*, 185; he gives a number of references; others can be found in W. G. Kümmel, *Promise*, 127, n. 79.

[318] *Kingdom*, 186.

[319] See J. Dupont, "Semence," 387-88; R. Bultmann, *History*, 418; W. G. Kümmel, *Promise*, 128.

growth toward maturity does express is the intimate link between Jesus' ministry and the full manifestation of the kingdom. There is a continuity between what is happening before the eyes of the disciples and crowds and the definitive establishment of the kingdom. What is taking place now is God's work, and what God has determined to achieve will be achieved without fail. The kingdom is thus already active, already present;[320] this presence is, of course, different from its glorious future coming, yet it is impossible to destroy it or to prevent it from reaching its goal, for the result is implicit in its beginnings. We must not identify this beginning with the small band of disciples,[321] much less imagine that the church is the growing kingdom of God.[322]

When we come to the precise problem which the parable was to solve, we can do little more than guess. One element must be stressed in this connection: Jesus is not merely affirming that the kingdom is coming; no one in his audience had doubts about that. What Jesus is affirming is that his coming and activity are intimately linked with the glorious manifestation of the kingdom in the future, that his ministry is the first step of its arrival. That there were doubts about this point is shown by such statements as Mt 11:6, "Blest is the man who finds no stumbling block in me,"[323] or Lk 12:8, "I tell you, whoever acknowledges me before men— the Son of Man will acknowledge him before the angels of God."[324] Jesus' activity was so unlike the expected kingdom, so unlike anything that might give the impression of being a preparation for the kingdom, that he had to emphasize the link between his coming and the arrival of the kingdom as well as the fact that the manner of this arrival is inscrutable and ineffable. The kingdom which *Jesus* is announcing *is* coming; it is already present in a hidden manner, the manner of its coming being so unexpected and humanly inexplicable that there is danger of missing its call. The parable

---

[320] See the opinions above of R. Schnackenburg, J. Gnilka, J. Dupont; see also N. A. Dahl, "Parables," 150; E. Percy, *Botschaft*, 204-6; G. E. Ladd, *Kingdom*, 187; G. Lundström, *Kingdom*, 234-35; J. Jeremias, *Parables*, 152.

[321] As does Jeremias, *Parables*, 149; for a criticism of Jeremias on this point, see R. Bultmann, *History*, 418.

[322] For reference to a recent example of this identification, see J. Dupont, "Semence," 370, n. 11.

[323] For its authenticity, see R. Bultmann, *History*, 23, 126; W. Grundmann, *Das Evangelium nach Matthäus* (*Theologischer Handkommentar zum Neuen Testament* 1; Berlin: Evangelische Verlagsanstalt, 1968) 304.

[324] For a discussion of the original form of this, likely authentic, logion, see N. Perrin, *Rediscovering the Teaching of Jesus* (*NT Library*; London: SCM, 1967) 186-91; J. Jeremias, "Die älteste Schicht der Menschensohn-Logien," *ZNW* 58 (1967) 168-70.

is basically an appeal to part with preconceptions and to submit to the only way in which God is setting up his kingdom, i.e., through Jesus.

(C) Mark's Understanding of the Parable

Since the parable shows no redactional intervention, we must look for Mark's understanding of it in the context. The context is the entire parable chapter, and particularly the evangelist's editorial insertions into it. Looked on as Jesus' teaching, it is no mere illustration or instruction, but an appeal to man's mind and heart to submit to its message and to the divine power radiating from it. It demands that it be understood—understood in the Marcan sense of the word, i.e., perceived in such a way that it transforms the life of the listener and the reader. This demand is brought out especially by vs. 13, which expresses two thoughts: first, the rebuke, and thus an implicit exhortation and warning to overcome the blindness characteristic of the Twelve before the resurrection of Jesus, but also of the minds and hearts of the community; secondly, that the other parables of the chapter should be read in the light of the explanation which follows in vss. 14-20.[325] Though there are no redactional insertions in vss. 14-20, with the possible exception of the phrase "persecution because of the word,"[326] vs. 13 shows clearly that Mark considers the explanation to be valid for other parables besides that of the Sower.

But what bearing does the explanation of the Sower have on the parable? It is to be noted, first of all, that the sower is not identified. Of primary concern is the fate of the seed. It is the seed which is interpreted as the Word, and its failures and success are fully described. The sower is there for the sake of the story; his feelings of frustration or elation are not recorded. Vss. 14-20 mirror the dedication to, and dependence on, the Word of the people who created and preserved them. It describes the fate of the Word in the time of the church.

Applying the explanation of the Sower to this parable, we may assume that Mark is interested primarily in the destiny of the seed, i.e., the word which is being proclaimed and is irresistibly and without human aid forcing its way toward the final manifestation of the kingdom. This word is the teaching of Jesus, an expression of that activity by means of which the divine world breaks into the world of men. The power residing in it is divine, and it cannot miss its goal, no matter how incalculable and mysterious its way toward that may seem to be. There is thus little doubt that, for Mark, this

---

[325] See C. E. B. Cranfield, *Mark*, 161; G. H. Boobyer, "Redaction," 66.

[326] See J. Dupont, "Semence," 388, n. 43; "Les paraboles du sénevé et du levain," *NRT* 89 (1967) 908.

parable is an excellent expression of the inner dynamism of revelation. The word is the light which has come in order that it may be seen by all. The community, having been given the mystery of the kingdom, is the product of this word. Its inner dynamism keeps the community and guides it ever further along its way toward the unfolding of what it already possesses. For that reason the community can rest calm and assured.[327]

Dupont[328] has identified the sowing in vs. 26 with the work of Jesus, and the growth in vss. 27-28 with the time of the church. What Mark is trying to say is, according to him, the following: Jesus had sown; now he seems to be absent and disinterested, but he will assuredly come at the time of the harvest. But this suggestion has little to commend it. Dupont's farmer seems, first of all, to be playing a double role; in vs. 27a he represents the calmly and trustfully waiting community, while in vss. 26, 27c and 29 he is identified with Jesus sowing in the past, apparently absent now, but coming again in the future. Secondly, the sower is not identified in vss. 14-20. Thirdly, vs. 27c lends itself poorly to a description of the risen Christ, as poorly as it does to a description of God in the pre-gospel stage of the parable. The rebuke administered to the disciples in vs. 13, the absence of the reference to knowledge in vs. 11, and the entire tendency of the Gospel to attribute the failure to understand to hardness of heart seem to indicate that vs. 27c, as Mark understands it, describes the community.[329] The community has, indeed, reached a degree of understanding, but this understanding is by no means perfect; even after Easter it cannot be said to know as it should know. It is still endangered by Satan and exposed to outer and inner temptations. It is perplexed by the ways of the kingdom; it is tempted to prefer its own, all too human fancies about it. In the face of persecution and "anxieties over life's demands, and the desire for wealth, and cravings of other sorts" (4:19), the unfathomable approach of the kingdom may become too much to bear. Yet the parable calls for confidence: no matter how dark the horizon or how strange the ways of God, he has decided to bring about his kingdom through the word of Jesus which the community has heard and accepted. It can therefore calmly and courageously brave the outer dangers and the inner uncertainty. Its own powers and calculations avail nothing; God alone can and will bring about the harvest.

Luke most likely omitted this parable because of its insistence on the all-sufficiency of God's power and the uselessness of human efforts. His redactional insertions and the rearrangement of pericopes with respect to Mark

---

[327] See J. Dupont, "Semence," 389; R. Schnackenburg, *Markus*, 115.

[328] "Semence," 389.

[329] It is thus possible that Baltensweiler's interpretation of the parable has not missed the mark completely.

show his emphasis on putting into practice the word which has been heard. In 8:13 (par Mk 4:17) he adds a reference to faith and temptation; in 8:15 (par Mk 4:20) he inserts a reference to holding fast to the word "in a spirit of openness." Mk 3:31-35 is placed by Luke after the logia on light and measure; to the text of Mark he adds a reference to doing the word (Lk 8:21 par Mk 3:35). But this episode is intended by him to close his parable section.[330] Matthew omits the parable most likely because vs. 27c hardly agrees with his emphasis on understanding.[331]

(3) Mark 4:30-32

He went on to say: "What comparison shall we use for the kingdom of God? What image will help to present it? [31]It is like a mustard seed which, when planted in the soil, is the smallest of all the earth's seeds, [32]yet once it is sown, springs up to become the largest of shrubs, with branches big enough for the birds of the sky to build nests in its shade."

(A) THE TEXT OF MARK 4:30-32 AND LUKE 13:18-19

For reasons which will appear later, only Mark's and Luke's versions of this parable will be considered.

(1) Despite the variety of readings contained in different MSS there is a fair degree of unanimity among the critical editions with regard to Mk 4:30-32. Disagreements concern the following points:

(a) the case of *kokkos* in vs. 31: Nestle, the United Bible Societies edition, Merk, Wescott-Hort, Tischendorf, and Tasker read the dative, while von Soden and Vogels give it in the accusative case.

(b) the form of the infinitive closing the text: Westcott-Hort, following the Codex Vaticanus, prefer *kataskēnoin* to *kataskēnoun*, the reading chosen by all the others.

To these may be added two variants which Merk considers as equally probable as the ones given in his text:

(a) in vs. 30 *tini* against *pōs*;

(b) in vs. 32 *auxei* against *anabainei*.

It would seem that the dative *kokkō* is preferable as a *lectio difficilior*,

---

[330] See J. Dupont, "Chapitre," 808-10. G. Harder ("Saat," 69-70) and H. Baltensweiler ("Saat," 69) suggest that the parable was missing in *Urmarkus*. Such a suggestion is as uncertain as the *Urmarkus* theory itself. B. H. Streeter (*The Four Gospels: A Study of Origins* [London: Macmillan, 1956] 171-72) attempts to explain the omission in Mt and Lk. But his suggestions (homoioteleuton is one of them) are equally unconvincing.

[331] See J. Dupont, "Chapitre," 815-16.

which is also better attested. *Kataskēnoun* is to be accepted against the sole authority of B.[332] *Tini* of vs. 30 is found, among other MSS, in Codex D which makes many attempts at producing a smoother text.[333] This tendency of D, and the suspicion that *tini* was borrowed from Lk 13:18, favor the reading *pōs*. *Auxei* in vs. 32 seems to be another attempt to avoid an awkward reading by borrowing from Lucan and Matthean parallels.

(2) In Lk 13:18-19 Merk, von Soden, and Vogels read *dendron mega* against *dendron* of Nestle, the United Bible Societies edition, Westcott-Hort, Tischendorf, and Tasker. Merk feels, further, that the omission of *eis* in vs. 19 (*egeneto eis dendron*) is as probable as its retention.

Mega is well attested in MSS; yet because it is not found in D, Sinaiticus, and some early versions speaks in favor of its exclusion from the text. It may have crept into the Lucan version under the influence of its Marcan and Matthean parallels. The omission of *eis* in D, which is known for its Semitisms,[334] and some early versions would favor its rejection. Still, the MS evidence seems too weak to warrant it.

(B) Discovering the Meaning of Jesus' Parable

It is commonly believed that Matthew's parable of the Mustard Seed (13:31-32) is not an independent witness but a conflation of Mark and Luke.[335] We shall consider the Marcan and Lucan versions, attempt to arrive at their sources, discuss the relationship between them, and the meaning of the parable on Jesus' lips.

(1) In discussing the parable of the Sower, A. Jülicher[336] was skeptical about the possibility of recovering its source, though he has no doubt that such a source did exist. Since his day, however, new insights have been gained, and many have ventured there where he refused to tread. We shall attempt to detect the traces of Mark's editorial activity in his parable of the Mustard Seed.

The manner in which the introductory question in vs. 30 is formulated is

---

[332] See M.-J. Lagrange, *Marc*, 116.

[333] See A. Jülicher, *Gleichnisreden* 2, 573-74.

[334] *Ginesthai eis*, however, is classical as well as Semitic; see E. Klostermann and H. Gressmann, *Das Lukasevangelium* (HNT 4; Tübingen: Mohr, 1919) 507.

[335] See R. Bultmann, *History*, 172; A. Jülicher, *Gleichnisreden* 2, 572; B. H. Streeter. *Gospels*, 246-47; V. Taylor, *Mark*, 269; J. Dupont, "Sénevé," 903. A. Schlatter (*Das Evangelium des Lukas* [Stuttgart: Calwer, 1960] 542-43) defends the thesis that Lk depends on Mt and Mk; he attempts to show that the Lucan form can be fully explained on the supposition that Lk reworked the Mt form of the parable. Yet his explanation of the manner in which Luke ought to have produced the opening question is too artificial to be convincing.

[336] *Gleichnisreden* 2, 514-15.

Semitic;[337] it imitates elevated Semitic language[338] and bears close similarity to the introductory formulae of rabbinic parables.[339] The second member of the verse is a *hapax legomenon*,[340] and as such it is rather difficult to attribute to Mark, whose Greek, though adequate, is not versatile. It is, moreover, doubtful that Mark would wish to compose such an elaborate introduction to a parable which is the last of the series;[341] he has already indicated the importance which he attributes to Jesus' teaching in parables by composing 4:1,2b. But the strongest argument in favor of the opinion that Mark reproduces his source is the double question which is also found in the Lucan parallel.[342] For these reasons it is difficult to admit with F. Grant[343] the possibility that the phrase "the kingdom of God" was added by Mark, or to agree with E. Lohmeyer and W. Grundmann[344] that the question itself is a composition of two questions which had been answered by two forms of the parable which, in their turn, were conflated into the version recorded by Mark.

The phrase *hōs kokkō sinapeōs* in vs. 31 comes most likely from the source. The numerous variants indicate that scribes found the dative strange. Mark, whose Greek is tolerably grammatical, can hardly be made responsible for it. What we seemingly have here is a double rendering of the rabbinic *lᵉ*—by means of *hōs* and by means of the dative. This may speak in favor of the phrase originating in the first stages of the community's translating its message from Aramaic into Greek.

The phrase *hōs hotan sparē* also seems to stem from the source. If Mark had composed or rephrased it, he would have harmonized the voice of the verb with that of the other two parables (4:3-4, *speirai, speirein*; 4:26, *balē*). *Epi tēs gēs* appears to be a favorite Marcan phrase. It occurs in three passages which are certainly or probably redactional (2:10; 4:1; 6:53). The word *gē* occurs 19 times in Mark; with *epi* 13 times, and with *epi* and the genitive 9 times. A comparison with Luke (*gē*—26 times; with *epi*—11 times; with *epi* and the genitive—5 times) and Matthew (*gē*—42 times; with *epi*—16 times; with *epi* and the genitive—10 times) is instructive.

---

[337] See V. Taylor, *Mark*, 269; M.-J. Lagrange, *Marc*, 115; J. Jeremias, *Parables*, 100-1.

[338] See F. Hauck, *Das Evangelium des Markus* (*Theologischer Handkommentar* 2; Leipzig: Deichert, 1931) 59; J. Schmid, *Mark*, 106.

[339] See *Str-B* 1, 653-55; 2, 7-9; F. Hauck, *TDNT* 5, 744-61.

[340] M.-J. Lagrange, *Marc*, 115.

[341] See A. Jülicher, *Gleichnisreden* 2, 570; C. Masson, *Paraboles*, 50.

[342] A. Jülicher, *Gleichnisreden* 2, 570.

[343] "Mark," 707.

[344] E. Lohmeyer, *Markus*, 88; W. Grundmann, *Markus*, 100.

Matthew avoids Mark's *epi tēs gēs* in five of his parallels (13:2,31; 14:23; 15:35; 26:39) and retains it in one (9:6). Moreover, the Marcan *epi* has no particular tendency to employ the genitive of other nouns; it is found with the genitive 21 times, with the dative 17 times, and with the accusative 34 times.[345] One might be inclined to attribute *epi tēs gēs* to Mark's editorial activity, but a number of considerations prevent us from doing so. That there was some reference to the ground which received the seed is shown clearly by Luke's *kēpos* and Matthew's *agros*. But it is not likely that either of these terms was found in the original form of the parable. Luke's "garden" is an adaptation to a hellenistic environment,[346] while "field" is a favorite term of Matthew.[347] This fact in itself would not be conclusive against its genuineness, were it not for the clearly derivative character of the Matthean parable. If one is tempted to look upon the "cultivated earth" of the *Gospel of Thomas* (§ 20) as a genuine echo of the pre-synoptic oral tradition, it should be pointed out that the term expresses the gnosticizing tendency of the gospel.[348] Mark's phrase, on the other hand, probably translates the Aramaic *b'r'*.[349] The LXX frequently renders the Hebrew *'l-h'rṣ* or *b'rṣ* by *epi tēs gēs*. We conclude, then, that Mark found this phrase in his source. C. Masson's suggestion[350] that Mark, considering "garden" or "field" too ordinary a ground to receive the word of God, changed one or the other into "earth," rests too exclusively on one particular interpretation of the parable to be convincing.[351]

Mikroteron on pantōn tōn spermatōn tōn epi tēs gēs is probably Mark's editorial addition.[352] It is an obvious insertion into an otherwise smoothly

---

[345] The precise classical distinctions in the use of cases with *epi* are no longer present in NT Greek; see Blass-Debrunner-Funk, *Grammar*, § 122-23; Bauer-Arndt-Gingrich, *Lexicon*, 285-89. It would thus serve little purpose to investigate the (classical) grammatical correctness of each case of *epi tēs gēs*.

[346] J. Jeremias, *Parables*, 27, n. 12.

[347] J. Jeremias, *Parables*, 83, n. 63.

[348] See E. Haenchen, *Die Botschaft des Thomas-Evangeliums* (*Theologische Bibliothek Töpelmann* 6; Berlin: Töpelmann, 1961) 45-46; H. Montefiore and H. E. W. Turner, *Thomas and the Evangelists* (*SBT* 35; London: SCM, 1962) 34-35, 53; W. Schrage, *Das Verhältnis des Thomas-Evangeliums zur synoptischen Tradition und zu den koptischen Evangelienübersetzungen* (*BZNW* 29; Berlin: Töpelmann, 1964) 65-66.

[349] See M. Black, *Approach*, 165.

[350] *Paraboles*, 45-46.

[351] How this type of argument can be used in ways leading to opposite results is shown by A. Plummer (*A Critical and Exegetical Commentary on the Gospel according to St. Luke* [*ICC*; Edinburgh: Clark, 1956] 344) who considers Luke's "garden" to be an image of Israel.

[352] C. H. Dodd, *Parables*, 142; V. Taylor, *Mark*, 269; C. E. B. Cranfield, *Mark*, 170; see also C. Masson, *Paraboles*, 45-46; E. Klostermann, *Markus*, 44.

running presentation, and it changes the gender from the masculine to the neuter in midstream. It recalls *katharizōn panta ta brōmata* of 7:19 which is Mark's insertion and refers to a subject (Jesus) who is only subsequently implied in the verb *elegen*—as in this case *on* seems to refer to the following *spermatōn*. Very likely Mark composed the phrase on the model of *meizon pantōn tōn lachanōn* in the following verse.

The second *hotan sparē* found at the beginning of vs. 32 was obviously called for by the editorial insertion of the previous phrase.[353] V. Taylor[354] considers *anabainei* to be a strange word to describe growth and thinks that it is due to the influence of the Aramaic original. There are, however, good reasons for taking it as redactional. The verb seems to be another favorite of Mark. We meet it in three redactional constructions (3:13; 10:32,33);[355] it is also found in 4:7,8 where it expresses the contrast between the growth of thorns and that of good fruit. Moreover, the similarity between 4:8 and 4:32 is striking:

> *kai edidou karpon anabainonta kai auxanomena kai epheren,*
> *anabainei kai ginetai meizon . . . kai poiei.*

It would thus be legitimate to conclude that Mark is responsible for the presence of *anabainei*. Confirmation may come from the fact that, in vs. 32, *anabainei* disturbs the two-membered structure which is noticeable if we omit the last phrase of vs. 31:

1. *homoiōsōmen . . . thōmen,*
2. *kokkō . . . sparē,*
3. *ginetai . . . poiei,*
4. *skian . . . kataskēnoun*

Ginetai meizon pantōn tōn lachanōn must come from Mark's source. The verb is found in the Lucan parallel, but *lachanon*, unlike the Lucan *dendron*, does not of itself convey the notion of great size, which is clearly essential to the thought of the parable.

We suspect that *megalous* is also Mark's own. The word itself is by no means characteristic of the Second Gospel (16 instances against Matthew's 20 and Luke's 25); it is nevertheless remarkable that 8 of the 16 instances occur in Mk 4:30—5:43.[356] It is of particular interest to note that "great"

---

[353] M. Black (*Approach*, 165) thinks that *hotan sparē* destroys the antithetic parallelism.

[354] *Mark*, 270.

[355] *Anabainō* occurs in Mk 1:10; 3:13; 4:7-8; 4:32; 6:51; 10:32-33; 15:8.

[356] Five of these instances are found in Mk 4:30-32 and 4:35-41; of the five, Matthew preserves two and Luke none.

is the word adopted to present the contrast in the story which immediately follows the parable of the Mustard Seed: "a great storm" (vs. 37), "great calm" (vs. 39), "great fear" (vs. 41).

If the above discussion has any validity, the parable of the Mustard Seed in the source utilized by Mark went like this:

1. pōs homoiōsōmen tēn basileian tou theou
ē en tíni autēn parabolē thōmen;
2. hōs kokkō sinapeōs,
hos hotan sparē epi tēs gēs
3. ginetai meizon pantōn tōn lachanōn
kai poiei kladous,
4. hōste dynasthai hypo tēn skian autou
ta peteina tou ouranou kataskēnoun.

The composition of the last phrase (no. 4) is too beautiful to be of Mark's making. F. Hauck[357] justly speaks of its poetic value. The succession of accentuated a- and u-syllables is striking.

(2) Even a superficial glance at the Marcan and Lucan versions of the parable discovers considerable diversity between them. The first, and likely the most important, difference lies in the fact that Luke gives a narrative, while Mark presents a typical occurrence. A further difference is found in Lucan exaggeration; while Mark allows the seed to grow into "the greatest of all shrubs," and it thus remains within the bounds of reality, Luke makes it into a tree. One would expect the opposite: to judge from his prolix form of the parable, Mark is more at pains to point out the disproportion between the beginning and the final stages of the seed; for that purpose "tree" would be more convenient than "shrub." Furthermore, Lucan imagery has already been transposed into hellenistic conditions; in Palestine mustard was a field plant,[358] but the Greeks grew it in their gardens.[359]

The Lucan parable is generally held to be independent of Mark, and to have come from Q.[360] The fact that Lk 8:4-18 follows the order of Mk 4:1-25, while the parable of Mk 4:30-32 is found in Lk 13:18-19, confirms the opinion that Luke in this passage is writing independently of Mark.

---

[357] *Markus*, 59.
[358] *Str-B* 1, 669.
[359] J. Jeremias, *Parables*, 27, n. 12.
[360] See F. W. Beare, *Records*, 115; R. Bultmann, *History*, 172; C. H. Dodd, *Parables*, 142; W. Grundmann, *Das Evangelium nach Lukas* (*Theologischer Handkommentar zum Neuen Testament* 3; Berlin: Evangelische Verlagsanstalt, 1961) 281; M.-J. Lagrange, *Évangile selon Saint Luc* (Paris: Gabalda, 1921) 385; W. Manson, *The Gospel of Luke* (*Moffat NT Commentary*; London: Hodder & Stoughton, 1955) 165; A. E. J. Rawlinson, *Mark*, 57; V. Taylor, *Mark*, 269.

That the Lucan version stems from the source common to Mt and Lk is best indicated by the parable of the Leaven which is linked to that of the Mustard Seed in both Gospels. The agreement of the two evangelists on its wording is almost complete. It is reasonable to assume that Matthew reproduces Q for the simple reason that he has no Marcan parallel with which to conflate it, as he does in the case of the parable of the Mustard Seed. Consequently it may also be assumed that Luke records the Q version of the Mustard Seed parable as faithfully as that of the Leaven.[361] *Kēpos*, found nowhere else in the Third Gospel, very likely comes from Q.[362] *Anthrōpos*, used instead of the usual Lucan *anthrōpos tis*, leads to the same conclusion.[363] There are no characteristic Lucan expressions in the passage.[364] Luke thus follows his customary procedure: "He preferred, where there was redundance in his sources, to omit radically rather than to conflate."[365] We may conclude, then, that Luke's version of the parable comes from Q, and that there is no direct dependence of Luke on Mark or Marcan source.

(3) There can be no doubt that the Marcan and Q versions of the parable derive from the same origin. Both speak of the mustard seed, of its being sown, and of its growth into a great plant which they illustrate by referring to OT imagery. They must have begun diverging while still in the oral stage of tradition. The difference in their vocabulary and in their OT quotations point in that direction; Mark's OT reference seems to consist of allusions to a number of OT texts,[366] while Luke's text is closer to Dan 4:21 (Theodotion).[367]

There is general agreement among exegetes that the main point of Marcan version lies in the contrast between the insignificance of the seed and the size of the full-grown plant.[368] The message of the Q version is

---

[361] See W. Grundmann, *Lukas*, 281.

[362] See A. Jülicher, *Gleichnisreden* 2, 573.

[363] See A. Plummer, *Luke*, 344. Cf. Lk 10:30; 12:16; 14:2,16; 15:11; 16:1,19; 19:12. Lk 14:16 and 19:12 belong to Q; other passages occur in material peculiar to Luke. Mt does not reproduce the Lk phrase in his parallels.

[364] See J. C. Hawkins, *Horae*, 16-23, 27-29.

[365] E. Hoskyns and N. Davey, *The Riddle of the New Testament* (London: Faber & Faber, 1958) 95; see also B. H. Streeter, *Gospels*, 246.

[366] Dan 4:21; Ezek 17:23; 31:6; Ps 103 (104):12. In Theodotion the term for birds is *ta ornea*; the LXX has *peteina*.

[367] A. Suhl (*Funktion*, 132, 154-55) places the OT allusions of Mk 4:32 under the heading, "Die Zitate des Redaktors Markus." But it is impossible to imagine that Luke followed Q in his parable and Mark in his reference to the OT.

[368] See A. Jülicher, *Gleichnisreden* 2, 576; R. Bultmann, *History*, 412; J. Jeremias, *Parables*, 147; F. W. Beare, *Records*, 114; C. E. B. Cranfield, *Mark*, 169; D. E. Nine-

disputed. Many hold that its principal point is likewise the contrast.[369] Others, however, disagree. C. H. Dodd, in keeping with his theory of realized eschatology, feels that the message should be sought in the final result of the previous growth. This final result is the ministry of Jesus: "Jesus is asserting that the time has come when the blessings of the Reign of God are available for all men."[370] He is followed by V. Taylor.[371] Others consider the growth of the plant to be the central point.[372] C. E. B. Cranfield's comment, however, is correct:

> The contrast between the smallness of the seed and the largeness of the plant cannot easily be pushed aside. Quite apart from the additional words in Mk the idea is present, for mustard seed was proverbial for its smallness (cf. Mt 17:20; Lk 17:6; and see further S.B. I 669). Moreover in Lk the hyperbolic *dendron* adequately emphasizes the contrast. This contrast is surely the key feature.[373]

The reference to the OT which serves to stress the greatness of the final product confirms this view.

On the question of greater originality opinions are again divided. Some[374] think that the Marcan form is more primitive, others[375] feel that the Lucan is. Marcan vocabulary certainly is closer to the Palestinian soil. The Lucan "tree," whether it is the result of a natural tendency of the oral tradition

---

ham, *Mark*, 144; A. E. J. Rawlinson, *Mark*, 57-58; J. Schmid, *Mark*, 106; J. Schniewind, *Markus*, 69-70.

[369] See J. Gnilka, *Verstockung*, 78; J. M. Creed, *The Gospel according to St. Luke* (London: Macmillan, 1965) 182; F. Hauck, *Marcus*, 59; E. Klostermann and H. Gressmann, *Lukas*, 507; K. H. Rengstorf, *Das Evangelium nach Lukas* (*NTD* 3; Göttingen: Vandenhoeck & Ruprecht, 1949) 168; B. H. Streeter, *Gospels*, 247; C. Masson, *Paraboles*, 46; F. C. Grant, "Mark," 707 (he thinks, however, that the original subject-matter of the parable was the good news).

[370] *Parables*, 142; for a criticism of this view, see W. G. Kümmel, *Promise*, 129-31.

[371] *Mark*, 268-69.

[372] See M.-J. Lagrange, *Marc*, 116; E. Lohmeyer, *Markus*, 88; A. Jülicher, *Gleichnisreden* 2, 576 (the growing number of believers constitutes the growth of the kingdom); W. Manson (*Luke*, 166) gives two possibilities: Jülicher's opinion or growth in faith. E. Grässer (*Parousieverzögerung*, 142) thinks that the Q form has been de-eschatologized, and that Luke's chief interest is the spread of the gospel by the missionary church. Dupont ("Sénevé," 901) is of the opinion that the Q version proclaims the certainty of growth.

[373] *Mark*, 169-70; see also E. Jüngel, *Paulus*, 153.

[374] W. G. Kümmel, *Promise*, 130-31; R. Bultmann, *History*, 412-13; J. Jeremias, *Parables*, 146, n. 68.

[375] A. Jülicher, *Gleichnisreden* 2, 571; C. Masson, *Paraboles*, 46; E. Klostermann, *Markus*, 44; E. Jüngel, *Paulus*, 152; K. L. Schmidt, *Rahmen*, 132; R. Schnackenburg, *God's Rule*, 154; J. Dupont, "Sénevé," 900; for further references, see W. G. Kümmel, *Promise*, 130, n. 91.

to exaggerate or of the influence of the OT quotation,[376] indicates a certain departure from the reality which Mark's "shrub" has preserved. "Garden" stems from the adaptation of the parable to hellenistic environment. The greater indefiniteness of Mark's allusions to the OT also seems to be more primitive and more in keeping with the freedom of the oral tradition.

When we consider the form of the parable, however, the opposite conclusion suggests itself. The great majority of synoptic parables present someone doing something instead of a description of what happens to an object; in other words, it is the active voice which predominates. Mark's Sower and Seed Growing Secretly belong to the category of parables where the active voice predominates. It may be that the passive voice in this parable has been brought about by the tendency to allegorize—God is sowing in the world. This development must have taken place in its pre-Marcan stage.

(4) If our conclusion about the principal message of the Q version of the parable is correct, there is every reason to think that the original form concerned the contrast between the insignificant beginnings and the magnificent result of the growth. We should refer, however, to the opinion of N. A. Dahl,[377] according to whom the message is

not the greatness of the coming Kingdom; that was described already in the Old Testament. Neither is the message to be found in the certainty that the Kingdom will come; that was a fact which no pious Israelite doubted. And it is not said that the Kingdom is to come in the immediate future. That may be the meaning, but in this parable the duration of the time of growth has no importance.

The problem which the parable was intended to solve was the incongruity between the glorious kingdom expected by the contemporaries of Jesus and the humility and apparent insignificance which characterized his ministry.

The parable gives the answer: Look at the mustard seed; in spite of its smallness, a great plant, providing shelter for the birds, grows out of it. The apparent smallness and insignificance of what is happening does not exclude the secret presence of the coming Kingdom—in the seed stage, so to say.

E. Percy[378] agrees with Dahl, adding that the community of God which accepts Jesus' message is the kingdom *in nuce*.

Many elements of this interpretation seem to be correct, particularly the insistence that the parable does not contain a general promise of the kingdom but refers to the concrete form of Jesus' preaching and other activity, pro-

---

[376] See J. Jeremias, *Parables*, 31; R. A. C. Leaney, *A Commentary on the Gospel according to St. Luke* (*Black's NT Commentaries*; London: Black, 1958) 207.

[377] "Parables," 147-48.

[378] *Botschaft*, 208-11; see also O. Kuss, "Sinngehalt," 652-53.

claiming that it is this activity which is introducing the kingdom. Another element is its stress on the intimate link between Jesus' ministry and the kingdom. To say, however, that the main point of the parable lies in the growth, and not in the contrast, is to strain the evidence of the texts, Lucan as well as Marcan. This parable, unlike that of the Seed Growing Secretly, wastes no words on the growth of the plant. It presents only its initial smallness and its great size at the end. The remarks made by Cranfield lose none of their validity. A good number of exegetes have accepted Dahl's contention that the parable implies the hidden presence of the kingdom but remain of the opinion that the contrast is its main point.[379] We agree with their view. It would be false to force all parables in which growth is mentioned or presupposed into the same mold. This is done by J. Jeremias[380] who considers the contrast to be the message of the Sower, the Seed Growing Secretly and the Mustard Seed; it is also done by Dahl[381] who thinks that the growth, and not the contrast, is the main point of all three parables. Each parable must be considered on its own merits; the fact that a certain motif occurs in all three is no reason to think that their message is the same.

What was the purpose of the parable? Jeremias classifies it under the heading "The Great Assurance."[382] "With the same compelling certainty that causes a tall shrub to grow out of a minute grain of mustard seed ... will God's miraculous power cause ... (the) small band to swell into the mighty host of the people of God in the Messianic Age, embracing the Gentiles."[383] We accept neither Jeremias' idea of the Semitic concept of growth nor his identification of the incipient kingdom with the small band of disciples. With such reservations, his opinion is shared by many exegetes. The parable is intended to give encouragement to those who are in danger of being overcome by the seeming insignificance of what is happening.

F. Mussner[384] has called attention to two Qumran texts which resemble our parable, 1QH 6:14b-17; 8:4-9.

> "They ... shall cause a shoot to grow
> into the boughs of an everlasting Plant.
> It shall cover the whole [earth] with its shadow ... (IQH 6: 14b-15).

---

[379] J. Jeremias, *Parables*, 152-53; R. Schnackenburg, *God's Rule*, 155-56; J. Gnilka, *Verstockung*, 78-79; G. E. Ladd, *Kingdom*, 230-31; E. Jüngel, *Paulus*, 154.

[380] *Parables*, 146-53.

[381] "Parables," 147-50, 152-54.

[382] *Parables*, 146.

[383] Ibid., 149.

[384] "1 Q Hodajoth und das Gleichnis vom Senfkorn (Mk 4, 30-32 Par)," *BZ* 4 (1960) 128-30.

The Qumran community is considered to be a small plant now, but it will become an imposing tree. The purpose of these passages seems to be encouragement for those who might become dispirited by the present insignificance of the community.

However, there may even be a more fundamental issue involved than that of encouragement and consolation. As in the case of the Seed Growing Secretly, the ultimate problem was not so much encouragement as faith in what Jesus was proclaiming. Dahl[385] says: "The Baptist and his disciples were probably not the only ones who asked: 'Are you he who is to come or shall we look for another?' The parable gives the answer. . . ." It is not a defense of the smallness or an attempt to mitigate the impact of the apparent insignificance of Jesus' coming,[386] but an appeal to men to realize that the only way open to them to enter the kingdom is by means of listening to and accepting Jesus' proclamation of it. In this feature of the parables, and not in attempts to identify Jesus with the sower, we should seek the christological dimension: there is no other kingdom than the one which Jesus is bringing. There are no other authoritative words about it besides the ones which Jesus is speaking; there is no other way into it apart from the one Jesus is pursuing.

(C) Mark's Understanding of the Parable

Two firm data enable us to see how Mark understood the parable. One is the contrast between the initial and final conditions of the seed; not only does he preserve the contrast found in his source, but he emphasizes it by his redactional insertions into the description of the beginning as well as the end of the process of growth. The other is the interpretation of the seed as the word, an interpretation which is pre-Marcan, but which he makes his own.

What is being contrasted? A good number of interpreters consider the parable in Mark to be an allegory of the church and its growth among the Gentiles at the time of the writing of the Gospel.[387] Some think that this was already the meaning in the source used by Mark.[388] According to J. Dupont, "on a bien l'impression de trouver là une description de l'expan-

---

[385] "Parables," 148; see also G. Bornkamm, *Jesus*, 72-73.

[386] E. Jüngel, *Paulus*, 154.

[387] W. G. Kümmel, *Promise*, 130-31; S. Schulz, *Botschaft*, 154; O. Kuss, "Sinngehalt," 653.

[388] E. Grässer (*Parousieverzögerung*, 142) thinks that the parable is an allegory of the church in all the Gospels; M. Karnetzki ("Die galiläische Redaktion im Markusevangelium," *ZNW* 52 [1961] 251-52) is of the same opinion.

sion chrétienne à la fin de l'époque apostolique. La parabole se présente ainsi comme une affirmation optimiste de la puissance du message apporté au monde par Jésus; son développement est inéluctable."[389] The parable is thus to be seen as a piece of church history.[390] The contrast described in it is between the preaching of Jesus, which was effectively limited to a narrow circle of disciples, and the preaching of the church which is spreading throughout the world. The birds nesting in the shade of the full-grown shrub are the Gentiles who are already being gathered into the church—this thought is derived from the allusions to OT texts in vs. 32.[391]

We consider this opinion to be incorrect. Nothing in the parable itself or in its context suggests that Mark identifies the kingdom with the church. The explanation of the Sower interprets the seed as the word, but it does not identify the community with the kingdom; neither for Jesus, nor for Mark is the band of disciples or the community the kingdom *in nuce*. "The kingdom of God is, indeed, active in this world and in the church; but it is not a visible reality and firm institution like the church. Neither is it subject to earthly developments as the church is in the course of her history."[392] Mark asserts that the good news must be preached to all nations before the end comes (13:10), but preaching is not the same as the definitive gathering of these nations within the kingdom. Mark is, moreover, keenly aware of the tribulations which are to remain the lot of the community until the coming of the Son of Man (cf. 13:5-23),[393] as well as of the imperfections which plague its life. To say with A. Suhl[394] that vs. 32 refers to the present, implying that pagans can enter the church if they wish to, hardly does justice to the strong contrast which Mark sets up between the two conditions of the seed. The word of Jesus, now proclaimed by the community, is still seemingly powerless; its results outwardly are insignificant. The first three classes of people described in vss. 14-20 do not convey the impression of irresistible power and luxuriant growth. The contrast, then, is not between the time of Jesus and that of the church, but between the time of the church and the end, the period of the hidden kingdom and that of the kingdom in power. Only at the end will the outward tribulations and inner temptations of the community cease. Only then will the word, proclaimed now, yield its ripe fruit and the nations find their definitive rest within the shade of the kingdom. Mark is fully aware that

---

[389] "Sénevé," 908-9.
[390] A. Suhl, *Funktion*, 155-56.
[391] See T. W. Manson, *Teaching*, 133, n. 1; A. Jülicher, *Gleichnisreden* 2, 576.
[392] R. Schnackenburg, *Markus*, 117 (my translation).
[393] Ibid.
[394] *Funktion*, 156.

the present smallness and insignificance contain the seeds of glory, but he is also aware that this glory is a future reality. Because it lies in the future, the present can appear as its very opposite. By means of the parable he is calling on his readers to remain true to the word which they have received. For it cannot but lead to that reality of which it is the hidden presence.

(4) Concluding Remarks

The parable chapter is, apart from ch. 13, the only lengthy discourse of Jesus within the Second Gospel. The solemn opening in vss. 1-2 tells us that Mark is presenting Jesus' teaching to his readers. This teaching is no mere instruction; rather, it is a manifestation of the divine eschatological power already at work in the world. The divine power is challenging the forces of darkness, encountering its opposition, yet is totally assured of its final victory. Vs. 11 tells us of the primary content of this teaching: it is the mystery of the kingdom of God. It not only speaks of the kingdom; it makes it mysteriously present. It introduces into the world the energies of the kingdom which infallibly strive toward its completion at the end. Once present, it cannot but tend toward the goal of which it is already a realization. Vss. 21-22, which are a redactional insertion, stress in particular that the purpose of what is taking place now, the purpose specifically of Jesus' teaching, is the final manifestation of the powers of the kingdom which have already been released and of the glory which is already discernible for those who have eyes to see. A history, if we can call it such, of this dynamic process toward manifestation can already be traced. Vs. 10 and vs. 34 express not only Mark's awareness, but also his emphasis, of the difference between the community's present and Jesus' past: what was reserved for the narrow circle of disciples before the resurrection of Jesus is now being proclaimed from the housetops to everyone who is willing to listen. The content of the proclamation is this: it is by Jesus, and by him alone, that the kingdom is being brought into the world. His words and deeds, and the subjection to the eschatological power which emanates from them, are the only means available to men for entering into the kingdom. The ten references to "hearing" which occur in the parable chapter clearly indicate the importance of what is being taught and the urgency of the message.

The irruption of the kingdom into the world cannot but bring about now what it will bring about definitively at the end; even now it gathers the elect into the true family of Jesus, gives them insight into God's eschatological plans and makes them participate in the power which is carrying out these plans. But it also condemns, blinding the eyes and stopping the ears of those who will not accept Jesus as the one in whom God's spirit is at work. The irrevocable division of the end-time is being anticipated every day. The teaching of Jesus is thus creating the community, setting its ultimate destiny before its eyes, giving it firm hope that its present acceptance of him as the Son of God and Messiah will find its confirmation at his final coming "in the clouds with great power and glory." Another product of Jesus' teaching, however, is the synagogue of Satan to the members of which his words are a source of hardening and impenitence.

The present kingdom is a hidden one. The community is still waiting for its coming with power. Its members are still in danger of being untrue to the word which they have accepted. Their need of consolation and exhortation has not ceased. The demand for repentance and faith, and the warning that they heed what they hear have lost none of their validity and urgency. The community knows that Jesus is the Messiah and the Son of God in whom the kingdom is already exercising its powers, and it proclaims these truths. But it does not know as it should know; defects of necessity disfigure its life, and it is subject to doubts and discouragement. The teaching of Jesus has not yet produced its definitive result.

The three parables depict the destiny of Jesus' teaching. The first one devotes most of its words to the preaching of the word, in the past as well as in the present. It is the word, not the sower, which is the center of attention. The community knows that the words of Jesus did not produce the desired result in all who heard them, and that its own proclamation all too often falls on deaf ears. But it refuses to be discouraged, for it knows that the word, being the divine eschatological manifestation, cannot fail. The very existence of the community is a witness to the victorious power of Jesus' teaching. This unfailing power is the main point of the second parable. Once proclaimed, the word of Jesus begins its way toward its final goal. The hidden kingdom brought into the world by the word belongs to God alone; man does not understand its way and can do nothing to change its course. The purpose of the last parable is to present above all the final goal of the now hidden kingdom. No matter how insignificant Jesus' words and works may have appeared, and despite the humble appearance of the community, the beginning of the end has arrived and the end itself cannot but come. Then all the nations will be gathered into the kingdom, and the community will enjoy the reward of its acceptance of Jesus' word. God's plan will have reached its completion.

Chapter III
THE ETHICAL DEMANDS OF THE KINGDOM

Since the kingdom remains hidden during the time of Jesus and the community, it remains possible to resist the Word. The community itself is threatened by outward and inward dangers. This hiddenness is not to last forever, for it is in the very nature of the kingdom to become manifest. In Chapter III we investigate the demands which the kingdom places on the members of the community. Of the passages to be considered, Mk 10:14-15, 23-25 demand a fairly lengthy treatment; we shall try to discover the way in which the logia found their way into their present context, their original meaning, and the message which Mark is conveying by means of them. Mk 9:47 and 12:34 will be given shorter treatment since the material which they contain is of lesser value for the purposes of our investigation.

(1) Mark 10:14-15

People were bringing their little children to him to have him touch them, but the disciples were scolding them for this. ¹⁴Jesus became indignant when he noticed it and said to them: "Let the children come to me and do not hinder them. It is to just such as these that the kingdom of God belongs. ¹⁵I assure you that whoever does not accept the kingdom of God like a little child shall not take part in it." ¹⁶Then he embraced them and blessed them, placing his hands on them. (Mk 10:13-16)

(A) THE ORIGINAL ISOLATED CHARACTER OF MK 10:15 AND THE QUESTION OF ITS GENUINITY

(1) The textual tradition of the logion in Mk 10:15 is singularly unanimous; critical editions note no variants.

E. Klostermann[1] summarizes various views about the origin and function of the verse: some look on it as an expansion of the previous verse, others as an independent saying inserted into a previously existing context, still others as the nucleus from which the entire story 10:13-16 had sprung. The most probable opinion seems to be the one which looks upon the logion in vs. 15 as originally independent, which only later came to be linked with its present context. E. Percy[2] who disagrees with this

---

[1] *Markus*, 100.
[2] *Botschaft*, 35.

opinion characterizes it as common. The main argument in its favor is the fact that the story in which it is now found deals with Jesus' attitude toward children, while the saying itself treats children as symbols.[3] Its Matthean variant in 18:3 should probably not be considered as an independent tradition.[4] Matthew simply did what some commentators would still wish to do, i.e., change the positions of Mk 9:37 and 10:15.[5] Jn 3:3,5, however, seems to be a genuine variant.[6]

Percy's reasons[7] for considering this saying as a natural part of the pericope are as follows: vs. 14 demands an explanation, which vs. 15 supplies; the thought of vs. 15 seems to be Jesus' own; the situation in which it arose demands children. But the second and third reasons are hardly convincing. The fact that the thought of vs. 15 is in all likelihood Jesus' own does not prove that the pericope, vss. 13-16, was from the very beginning the place where this thought was preserved by the tradition. It is not at all certain that the situation in which the logion arose demanded children. Even if such were the case, the event of which this pericope is a reflection was surely not the only occasion at which children were present. His first argument is a good explanation why vs. 15 was inserted into the pericope. The thrust of the story, and of vs. 14 in particular, is, however, so dissimilar from that of vs. 15 that it is difficult to imagine that these two verses formed an original unity.

Mention should be made of J. Sundwall's reconstruction of the history of 10:13-16 and 9:36-37.[8] He thinks that vs. 15 was originally an independent logion, but it became part of a catena of sayings, consisting of Mk 10:14; 9:37; 10:15. Out of this catena an apophthegm was created which consisted of Mk 10:13,14,15,16 (=9:36); 9:37. Mark would have divided

---

[3] It is considered as an originally independent logion by R. Bultmann (*History*, 32), D. E. Nineham (*Mark*, 269), E. Schweizer (*Mark*, 206), F. W. Beare (*Records*, 193), J. Schmid (*Mark*, 188), W. Grundmann (*Markus*, 205), and E. Lohmeyer (*Markus*, 202). C. E. B. Cranfield (*Mark*, 324) is uncertain, and E. Klostermann (*Markus*, 100) does not voice an opinion.

[4] R. Bultmann, *History*, 32; see below.

[5] E.g., V. Taylor, *Mark*, 406; D. E. Nineham, *Mark*, 269.

[6] See E. Schweizer, *Mark*, 206; E. Lohmeyer, *Markus*, 206; J. Jeremias, *Infant Baptism in the First Four Centuries* (London: SCM, 1960) 51-52. The Johannine version of the saying is an effective argument against F. A. Schilling's interpretation: "Here it [i.e., the kingdom] is like a child, normally a being fresh, at the beginning of life, in need of affection" ("What Means the Saying about Receiving the Kingdom of God as a Little Child (*tēn basileian tou theou hōs paidion*)? Mk x.15; Lk xviii.17," *ExpT* 77 [1965-66] 57). According to him, the kingdom is here compared to a child as it is elsewhere compared to a seed.

[7] *Botschaft*, 35.

[8] *Zusammensetzung*, 64.

the apophthegm into two independent stories, in one of which 10:16 forms the ending, and in the other (9:36) the beginning.

This reconstruction is at best hypothetical. There is little doubt that verbal correspondences played a role in the arrangement and sequence of logia collections. But Sundwall tries to explain too much by means of this device. There is, further, a considerable difference between 10:16 and 9:36; the theme of the child being placed in the midst of the disciples, found in 9:36, is quite foreign to the message of 10:16.[9] He seems, furthermore, to proceed from the assumption that the oral tradition contained originally Jesus' logia alone, and around these logia it constructed ideal stories. On the character of 10:13-16 more will be said later. Here we may mention only the growing awareness that Jesus' manner of acting was as important to the primitive church as were his words.[10] Mk 10:16 is a great deal more than a mere illustration of the previous verse;[11] it is a manifestation of the character of the kingdom in deed, as vss. 14-15 are in word.

Hence it is probable that Mk 10:15 circulated originally as an isolated saying. Later it was joined to the story in which it is now found. There is no reason to think that it was Mark who first effected this combination.

(2) The logion is looked upon as genuine by the majority of today's commentators.[12] The main reasons for this opinion are stated by R. Bultmann;[13] the logion contains something characteristic and new, which is beyond popular piety and rabbinic or apocalyptic thought. To this E. Schweizer[14] adds that it gives voice to the tension between the future and the present kingdom, a tension characteristic of Jesus' proclamation. He also observes that the child plays no significant role in the primitive church.

Others, however, hesitate to accept the saying as genuine, either in its entirety or in its wording. E. Lohmeyer, who remarks that in vs. 14 Jesus speaks as the divine Wisdom of Proverbs 8 or as the glorified Christ of Rev 22:27, thinks that vs. 15 is a community rule, expressing the meaning of Mk 10:13-16.[15] J. Dupont[16] wonders whether the catechetical interests

[9] See R. Schnackenburg, "Mk 9," 198.
[10] See G. Bornkamm, *Jesus*, 80-81; J. M. Robinson, *A New Quest of the Historical Jesus* (*SBT* 25; London: SCM, 1966) 48-58.
[11] As some commentators still seem to think: see D. E. Nineham, *Mark*, 268; E. Schweizer, *Mark*, 207. It is considered as more than an illustration by V. Taylor (*Mark*, 424), W. Grundmann (*Markus*, 207-8), E. Lohmeyer (*Markus*, 203-5).
[12] See E. Percy, *Botschaft*, 35; D. E. Nineham, *Mark*, 269; V. Taylor, *Mark*, 424; also C. G. Montefiore, *Gospels* 1, 238.
[13] *History*, 105.
[14] *Mark*, 206; cf. J. M. Robinson, *New Quest*, 121.
[15] *Markus*, 205-6.
[16] *Les béatitudes* (Bruges: Abbaye de Saint-André, 1954) 154, n. 1.

of the primitive church may not have influenced its present form, but in the end seems to opt for its genuinity. R. Schnackenburg[17] doubts the genuinity of the phrase "to receive the kingdom," for nowhere else in NT is the kingdom spoken of as a present gift. He suggests that it grew out of the phrase "to receive the word of God," which forms part of the missionary terminology of the primitive church. A reference to the content of the message would thus have replaced the reference to the message itself.

Another disputed point is the relation between Mk 10:15 and Mt 18:3. J. Jeremias,[18] who is followed by J. M. Robinson,[19] considers the Matthean form as more Semitic and therefore closer to the original words of Jesus. We have already referred to R. Bultmann's view to the contrary. Percy,[20] likewise disagrees with Jeremias, because Mark's form is clearer than that of Matthew, who must explain it in the following verse;[21] further, the Matthean form can be explained as derived from that of Mark, but not vice versa. Matthew felt that a saying of Jesus setting a child as an example was missing in Mk 9:36-37 and to fill the gap he took Mk 10:15, adapting it to the new context. Percy also expresses doubt about the correctness of Jeremias' retranslation of Mt 18:3 into Aramaic.

Mk 10:15 is substantially a genuine saying of Jesus. The modifications which the original form underwent are discussed in the following subsection.

(B) The Meaning of Jesus' Logion

(a) *"To Enter the Kingdom"*

(1) G. Dalman[22] and P. Billerbeck[23] furnish the rabbinic parallels to this phrase: "to come, or to enter, or to inherit the future aeon, or the life of the future aeon." If their choice of parallels is correct, the conclusion to be drawn is that "to enter the kingdom" refers to the future eschatological fulfilment of God's reign. In his classic article "Die Sprüche vom Eingehen in das Reich Gottes,"[24] H. Windisch traced the prehistory

---

[17] *God's Rule*, 142; see also W. G. Kümmel, *Promise*, 126, n. 77.
[18] *Parables*, 190, n. 75, 76.
[19] *New Quest*, 121.
[20] *Botschaft*, 36, n. 5.
[21] Others agree that Mt 18:4 is the evangelist's own explanation of the previous verse: T. H. Robinson, *The Gospel of Matthew* (*Moffat NT Commentary*; London: Hodder & Stoughton, 1951) 152; J. Dupont, *Béatitudes*, 152.
[22] *Words*, 116-17.
[23] *Str-B* 1, 252-53, 829.
[24] *ZNW* 27 (1928) 163-92.

of the phrase and arrived at the conclusion that these NT "toroth of entry" refer to the definitive coming of the kingdom in the future. Contemporary exegetes agree with him in this respect.[25]

A discordant note, however, is sounded by T. W. Manson who claims that "in the mind of Jesus, to become a genuine disciple of his and to enter the Kingdom of God amounted to much the same thing."[26] G. E. Ladd likewise disagrees with the consensus. According to him, "it is arbitrary to insist that all sayings about entry into the Kingdom are eschatological unless it is established that the eschatological concept exclusively dominated Jesus' thinking; and this is precisely the question at issue."[27] Manson's procedure in establishing Jesus' concept of the kingdom has been criticized by N. Perrin[28] and G. Lundström.[29] As for Ladd's view, it should be pointed out that even though for Jesus the kingdom is really somehow present in his acts and words, that does not automatically determine the meaning of a phrase. We cannot go into a detailed examination of Lk 16:16/Mt 11:12 which he adduces in favor of his thesis; the saying is discussed in every work which deals with Jesus' concept of the kingdom. The other sayings which, according to Ladd, speak of the possibility of entering the kingdom in the present, i.e., Mt 21:31; 23:13; 11:11; Lk 11:52,[30] can just as easily be interpreted in a sense contrary to that of Ladd.[31] In this connection J. M. Robinson's study of the present-future tension found in the very structure of many of Jesus' sayings should be mentioned;[32] the entry sayings are of paramount importance in his discussion.

(2) Before we continue, a word must be said about the relationship between eschatology and ethics in Jesus' message. What role is being played in Jesus' ethical thought by the central theme of his proclamation, the good news of the approaching kingdom? That eschatologically and sapientially

[25] See R. Schnackenburg, *The Moral Teaching of the New Testament* (Freiburg: Herder, 1965) 20; *God's Rule*, 161, 227; J. Theissing, *Die Lehre Jesu von der ewigen Seligkeit* (Breslau: Müller & Seiffert, 1940) 75, 78-81; E. Lohmeyer, *Markus*, 204; G. Lundström, *Kingdom*, 236; N. Perrin, *Kingdom*, 183-84, 192; R. H. Fuller, *The Mission and Achievement of Jesus* (SBT 12; London: SCM, 1963) 29-30; W. Trilling, *Das wahre Israel: Studien zur Theologie des Matthäus-Evangeliums* (StANT 10; München: Kösel, 1964) 107.
[26] *Teaching*, 206; see also *Ethics and the Gospel* (London: SCM, 1966) 65-68.
[27] *Kingdom*, 193.
[28] *Kingdom*, 95-97.
[29] *Kingdom*, 111-13.
[30] *Kingdom*, 193.
[31] See R. Schnackenburg, *God's Rule*, 161, 227; N. Perrin, *Kingdom*, 184.
[32] *New Quest*, 121-25.

motivated ethical sayings of Jesus exist side by side in the gospels has long been acknowledged.[33] A historical conspectus of the struggle of the eschatological view of the kingdom against the earlier liberal views and of its eventual victory over these, as well as of the discussion that followed the first presentation of the interim ethics theory, is out of place here.[34] We shall attempt, however, to summarize and briefly comment on the views of three contemporary exegetes: A. N. Wilder, G. Bornkamm, and R. Schnackenburg. These scholars seem to represent the main tendencies in the present discussion of the problem.

Wilder has set forth his understanding of Jesus' ethics in his book, *Eschatology and Ethics in the Teaching of Jesus*.[35] The tension between the present time of salvation and the future fulfilment, and consequently between the sapiential and eschatological elements in Jesus' ethical teaching, resolves itself for Wilder in favor of the former. He recognizes clearly the reality of the future eschatological event in Jesus' proclamation. Yet he points out that Jesus stripped away the Jewish apocalyptic imagery until the thought of impartial judgment remained almost alone. Even this "apocalyptic event in the future is secondary to and derivative from the judgment inherent in the offered time of salvation" (p. 179); it is "essentially of the character of myth" (p. 182). "The radical character of Jesus' ethics does not spring from the shortness of time but from the new relation to God in the time of salvation" (p. 161). The extreme demands which Jesus places upon his disciples are not an interim ethic, nor are they meant to be permanent rules of conduct; they are concrete demands placed on individual men at a time of crisis, a time when Israel was deciding whether or not to be faithful to its vocation.[36]

One cannot but agree with Wilder's emphasis on the present time of salvation with its radical ethical demands, as well as with his opposition to the theory of interim ethics. His emphasis on the present, however, seems to be one-sided. Though Wilder does not deny the reality of the future

---

[33] See H. Windisch, *Der Sinn der Bergpredigt* (Untersuchungen zum Neuen Testament 16; Leipzig: Hinrichs, 1937) 20-24; H. Schürmann, "Das hermeneutische Hauptproblem der Verkündigung Jesu," *Traditionsgeschichtliche Untersuchungen zu den synoptischen Evangelien* (Düsseldorf: Patmos, 1968) 13-35, particularly 15-26.

[34] A good survey of this history can be found in N. Perrin's and G. Lundström's books with the identical title, *The Kingdom of God in the Teaching of Jesus*.

[35] Published in New York: Harper, 1950.

[36] Ideas which bear great similarity to those of A. N. Wilder are expressed by H. Schürmann in "Eschatologie und Liebesdienst in der Verkündigung Jesu," *Kaufet die Zeit aus: Beiträge zur christlichen Eschatologie: Festgabe für Th. Kampmann* (ed. H. Kirchhof; Paderborn: Schöningh, 1959). Schürmann seems to be more reserved on this matter in his later article, "Hauptproblem."

sanction in the thought of Jesus, he nonetheless seems to empty it of significance by branding it as myth. This future may be a myth to many of our contemporaries, but it is difficult to imagine that it was a myth to Jesus.[37] Wilder, furthermore, seems to identify the future as such with its imminence. Interim ethics, which he rejects, slip in by the back door in his restricting the validity of Jesus' extreme demands to a historical period of crisis. Are we to think that the evangelists have misjudged the thought of Jesus each time that they extend to all Christians a radical demand addressed by Jesus to an individual contemporary? Is Wilder's statement that "Jesus' appeal was in its general aspect a summoning of all to a total response of obedience to the Father" (p. 183) the only criterion which we are to use for separating the commands for the time of crisis from those which are permanently valid?

G. Bornkamm, in his book *Jesus of Nazareth*, points to Jesus' sober presentation of the future and the absence of Jewish particularism in his image of judgment. The victory of the kingdom is already being won in Jesus' words and actions. These words and acts are calling for a complete turning to God, for in view of the coming judgment everything pales into insignificance. Bornkamm rejects a number of attempts to solve the tension between the present era of salvation and that of the final fulfilment: Jesus was neither a realized eschatologist, nor was he a mere apocalypticist. The present and the future intertwine and interpenetrate each other: "The future of God is *salvation* to the man who apprehends the present as God's present, and as the hour of salvation" (p. 93). By hearing and accepting the word of God, men are placed in a new condition in which "the world and its possibilities end, and the future of God begins" (p. 109). While Bornkamm clearly recognizes the tension between the present and the future and does not, verbally at least, try to suppress one or the other in order to arrive at a simple and clear theory, one wonders whether N. Perrin's comment on R. Bultmann is not equally applicable to Bornkamm: "It is difficult to determine how far Bultmann regards Jesus as the author of this existentialist understanding of eschatology."[38] Bornkamm's statement on p. 109, quoted above, seems to be little more than a play on words. Jesus' proclamation of the future undoubtedly called for a radical decision in the present, but it seems quite impossible to reduce this future to a mere factor in man's personal decision. A reaction against the place- and time-bound interim ethic does not justify flight to a virtual abolition of the time-element in the thought of Jesus.

[37] See G. E. Ladd, *Kingdom*, 285; N. Perrin, *Kingdom*, 156.
[38] *Kingdom*, 116.

In various books and articles, and particularly in *God's Rule and Kingdom* and *The Moral Teaching of the New Testament*, R. Schnackenburg has insisted on the primacy of the eschatological kingdom as well as the reality of the present age of salvation in Jesus' proclamation of the good news. When we speak of the kingdom as already present we should characterize it as preparatory, hidden, or dynamic to distinguish it clearly from its full realization which is yet to come. The ultimate reason for Jesus' radical ethical demands is not a generic will of God but the nearness of his eschatological kingdom. This nearness as the basis of ethical demands should not, however, be conceived as closeness in time but, primarily, as productive of a new relationship between God and man through the work of Jesus.

The same view is voiced by E. Neuhäusler in his book *Anspruch und Antwort Gottes*.[39] He notes that Mt 5:45, according to which the heavenly Father makes his sun rise on the just as well as on the unjust, could easily be misunderstood as a general ethical maxim, if we forget that this same Father is to carry out the judgment which is approaching. The word about the eunuch (Mt 19:12), on the other hand, seems to be eschatologically colored to such a degree that it can have no more than an analogous value for a believer whose expectation of the end is less intense. The two classes of sayings find their unity in the words and works of Jesus and thus ultimately in his person. It is Jesus who determines the believer's existence. Not only will the future judgment of God punish and reward, but the poor are pronounced blessed in the present—they already have a share of the future happiness now.

This view, however, of Jesus' sapientially and eschatologically motivated ethical demands juxtaposes them instead of attempting to harmonize them on their own level; it rests content with deducing their ultimate cohesion from the mission of Jesus. Yet our desire to reduce disparate sayings to a mentally satisfying unity may run the danger of oversimplification and of sacrificing totality for intellectual consistency.

(b) *"To Receive the Kingdom"*

(1) Apart from Mk 10:15 and its Lucan parallel, the kingdom of God is not spoken of as a present gift in the NT. G. Dalman[40] and T. W. Manson[41] have suggested that the phrase is equivalent to the rabbinic

---

[39] *Zur Lehre von den Weisungen innerhalb der synoptischen Jesusverkündigung* (Düsseldorf: Patmos, 1962) 37-42.
[40] *Words*, 124-25.
[41] *Teaching*, 135, n. 1.

expression "to take (*qbl*) upon oneself (the yoke of) the kingdom," an expression which describes obedience to the Law. W. Pesch,[42] however, has pointed out that, if this were the case, the verb used would have to be *airein*, not *dechesthai*, for in the LXX the latter verb translates *qbl* only in those cases in which it describes reception (of a gift). In the intertestamental literature there are texts in which the verb "to receive" has the future aeon as its object.[43] However, no such instances are to be found in the NT,[44] where *dechesthai* is used primarily in the sense of receiving a guest[45] or as a technical term for accepting Christian message.[46] In the latter sense "to receive the word of God" stands practically for "to believe."[47]

In the OT the verb *dechesthai* is frequently found in sapiential contexts where it translates the Hebrew verb *lqh*. It occurs in prophetic books, where it has *paideia* (*mwsr*)[48] or *logos theou*[49] as its object. In the Book of Proverbs it occurs 10 times;[50] in every instance but one (1:9) it describes the reception of wisdom. Its direct objects are instruction (*paideia, prosthesis*), words (*logoi*), knowledge (*gnōsis*), or commands (*entolai*). The Hebrew verb which it translates is, as in the prophetic books, *lqh*.[51] From this OT evidence *dechesthai* is to be seen as a technical term which describes a willing and understanding acceptance of wisdom in its various manifestations.[52]

In other Wisdom books the term *dechesthai* does not occur as frequently as in the Proverbs, at least in the sense of receiving wisdom. Neither, for that matter, does the Hebrew verb *lqh* in the same sense. In Sirach it is found seven times; in three instances it carries the meaning of accepting wisdom.[53] In Job it is found five times, once with instruction as its object.[54]

---

[42] *Der Lohngedanke in der Lehre Jesu verglichen mit der religiösen Lohnlehre des Spätjudentums* (Münchener Theologische Studien 1, Hist. Abt. 7; München: Zink, 1955) 55.

[43] 2 Esdras 7:14 ("to enter" and "to receive" have their roles reversed in comparison with Mk 10:15); 2 Apoc. Bar. 14:13; 51:3.

[44] See E. Percy, *Botschaft*, 35.

[45] This is particularly true of the Gospels.

[46] This is found primarily in the Acts.

[47] W. Pesch, *Lohngedanke*, 55.

[48] Jer 2:30; 5:3; 7:28; 17:23; Zeph 3:2,7; see W. Grundmann, *TDNT* 2, 50-51.

[49] Jer 9:19.

[50] 1:3,9; 2:1; 4:10; 9:9; 10:8; 16:17; 21:11; 30:1 (Rahlfs); 24:22a (Rahlfs).

[51] Except in 30:1, which has no Hebrew equivalent; the verse is interesting: *dexamenos autous* (i.e., *tous emous logous*) *metanoei*.

[52] See W. Grundmann, *TDNT* 2, 52.

[53] 6:23; 51:16,26; in places for which we possess the original text (2:4?; 41:1; 50:12) it translates the piel of *qbl*.

[54] 4:12; in the Hebrew text *lqh* with the same meaning occurs twice (4:12; 22:22).

In Qoheleth neither *dechesthai* nor *lqh* appear. In the book of Wisdom it occurs twice in the sense of hospitality (12:7; 19:14).

In particular, there is a frequent juxtaposition of the verbs *didōmi* and *dechomai* with wisdom as their object:

> That men may . . .
> receive (*dexasthai*) training in wise conduct,
> in what is right, just and honest;
> that resourcefulness may be imparted (*dō*) to the simple,
> to the young man knowledge and discretion (Prov 1:2-4).

Other instances of the same combination occur in Prov 2:1,6; 9:9; Sir 51:16-17. The corresponding Hebrew verbs are *ntn* and *lqh*.[55]

In the intertestamental literature the verb does not seem to be much used with wisdom as its object. There are, however, frequent instances of what may be regarded as counterpart expressions, viz., those which describe the bestowal of wisdom. Some examples from *1 Enoch*:

And then there shall be bestowed on the elect wisdom (5:8).

. . . such wisdom has never been given by the Lord of
Spirits as I have received (37:4).

Now three parables were imparted to me (37:5).

And I asked the angel who . . . showed me what was hidden:
"What are these?" And he said to me: "The Lord of
Spirits hath showed thee their parabolic meaning
(lit. 'their parable')" (43:3-4).

. . . Enoch gave me the teaching of all the secrets . . .
which had been given to him (68:1).

. . . I have given the books concerning all these:
so preserve . . . the books . . . and . . . deliver them
to the generations of the world (82:1).

I have given wisdom to thee and to thy children . . . (82:2).

Who has given understanding and wisdom to all that moves
. . . ? (101:8) [the answer implied is God].

In *2 Esdras* we find similar statements:

O my soul, drink thy fill of understanding,
and, o heart, feed on wisdom! (8:4).

---

[55] Sir 51:17 has *ntn*; for 51:16 *dechomai* the original text is not available.

... the Most High hath revealed many secrets unto thee" (10:38).

... deliver the secret to the wise [i.e., the secret which God will give to him] (14:25-26, 44).

Two other statements which throw light on this usage are found in *1 Enoch* 60:1 and *2 Esdras* 5:10.

In Qumran literature there are highly interesting parallels:

Blessed art [Thou, O my Lord],
who hast given to [Thy servant]
the knowledge of wisdom
that he may comprehend Thy wonders ... (1QH 11:27-28).

I, the Master, know Thee, O my God,
by the spirit which Thou hast given to me,
and by Thy Holy Spirit I have faithfully hearkened
to Thy marvelous counsel.
In the mystery of Thy wisdom
Thou hast opened knowledge to me ... (1QH 12:11-13).

[Blessed art Thou,] O Lord,
who hast given understanding
to the heart of [Thy] servant ... (1QH 14:8).

Ntn is the Hebrew verb which describes God's bestowal of his gift, or the spirit, of knowledge in all the Qumran passages quoted above. The same verb in the same context appears also in 1QH 10:27; 13:18-19; 16:11-12; 17:17; 18:27. Thus we meet in the Qumran texts and the intertestamental literature the *ntn—lqḥ* (*didōmi—dechomai*) complex which we have seen in the books of Proverbs and Sirach.

Hence it is likely that *dechesthai* in certain NT contexts retains the technical meaning which it had acquired in Wisdom literature. As it described an understanding and willing acceptance of wisdom in the OT, so in the NT it describes an understanding and willing acceptance of divine revelation. The transition from acceptance of human wisdom to that of divine revelation begins already in the OT.[56] The intertestamental literature brings this notion of the bestowal of divine revelation into sharper focus by insisting that true wisdom can come only to those to whom God chooses to reveal his mysteries. The idea of "mysteries" reserved to God alone is no monopoly of the intertestamental apocalyptic literature; it has a long prehistory in the religious thought of Israel, but it is not foreign to Wisdom

---

[56] See Job 28; Prov 8:22-31; Sir 24:3-12; 39:5-8; 51:13-16.

literature.[57] Thus it is not difficult to imagine that a sapiential concept and a sapiential technical term would pass into the NT through the medium of apocalyptic.

(2) Is there any reason to suppose that this sapiential nuance of *dechomai* is present in Mk 10:15? Since the kingdom is not spoken of as a present gift in the rest of the NT there are reasons to doubt the authenticity of the phrase "to receive the kingdom." Can we arrive at a reasonably likely surmise about the authentic words of Jesus, and about the causes which brought about a change of his words in the tradition or redaction? A comparison of this logion with that in 4:11 might throw some light on the question.

In our discussion of Mk 4:11 we attempted to show the likelihood, or at least the possibility, that the Mt-Lk form of the logion, which contains a reference to knowledge, is more original than that of Mark which does not contain such a reference. Mark's form of the logion is in keeping with his consistent refusal to attribute knowledge to the disciples. Though it is quite impossible to determine whether the absence of *gnōnai* is due to Mark or to his tradition, it must be admitted that this absence suits the evangelist's portrait of the disciples.

Though the term *dechesthai* occurs only once in the intertestamental literature with the nuance which it carries in the Wisdom literature (*1 Enoch* 37:4), we have noted the frequency of its counterpart, i.e., the divine bestowal of wisdom on men. This also has precedents in Wisdom literature.[58] We would suggest as a solution (close to that of Schnackenburg) that in the original form of the logion there followed after *dexētai*, a term which, in connection with the verb, suggested reception of knowledge. We can only guess what this term may have been; perhaps *ho logos tēs basileias* or more likely, if a comparison with 4:11 has any validity, *ta mystēria tēs basileias*. A reference to "the word" or to "mysteries" may have been dropped either in the course of oral transmission or by Mark himself. It is significant that the logion no longer attributes knowledge to the disciples or to other followers of Jesus (compare Mk 4:11; see above pp. 86-88). By producing, or reproducing, his form of the logion, Mark achieves two results: he avoids the suggestion that the disciples already possess knowledge, or that knowledge is an automatic consequence of following in the footsteps of Jesus. One may be a follower of Jesus, in other words, but that does not absolve one from striving for understanding. The other result

[57] See R. E. Brown, "Pre-Christian."
[58] Prov 1:4; 2:6; 9:9; Qoh 2:26; Sir 15:17; 43:33; 45:26; 51:17; Wis 7:7,15,17; 8:21; 9:4,17. In these passages we find *didōmi* with wisdom as its object; for the complex *didōmi-dechomai*, see above, p. 145.

that he achieves is that the kingdom is seen much more clearly as a present gift. Thus we have two tendencies at play in the logion which are evident in the rest of the Gospel. This suggestion about the sapiential background of the verb *dechomai* finds further support when we consider the next phrase of the logion.

(c) *"Like a Child"*

(1) Mark 10:15 was originally an isolated logion. Its synoptic parallels do not make any more evident than it does how we are to understand the phrase *hōs paidion*. Neither the NT nor the OT nor intertestamental literature nor rabbinic literature possesses a "tract" about children as understood by Jesus' contemporaries. We have to search for implications and insinuations in isolated references to children. Mt 18:4 may be helpful; but the contiguity of 18:3 and 18:4 is probably due to Matthew, and thus it may not necessarily indicate what was meant by the original logion. The fact that children, along with the poor and sinners, are treated as a privileged group with regard to the kingdom[59] gives us a hint. But there may be fine shades of meaning which escape us in such a general comparison. The possibility that Mk 9:37 was, in pre-Marcan tradition, closely followed by Mk 9:42[60] may throw some light on the question; but it does not permit us to draw final conclusions, because the pre-Marcan collection of sayings, if it existed, was undoubtedly a secondary formation. In any case, one may not identify "children" and "the little ones" *a priori*.

However, a rather general consensus exists on the meaning of the phrase among contemporary exegetes: The child was, to the Jews of Jesus' time, a prototype of insignificance, dependence, unimportance, helplessness, and immaturity; the child was looked upon as one who deserved no attention, who had nothing to offer, and therefore could make no claims. The child had to receive whatever it received as a pure gift.[61] Some view the matter differently (generally without producing proof).[62] Still others, though agree-

---

[59] See J. Dupont, *Béatitudes*, 141-64.

[60] E. Klostermann (*Markus*, 93) and R. Bultmann (*History*, 149-50) are of this opinion; R. Schnackenburg ("Mk 9," 187) disagrees with it.

[61] See G. Bornkamm, *Jesus*, 84; E. Schweizer, *Mark*, 207; W. Grundmann, *Markus*, 196; J. Schmid, *Mark*, 188; E. Lohmeyer, *Markus*, 205; J. Dupont, *Béatitudes*, 148-58; E. Percy, *Botschaft*, 36; W. Pesch, *Lohngedanke*, 56-57; J. Behm, *TDNT* 4, 1002-3; R. Schnackenburg, *Moral Teaching*, 30, n. 25, 257; C. E. B. Cranfield, *Mark*, 324; V. Taylor, *Mark*, 423; E. Klostermann, *Markus*, 94; E. Neuhäusler, *Anspruch*, 136.

[62] M.-F. Lacan ("Conversion et royaume dans les évangiles synoptiques," *LumVie* 9 [1960] 31): to be like a child means to be open. N. Walter ("Zur Analyse von Mc 10,17-31," *ZNW* 53 [1962] 211): trust in the father. M.-J. Lagrange (*Marc*, 247): trust. N. Perrin (*Teaching*, 146): "ready trust and instinctive obedience."

ing with the general opinion, add other possibilities of interpretation.[63] One wonders whether present-day exegetes have not overreacted against their predecessors' romantic idealization of children.

A full presentation of the common opinion is given by A. Oepke in *TDNT* 5, 636-54. The OT, according to him, is in general negatively disposed toward the child; the thought of a child's innocence is foreign to it. However, this innocence seems to be implied in at least one OT passage, Jon 4:11. This text may be isolated, but it is interesting because it deals with pagan children. Furthermore, Sir 30:1-13, to which Oepke refers as an indication of the low esteem for children, could just as easily be understood otherwise. We are not certain that the OT writers were as able or as willing as we are to make the distinction between a child's present uselessness and his promise for the future. The evidence which Oepke finds in Billerbeck is not as one-sided as he seems to imply. For instance, the saying of R. Dosa ben Archinos that it is a waste of time to chatter with children is by no means representative of the entire rabbinic thought—at least as portrayed by Billerbeck.[64]

J. Jeremias gives a different interpretation of the phrase.[65] He takes up T. W. Manson's suggestion that *abba* is a distinctively Christian manner of addressing God and is to be traced to Jesus himself.[66] He uses this to interpret Mk 10:15 and paraphrases the saying thus: "If you do not learn to say Abba, you cannot enter the Kingdom of God." He continues: "In favour of this interpretation . . . are its simplicity, and the fact that it is rooted in the heart of the gospel."[67] The difficulty with this interpretation is the original isolated character of the logion, which Jeremias himself affirms. The logion thus had to speak for itself, and convey whatever message it contained by the force of its own words. It remains to be proven that the first thought that came to the mind of a Jew in connection with

---

[63] R. Schnackenburg (*Moral Teaching*, 30, n. 25): simplicity and trust. W. Grundmann (*Markus*, 207): they can say "abba," they trust and obey. E. Neuhäusler (*Anspruch*, 136): spontaneity.

[64] *Str-B* 1, 607, 773-74, 780-81, 786; 2, 423, 528-29; 4, 468-69. It is interesting to note that E. Neuhäusler (*Anspruch*, 135) arrives on the basis of the same evidence at a quite different conclusion from A. Oepke (*TDNT* 5, 646-47) in the matter of the rabbinic view of children.

[65] *Infant Baptism*, 49 and *Parables*, 191. He is followed by N. Perrin (*Kingdom*, 192); and seemingly also by G. Lundström (*Kingdom*, 171-72). W. Grundmann (*Markus*, 207) gives Jeremias' opinion as one of the options.

[66] T. W. Manson, *Teaching*, 331. The thesis is further developed by J. Jeremias in "The Lord's Prayer in the Light of Recent Research," *The Prayers of Jesus* (*SBT*, 2/6; London: SCM, 1967) 82-107, and in "Characteristics of the *ipsissima vox Jesu*," ibid., 108-15.

[67] *Parables*, 191.

the child was the fact that he cries "Dad." It also remains to be proven that Jesus pronounced the logion in close temporal or contextual proximity of the "Our Father."[68]

(2) References to children in the Gospel do not help us to arrive at a clear notion of what Jesus meant when he uttered the phrase "like a child."[69] An examination of other NT passages containing *paidion* brings us no further. Turning to the word *teknon*, we find in Phil 2:22 a statement of Paul which may be of some significance; referring to Timothy, he says of him that "he was like a son at his father's side serving the gospel along with me." Paul seems to be implying that Timothy was faithful and obedient to him in the work of preaching the good news. There are other passages which either imply children's dependence (2 Cor 12:14) or demand that they be obedient (Eph 6:1; Col 3:20; 1 Tim 3:4,12; Tit 1:6; 1 Pet 1:14).

In the OT *paidion* occurs in some passages which may aid us to discern the undertones of the word. Describing the chaos which was to be caused by God's punishment of his people, Isa 3:5 says: "The child (*nʽr—paidion*) shall be bold toward the elder, and the base toward the honorable." The words and the context (vs. 4) clearly indicate that the normal condition of the child is thought of as one of obedience and subjection. In Isa 10:19 and 11:6-8 the inexperience, helplessness, and vulnerability of children are the basis of the images presented. However, the most significant, and the most numerous, instances of *paidion* and *teknon* occur in the book of Tobit.[70] Tobit's *paidion* reminds us of *bn*, a technical term of address for disciples of wise men in other Wisdom books. The suspicion turns into probability when we examine the LXX translation of *bn* in the book of

---

[68] See J. Dupont, *Béatitudes*, 153, n. 2.

[69] In 18:4 Matthew seems to feel that the previous verse needs an explanation; he shows thereby that the logion no longer carried a clear message by itself.

[70] Tobit is a sapiential book. Computing the number of times that *paidion* and *teknon* occur as an address used by an elder for a junior (Tobit, Raguel, Anna, Edna addressing Tobias, and occasionally Sara; sometimes Tobias with his children) is difficult because of diverse textual traditions. For convenience we have used E. Hatch and H. A. Redpath, *A Concordance to the Septuagint*. *Paidion* as an address occurs in S 30 times, in AB 10 times; *paidia* in S twice, in AB once; *teknon* in S 3 times, in B 8 times, in A 7 times. Of particular interest are the passages in which the MSS disagree; they give us some indication of the wording in the original (Aramaic or Hebrew): 7:17 BA *teknon*, S *thygatēr*; 10:12 B *teknon*, A *thygatēr*, S *paidion*; 10:13 BA *adelphē*, S *teknon*; 11:14 BA *huion*, S *teknon*: 12:1 BA *teknon*, S *paidion*; 14:3 BA *teknon*, S *huios*; 14:4,8,10 BA *teknon*, S *paidon* (*paidia*); 3:15 BA *paidion*, S *teknon*. From these variations we conclude that *teknon* and *paidion* are easily interchangeable, and that in all probability some cases of one as well as the other stand for an original *bn*.

Sirach. In Sirach the word *teknon* occurs 47 times. In 15 cases at least[71] it translates the Hebrew *bn*, while *huios* stands for *bn* in at least 21 instances.[72]

This practice of the wise men addressing their disciples as sons has a very natural background, viz., that of parents' educating their children. It is clearly seen in the book of Tobit and Sir 30:1-13.

He who disciplines (*paideuōn*) his son will benefit from him,
and boast of him among his intimates.
He who educates (*didaskōn*) his son will make his enemy jealous,
and shows his delight in him among his friends (Sir 30:2-3).

Ps 78, a wisdom psalm,[73] expresses the same theme on the subject of religious instruction:

I will open my mouth in a parable,
I will utter mysteries from of old.
What we have heard and know,
and what our fathers have declared to us,
We will not hide from their sons
 (*mbnyhm; apo tōn teknōn*) ;
we will declare to the generation to come
The glorious deeds of the LORD and his strength
and the wonders that he wrought (vss. 2-4).

How shall a young man (*nʿr; ho neōteros*) be faultless in his way?
By keeping to your words (Ps 119:9).[74]

The intertestamental literature kept the theme alive. The *Testaments of the Twelve Patriarchs*, received their basic structure from it. The narrower sapiential use of the address "son" is evidenced in *1 Enoch* 79:1; 82:1-3; 91:2. The rabbis too addressed their disciples as *tekna*.[75] The "children" of Mk 10:24 is likely a witness to this rabbinic practice in Jesus' own instruction of his disciples.

The phrase "like a child" in Mk 10:15 has to be understood against this background. The specific characteristic of the child is his lack of wisdom, his need of instruction, and education.[76] E. Lohmeyer's suggestion[77] that Jesus in Mk 10:14 speaks as does the divine Wisdom in Prov 8:4-36 harmonizes well with this view. "Like a child," or like a disciple of the wise

---

[71] All the passages cannot be checked because of the incomplete Hebrew text.
[72] The LXX of Proverbs is quite consistent: *bn* is always rendered by *huios*.
[73] See A. Weiser, *Psalms*, 539.
[74] See A. Weiser, *Psalms*, 740.
[75] See E. Lohmeyer, *Markus*, 214; W. Grundmann, *Markus*, 213.
[76] See E. Lohmeyer, *Markus*, 203.
[77] *Markus*, 205.

who admits his lack of wisdom as he asks to be instructed, the follower of Jesus must admit that he is unwise in matters of the kingdom, and that his only way into the kingdom consists in his acceptance of its mystery, with the proclamation of which Jesus has been entrusted by the Father. Mk 10:15 is thus a sapientially colored *torah* of entry. It expresses the same thought as the Q logion Mt 11:25/Lk 10:21. There "the wise and understanding" are contrasted to babies. Though this Q logion may not reproduce *ipsissima verba Jesu*, it undoubtedly stems from the oldest Jewish-Christian tradition[78] and reflects the thought of Jesus.

The Aramaic word which *paidion* translates was probably *bar*.[79] This is more likely than the suggestion made by M. Black[80] that it was *ṭly'* "servant." The word *ṭly'* was used to designate a young person or a young servant, not a child.[81] If this suggestion is correct, it may be easier to explain the divergence between Mk 10:15 and its probable Johannine variant in 3:3,5. "Son" brings the thought of birth and rebirth more readily to mind than does "child." In it there may also lie the explanation of the similarities and differences between Lk 10:16; Mt 10:40; Jn 13:20 and Mk 9:37—the logia which speak of receiving or rejecting disciples (children) in the name of Jesus. The original reference was to the disciples as the "sons" of their Master; some strands of tradition transmitted the logion according to the sense, others translated literally.

(3) Mk 10:14c, in the present form of the pericope, should be read in the light of vs. 15, for there is little doubt that vs. 15 forms the climax of the story. The kingdom belongs to those who resemble children.[82] Its original message, however, concerned children as such; the verse stated simply that children have a share in the kingdom of God.[83] The narrative framework in vss. 13 and 16 confirms this view, for it speaks of children being brought to Jesus and blessed by him. Catechetical interests of the community were likely responsible for a spiritualization of children, turning them into symbols of proper attitude in regard to the kingdom, and thus for the insertion of a logion which already contained this thought, i.e., vs. 15.[84]

[78] See W. Grundmann, *Lukas*, 214.
[79] Cf. the targum rendering of the sapiential *bny* as *bry* in Prov 2:1; 3:1, etc.
[80] *Approach*, 221.
[81] See J. Levy, *Wörterbuch über Talmud und Mischna* (Berlin-Wien: B. Harz, 1924), *ṭly*.
[82] See J. Dupont, *Béatitudes*, 149-50. It is commonly agreed that the phrase should be translated, "for to such belongs the kingdom," and not "of such consists the kingdom"; see C. E. B. Cranfield, *Mark*, 323; J. Theissing, *Seligkeit*, 7.
[83] See R. Bultmann, *History*, 32; E. Lohmeyer, *Markus*, 203.
[84] See J. Dupont, *Béatitudes*, 150.

Commentators often consider the pericope itself to be a pronouncement story.[85] E. Lohmeyer[86] disagrees and characterizes it as a biographical anecdote. Yet the very form of Jesus' statement in vs. 14 speaks in favor of its being the real center of attention, even before the insertion of vs. 15. The insistent demand contained in the words, "Let the children come to me and do not hinder them," speaks in favor of the more common opinion. It would be false, however, to draw the conclusion that vs. 16 is nothing more than a pictorial expression of the truth enunciated in vs. 14b. We would agree with V. Taylor who thinks that the action of Jesus in vs. 16 "is as significant as his words."[87] Jesus promises the kingdom in word and deed; his gesture manifests the meaning of the word and the word explains the sense of the gesture, for he has the power to decide who belongs to the kingdom.[88] But it is again going too far to say, as do Taylor and others,[89] "that in a true sense Jesus Himself is the Kingdom." Such a statement is somewhat superficial; the fact that in Gospel parallels a reference to Jesus can replace a reference to the kingdom does not allow us to conclude that one is identical with the other, but simply that the two are intimately linked. E. Percy's opinion that Jesus looked upon himself as the representative of the kingdom before men[90] is valid for the thought of the Second Evangelist as well; but there is no suggestion in Mark that he identified the one with the other.

In the Second Gospel Jesus represents the kingdom, i.e., he makes it present in a mysterious manner and acts on its behalf. This is particularly evident in 10:16, where Jesus embraces and blesses those to whom the kingdom belongs. Jesus, moreover, decides who is to enter the kingdom, and sets conditions of entrance (10:14-15,21,23-25). It is Jesus, in fact, who is the main condition of entry (10:14,21,28), since he by his destiny, his work, and his word leads the way to the kingdom. Man's attitude toward Jesus determines his attitude to the kingdom (8:38—9:1; 10:22). We feel that *lypoumenos* of 10:22 should be understood in the same objective sense as fear on the part of disciples and others; not as a psychological reaction, but as a theological condition of the man who refuses to follow Jesus.

---

[85] See R. Bultmann, *History*, 32, 55-56; J. Dupont, *Béatitudes*, 154, n. 1; D. E. Nineham, *Mark*, 268; W. Grundmann, *Markus*, 205-6; E. Schweizer, *Mark*, 207.
[86] *Markus*, 205.
[87] *Mark*, 424.
[88] See E. Lohmeyer, *Markus*, 203-4; W. Grundmann, *Markus*, 207-8.
[89] V. Taylor, *Mark*, 423; C. E. B. Cranfield, *Mark*, 323; K. L. Schmidt (*TDNT* 1, 588-90) points out that in Gospel parallels a reference to Jesus can replace a reference to the kingdom.
[90] *Botschaft*, 34-35.

Does vs. 14 give us authentic words of Jesus? R. Bultmann[91] does not think so. "The original unit, vv. 13,14,16, may well be an ideal construction, with its basis in the Jewish practice of blessing, and some sort of prototype in the story of Elisha and Gehazi (2 Kings 4,27) and an analogy in a Rabbinic story." The OT story to which Bultmann refers does have some similarity with this pericope; what is missing, however, is its salient point, viz., children. His recourse to the rabbinic story, presumably in order to account for the presence of children, is hardly convincing. The stories found on the page of Strack-Billerbeck to which he refers[92] bear little resemblance to the Gospel pericope. Hence we would rather agree with J. Dupont and E. Percy[93] that in vs. 14 at least we have to do with basically authentic tradition.

O. Cullmann has suggested[94] that the story played a role in early Christian discussions about infant baptism. He bases his argument on the verb *kōlyō*, which was used in early Jewish-Christian baptismal liturgies in connection with the question about possible impediments to baptism. This suggestion has found more or less wholehearted acceptance by some;[95] others consider it possible,[96] or merely refer to it;[97] still others have strong doubts about it.[98] If the theme of infant baptism did find its way into the pericope during the course of its transmission by the community, it remained secondary. The main message of the story concerns acceptance of the kingdom "like a child."

(C) MARK'S MESSAGE IN 10:15

But what meaning did the logion take on in the Second Gospel? We will consider this question under three headings: Mark's understanding of "receiving the kingdom," of "like a child," and of "entering the kingdom."

(1) How does Mark understand the verb "to receive," now that the reference to knowledge has been excised? The most natural notion, that of

---

[91] *History*, 32.
[92] *Str-B* 1, 808.
[93] J. Dupont, *Béatitudes*, 154, n. 1; E. Percy, *Botschaft*, 34; see also V. Taylor, *Mark*, 421-24.
[94] O. Cullmann, *Baptism in the New Testament* (*SBT* 1; London: SCM, 1950) 71-80.
[95] A. Richardson (*An Introduction to the Theology of the New Testament* [London: SCM, 1961] 360-61) accepts it. D. E. Nineham (*Mark*, 268) considers it "at least plausible."
[96] C. E. B. Cranfield, *Mark*, 323.
[97] R. Bultmann, *History*, 387; W. Grundmann, *Markus*, 206.
[98] E. Schweizer, *Mark*, 207-8; E. Haenchen, *Weg*, 347.

receiving a gift, should be mentioned first. That this notion can be attributed to Mark is confirmed by 4:11. There the evangelist describes the mystery of the kingdom as a gratuitous gift of God entrusted to those who are with Jesus. There is an echo of the same thought in 10:14b and 10:16. This, however, does not exhaust the implications of the verb in Mark. Mk 9:33-37 helps us to answer the question with greater precision.

There is almost universal agreement that 9:33-37, though containing traditional material, should be considered a Marcan construction in its present form.[99] Vs. 33 is most probably redactional:[100] "the house," "on the way" are characteristically Marcan expressions,[101] and so is the impersonal plural followed by a singular.[102] In vs. 34 there occurs "on the way" again, along with the characteristically Marcan *siōpaō*.[103] That the reference to the discussion about greatness in this verse should be attributed to Mark's redactional arrangement is shown by Mk 10:35-45 which, like this pericope, follows Jesus' prediction of the passion. The structure of 9:30-37 and 10:32-45 is the same, the themes and even the logia are remarkably similar in both passages. It is practically impossible not to perceive that the evangelist is pursuing the same purpose in both, by arranging the sequence of pericopes and verses. Vs. 35b was originally an isolated logion.[104] It differs from its variants in two respects: it is the only version of the logion which contains the contrast of "first-last," and it does not have a regular two-membered structure. We suspect that it was Mark who added the phrase "and the servant of all," a phrase which does not fit into the contrast of "first-last" and disturbs the parallelism of the logion. Mark probably added this phrase under the influence of 10:43-45, the passage in which the thought of service predominates: vs. 43, great—servant; vs. 44, first—slave of all; vs. 45, "not to be served but to serve." In the structure of Mk 8:27—10:52 the passages 9:33-37 and 10:35-45 play the same role. That Mark constructed vs. 35a is indicated by the term *hoi dōdeka*.[105] Vs. 36

---

[99] J. Sundwall, *Zusammensetzung*, 61; R. Bultmann, *History*, 65; R. Schnackenburg, "Mk 9," 186; V. Taylor, *Mark*, 403-5; J. Wellhausen, *Marcus*, 75; E. Schweizer, *Mark*, 191-92.

[100] See R. Bultmann, *History*, 65; J. Schreiber, *Vertrauen*, 162-63; F. Neirynck, "The Tradition of the Sayings of Jesus: Mark 9,33-50," *Concilium* 20 (1966) 68.

[101] See J. C. Hawkins, *Horae*, 12-13.

[102] See C. H. Turner, "Usage," *JTS* 26 (1924-25) 229.

[103] See J. C. Hawkins, *Horae*, 12-13. F. Neirynck ("Tradition," 68) considers vs. 34 to be redactional.

[104] For a thorough discussion of Mk 9:35 and its variants, see R. Schnackenburg, "Mk 9," 185-200; see also E. Schweizer, *Mark*, 191-92. The variants are: Mk 9:35; Mt 20:26-27 (Mk 10:43-44); 23:11; Lk 9:48c; 22:26.

[105] See C. H. Turner, "Usage," *JTS* 26 (1924-25) 339.

was either constructed by Mark in imitation of 10:16,[106] or he found it in his source but inserted the phrase "and taking him in his arms."[107] The only possible Marcan characteristic in vs. 37, another originally isolated logion, is the position of the verbs at the end of clauses.[108]

L. Vaganay[109] has suggested that an originally Aramaic source had already assembled the logia which are found in Mk 9:33-50. The only redactional interventions of Mark are, according to him, the omission of a logion whose position in the source was between the two halves of the present 9:35 and the insertion of the phrase "and taking him in his arms" in vs. 36. We need not discuss Vaganay's entire article; it is sufficient for our purpose to point out that the Aramaic word which he makes serve as the link between vss. 35 and 36 cannot fulfil the task assigned to it. He suggests that the word *ṭly'* has been translated in vs. 35 as *diakonos*, and in vs. 36 as *paidion*.[110] We have already mentioned that this Aramaic word does not mean "child," but "young person, young slave." It is thus a great deal more likely that Mark is responsible for the construction of 9:33-37, particularly in view of all the Marcan characteristics in the verses which form the framework of the two logia, vss. 35b and 37.

If it is commonly agreed that the pericope should be looked upon as a Marcan construction, there is less agreement about its unity. Some think we have to do with two scenes which should be kept separate, vss. 33-35 and 36-37.[111] Others treat it as one scene.[112] Whether one scene or two, we should assume a basic unity of thought behind 9:33-37, once we accept the proposition that Mark composed it. What leads many to doubt the unity of the pericope is the inability to see the connection between the question implied in vs. 34 and the answer supplied in vs. 37. But is seems that the question in vs. 34 is often misunderstood. It is precisely this misunderstand-

---

[106] R. Bultmann, *History*, 61-62.

[107] L. Vaganay, "Le schématisme du discours communautaire à la lumière de la critique des sources," *RB* 60 (1953) 217. J. Sundwall's opinion has already been discussed.

[108] See C. H. Turner, "Usage," *JTS* 29 (1927-28) 354-55. J. Schmid (*Mark*, 179) seems to think that even vs. 37 was composed by Mark out of two traditional sayings.

[109] "Schématisme."

[110] "Schématisme," 212-17. The same opinion is held by M. Black (*Approach*, 220-21); for a criticism of L. Vaganay's opinion, see F. Neirynck ("Tradition," 68) and A. Descamps ("Du discours de Marc IX, 33-50 aux paroles de Jésus," *La formation des évangiles: Problème synoptique et Formgeschichte* (*RechBib* 2; Bruges: Desclée de Brouwer, 1957) 157-58); see also below, p. 173.

[111] R. Schnackenburg, "Mk 9," 186; F. W. Beare, *Records*, 148.

[112] R. Bultmann, *History*, 65; C. E. B. Cranfield, *Mark*, 307; E. Schweizer, *Mark*, **192-93.**

ing which lies behind the frequently voiced suggestion that 9:37 and 10:15 should change places.[113] Yet the question in vs. 34 concerns not pride and humility, nor importance and insignificance, but domination and service.[114] This is clearly confirmed by vs. 35 which ends with "servant of all"—and the presumption is that the ending is the most important part of the saying. It is also confirmed by the longer variant of 9:35 found in 10:43-44, where greatness is set in opposition to service. And we must keep in mind the parallel function of 9:33-37 and 10:35-45: It is quite possible, and even likely, that Mark added the last phrase of vs. 35 in order to create the proper contrast to "the greatest" of vs. 34.[115]

If, then, vss. 34 and 35 speak primarily of service, it is not difficult to see the connection between them and vs. 37. In vs. 37 Jesus demands that to a child the same service be rendered as to the most honored guest conceivable.[116] Thus Mark's *dechomai*, besides expressing the thought of receiving something as a gratuitous divine gift, contains the idea of service and subjection to someone who is to be welcomed[117] as being sent by God himself.[118]

(2) If the above conclusion is correct, we can describe Mark's child as "one who is the last of all and expected to subject himself to others." There is a gradation and intensification of thought in 9:33-37. In vs. 35 the Twelve are told that their greatness consists in being servants of all. How widely their service must extend, and what radical forms it must take, is shown by the next two verses—they must serve even children. What Mark is saying is this: the greatest among you is the one who serves, the one who subjects himself even to those who are most clearly expected to be subject to others.

Objections to this notion of child are raised by A. Descamps,[119] for whom there is no similarity of thought between the servant of vs. 35 and the child of vs. 36, since the child evokes the thought of humility, not of service; and vs. 37 speaks of a mystical union between Jesus and children, not of service.

---

[113] V. Taylor, *Mark*, 406; E. Lohmeyer, *Markus*, 193; M. Black, *Approach*, 220. R. Bultmann (*History*, 149) thinks that vs. 36 is a most unsuitable introduction for vs. 37.

[114] See E. Lohmeyer, *Markus*, 193-94; E. Schweizer, *Mark*, 192-93.

[115] For a somewhat different suggestion, see J. Sundwall, *Zusammensetzung*, 60.

[116] We cannot understand R. Bultmann's remark referred to in note 113; Jesus' embrace is designed perfectly to indicate the union of Jesus and the children which vs. 37 expresses in words.

[117] See V. Taylor, *Mark*, 305.

[118] For the overtones of the Jewish *šālîaḥ* institution present in this verse, see *Str-B* 1, 590.

[119] "Discours," 154-56.

To the suggestion that the contrast of "first-last" in vs. 35 is responsible for the introduction of the child in vss. 36-37 he replies that vs. 37 speaks of devotion, not humility. He opts for a catchword combination or a pre-Marcan source to explain the sequence of vss. 33-35 and 36-37. But we should note that the child does evoke the notion of subjection; that the idea of mystical union seems to be most unlikely; and that devotion really means service. Mark is not interested in static virtues, such as humility, or a static absence of qualities, such as insignificance. Rather, his primary concern is action.

(3) There is little doubt that Mark considers the kingdom as an essentially future reality; he is still looking forward to a fulfilment of which the present experience is but a foretaste. Yet this foretaste participates in the character of the future kingdom: the mystery of the kingdom is a present gift. Mk 10:28-30ab points in the same direction: those who follow Jesus already share in the reward. This reward comes to them with persecution, indeed, but it is nonetheless a present reality. To follow Jesus is thus to enter an as yet hidden kingdom. The tension which existed between the two members of the original logion, the present acceptance of the mysteries of the kingdom and the future entry into it, has shifted somewhat. The tension in Mark is between the present acceptance of, and entry into, a hidden kingdom and the entry into a fully manifested and victorious kingdom which is fast approaching.

In conclusion, we would paraphrase Mk 10:15 as Mark understands it: Whoever does not joyfully subject himself to the hidden kingdom offered by a hidden Messiah, by accepting it as a free gift of God, will not be allowed to share in its present and future blessings.

(2) Mark 10:23-25

Jesus looked around and said to his disciples, "How hard it is for the rich to enter the kingdom of God!" [24]The disciples could only marvel at his words. So Jesus repeated what he had said: "My sons, how hard it is to enter the kingdom of God! [25]It is easier for a camel to pass through a needle's eye than for a rich man to enter the kingdom of God." [26]They were completely overwhelmed at this, and exclaimed to one another, "Then who can be saved?" [27]Jesus fixed his gaze on them and said, "For man it is impossible but not for God. With God all things are possible."

(A) Text and Composition of Mark 10:23-25

(1) It is not surprising that MSS present a rather disunited picture of this text. For there is a strange alternation between statements which

affirm the difficulty of entering the kingdom for the rich and those which affirm or imply the same difficulty for all. Various attempts have been made by copyists to put some order into the text. Codex Bezae and the Vetus Itala place vs. 25 immediately after vs. 23, undoubtedly to bring together the verses speaking of the rich, and those referring to everyone. But no critical edition accepts this sequence of verses. Other MSS, e.g., Alexandrinus and the Codex Ephraemi rescriptus, have vs. 24c speaking of the rich, or, to be more precise, of those who place their trust in riches. This reading is a rather obvious attempt to harmonize vs. 24 with vss. 23 and 25, as well as to soften the harshness of the entire passage. Despite the fact that this reading makes nonsense of vs. 26, it is accepted as the more likely by Vogels and Merk; it is considered doubtful, but still left in the text, by von Soden. We feel that the correct reading is the one found in, among others, Vaticanus and Sinaiticus and accepted by Westcott-Hort, Nestle, the United Bible Societies edition, and Tasker.

It is difficult to decide whether to follow Westcott-Hort which in vs. 26 reads *pros auton*, or all the other critical editions which read *pros heautous*. The reading preferred by Westcott-Hort is found in Vaticanus, Sinaiticus, and others. Yet the majority of MSS, Alexandrinus, Bezae and Koridethi among them, have the reading accepted by other critical editions.

(2) The sequence of verses is indeed strange. Vss. 23 and 25 express the same thought in different manners, whereas the statement of Jesus in vs. 24 and the disciples' reaction in vs. 26 clearly belong to each other. Vs. 27 is a fitting answer to both problems raised in vss. 23-26: the salvation of the rich as well as that of everyone. How are we to explain the formation of this text?

A confusing array of attempts has been made to sort out the obvious inconsistencies and lack of logic and good order. We can leave aside those attempts which continue the tradition of some copyists by rearranging verses,[120] and those which try to solve the problem by amputating the text.[121] To consider vs. 24 as a mere restatement of vs. 23,[122] for the sake of emphasis, is quite inadequate in view of vs. 26; the disciples' question in the latter implies that the circle of those who will be saved with great difficulty is very wide and not limited to the rich alone. F. W. Beare[123] is

---

[120] See, e.g., D. E. Nineham, *Mark*, 272, 275; A. E. J. Rawlinson, *Mark*, 140.

[121] This solution is offered by J. Wellhausen, *Marcus*, 81; E. Klostermann, *Markus*, 103; W. Grundmann, *Markus*, 209. For a criticism of such attempts, see N. Walter, "Analyse," 206-9.

[122] The opinion voiced by S. E. Johnson, *Mark*, 175.

[123] *Records*, 194. A similar suggestion is made by E. Best, "Uncomfortable Words: VII. The Camel and the Needle's Eye (Mk 10:25)," *ExpT* 83 (1970-71) 84: Mark has added vss. 26-27 to 23-25.

of the opinion "that the question and answer of vv. 26-27 are a secondary supplement to the pronouncement of v. 25, which may always have been coupled with that of v. 23 (as in Luke). There is certainly something artificial about the question: 'Who then can be saved?'" On the contrary, this artificiality disappears if we suppose that vss. 26-27 originally did not refer to vss. 23 and 25, but to vs. 24c. It is, moreover, much more likely that Luke's parallel (18:24-27) is a correction of Mark's text, not a more faithful reflection of Mark's source. Vss. 24c and 26 belong to each other; it is quite incorrect to regard one as primary and the other as a later addition —at least insofar as Mark or his source are concerned.

E. Lohmeyer's suggestion[124] that the pericope presents a Johannine method of conversation is unconvincing since such method is not evident in the rest of the Second Gospel. It is equally difficult to accept R. Bultmann's suggestion[125] that vs. 24c is a doublet of vs. 23b. If we must look on the two sayings as springing from one authentic word of Jesus, we would rather agree with J. Sundwall, N. Walter and S. Légasse[126] who consider vs. 24c to be an older form of vs. 23b; for it is easier to imagine that a difficult word of Jesus affecting everyone would in the process of transmission be limited to the rich alone. Contrary to R. Bultmann,[127] we feel that Mark more likely found vss. 24, 26-27 already joined to 10:17-22 and added vss. 23 and 25.

But we must give closer attention to the opinions of N. Walter[128] and S. Légasse.[129] Though they agree on a number of important points, there are enough divergences between them to justify a separate treatment. Walter thinks that originally Mk 10:17-22 was a narrative of a call to

[124] *Markus*, 213.
[125] *History*, 105.
[126] *Zusammensetzung*, 66. It is difficult, however, to accept Sundwall's opinion on the formation of vss. 23-27 (it is a Marcan construction; vss. 23b, 25 were bound with 10:15 in the pre-Marcan tradition by means of the catchword-phrase "to enter the kingdom of God," but they were separated from 10:15 by Mark who found another set of catchwords, viz., "having riches" in the story of the rich man). But J. Sundwall seems to be solving too many problems by means of catchwords. Besides, he follows the D reading to prove his point. For references to N. Walter and S. Légasse, see below, pp. 160-62.
[127] *History*, 22; so also E. Schweizer, *Mark*, 209-10. Schweizer's opinion that vs. 26 is redactional is effectively disproved by the evident connection between vss. 24c and 26, as well as by the singular usage of *sōzesthai* (see E. Lohmeyer, *Markus*, 214, n. 4).
[128] In "Analyse."
[129] *L'appel du riche (Marc 10,17-31 et parallèles): Contribution à l'étude des fondements scripturaires de l'état religieux* (VS, collection annexe 1; Paris: Beauchesne, 1966); "Jésus a-t-il annoncé la conversion finale d'Israël? (A propos de Marc x 23-27)," *NTS* 10 (1963-64) 480-87.

discipleship ending with a refusal. Tradition shifted the accent from the call to the discussion on the way to the kingdom. In the third stage the emphasis shifts again, this time to the negative result of the call. It was at this stage that vss. 22b, 23, 24a were added; these verses reflect the theology of poverty. They also furnish the story with a general application. In order to emphasize, as well as to soften, the message of vs. 23 Mark added vss. 24bc-27.[130] Mark understood the message of vss. 17-22 to be that of vs. 23; he added vs. 24bc on account of its similarity with vs. 23; to harmonize vs. 25 with vs. 23 he replaced the original *anthrōpon* with *plousion*, even though he thereby weakened the impact of vs. 26. But, being no anticapitalist, he wished, by means of vs. 27, to stress that there is possibility of salvation even for the rich. *Palin* in vs. 24b and *perissōs* of vs. 26 were also added by Mark to create a certain outward intensification of the argument.

In criticism of N. Walter's opinion, we would ask why Mark bothered adding vs. 24bc at all. If he did not hesitate to change vs. 25 to conform it to vs. 23, he must have realized that vss. 24bc and 26 needed much more drastic surgery if they were to convey the message which they were, according to Walter, designed to convey.

Like Walter, Légasse thinks that vs. 24c is the original form of the logion. It was imbedded in a pericope which consisted of 24bc, 25 (without *plousion*), 26, 27; the message of the pericope dealt with the impossibility of salvation without God's effective aid. At some stage of tradition vss. 24bc and 26a were separated from the rest of the pericope, and attached to the story of the rich man; but in order to conform the logion to the story, *hoi ta chrēmata echontes* was inserted into the logion in vs. 24c. At a later stage someone, wishing to restore the truncated pericope, added it in its entirety, but in vs. 25 changed the original *anthrōpon* or *tina* into *plousion*. To smooth the course of the story, Mark added *palin* in vs. 24b and *perissōs* in vs. 26.

This reconstruction of the prehistory of the pericope is admissible, though not highly probable in our opinion. Two serious objections can be raised against Légasse's method of argumentation. He thinks that the fear of the disciples in vs. 24a is out of place (and that therefore vs. 23 cannot be the original form of the logion) because the disciples have left everything and thus have nothing to fear for themselves. But this interpretation of the disciples' fear is quite false: fear in Mark is an expression of defective faith; it is a theological, not a psychological datum. Again in both Légasse's and Walter's procedure, there is a tendency to postulate, without proof, a

---

[130] He thinks that vs. 24c is more original than vs. 23b, and that vss. 24b-27 did not speak of the rich in its pre-Marcan form.

specific *Sitz im Leben Jesu* in order to escape unpleasant conclusions or to respond to an objection.[131]

We would rather suggest that Mark found 10:17-22 in the tradition substantially in the form in which he gives it,[132] but without vs. 22c, *ēn gar echōn ktēmata polla*. Vss. 24bc, 26, 27 had already been linked to the story in the pre-Marcan tradition. To this Mark added vss. 22c, 23, 24a, 25, *palin* in vs. 24b and *perissōs* in vs. 26a.

To show that this hypothesis is plausible, we begin by noting the formulae which echo one another:

1. 21 *emblepsas . . . eipen autō*
   23 *periblepsamenos . . . legei tois mathētais*
2. 22 *ho de stygnasas epi tō logō*
   24 *hoi de mathētai ethambounto epi tois logois*
3. 22 *echōn ktēmata*
   23 *hoi ta chrēmata echontes*

With regard to the first pair note the word *periblepsamenos* in vs. 23; though most likely borrowed from older tradition,[133] it is characteristic of Mark. *Legei*[134] as well as *tois mathētais autou*[135] could well be due to Mark's composition. Mark probably composed vs. 23a in conscious imitation of vs. 21a. Thus he laid a stone in the construction of the contrast between the disciples and the man who refused to follow Jesus. One objection which could be voiced against the Marcan composition of vs. 23a is that it does not mention Jesus being alone with the disciples, a feature which one would expect in the introduction to a scene which presents him giving special instruction to the Twelve. However, the same result is achieved by vs. 22 which tells us that the man had gone.

With regard to the second pair, it is difficult not to see an intended affinity between *stygnasas*[136] and *ethambounto*.[137] The two verbs seem to

---

[131] S. Légasse (*Appel*, 71-73) attempts to escape the conclusion that few will be saved. N. Walter ("Analyse," 211) is responding to E. Percy (*Botschaft*, 92, n. 1), who considers it unthinkable that Jesus would compare the difficulty of entering the kingdom with that of a camel going through a needle's eye.

[132] Vs. 17a is probably redactional; *mē aposterēsēs* of vs. 19 and vs. 21a are possibly so. See S. Légasse, *Appel*, 20, 42.

[133] R. Bultmann, *History*, 332.

[134] See M. Zerwick, *Markusstil*, 49-50, 52, 57-58; G. Minette de Tillesse, *Secret*, 167, 174, 179-80.

[135] See C. H. Turner, "Usage," *JTS* 26 (1924-25) 235-37.

[136] Bauer-Arndt-Gingrich (*Lexicon*, 779) gives "to be shocked, appalled" as the first meaning of the verb.

[137] *Thambeomai* is a verb found only in Mark; cf. J. C. Hawkins, *Horae*, 12.

convey the same meaning,[138] at least in the eyes of Mark. It is well nigh impossible to regard *stygnasas* as redactional, since Mk 10:22 is the only certain instance of the verb in NT; but it is quite possible to regard *thambeomai* as such. The presence of this verb speaks in favor of considering vs. 24a as redactional, for Mark stresses the ignorance and fear of the disciples. *De* is not characteristic of the Second Gospel, yet its presence can easily be explained as the result of an almost word for word repetition of vs. 22a in vs. 24a.

When we come to the third pair of phrases the similarity is again striking. *Chrēmata* could easily recall *ktēmata* because the phrase *chrēmata kai ktēmata* seems to have been current in the Greek-speaking world.[139] It is less likely that the evangelist inserted *hoi ta chrēmata echontes* in vs. 23, for the phrase betrays no Marcan characteristics. But he probably attached vs. 22c to the preceding story. Reasons to support this suggestion are not far to seek. There is, first, Mark's predilection for periphrastic constructions;[140] he shows, moreover, a strong tendency to form short sentences with *gar*.[141] Apart from this last statement the man's riches are barely implied in the story. This short statement at the end of the story gives the impression of being an afterthought, added in the interests of the following verse; it recalls *ēsan gar halieis* of 1:16.

We need not discuss the literary form of 10:17-22 and the likely stages of its development. As Mark found it, it undoubtedly already contained the discussion about the way to the kingdom as well as the refusal of the call to discipleship. This final note stands in sharp contrast to 10:16 which shows us Jesus lovingly receiving children, designating them thereby as already, in some manner, belonging to the kingdom. The record of the refusal would naturally lend itself to a generalizing addition pointing out the difficulty of entering the kingdom. This addition, pre-Marcan in all probability, consisted of vss. 24bc, 26, 27; it ends on a note very similar to that of 10:16 and 4:11. In all these verses the thought is present that it is God's initiative alone which can bring men into the kingdom. The disciples' astounded question in vs. 26 is in perfect harmony both with Jesus' state-

---

[138] See E. Schweizer, *Mark*, 213; J. Weiss, "Zum reichen Jüngling Mk 10,13-27," *ZNW* 11 (1910) 80-81.

[139] See H. G. Liddell and R. Scott, *Lexicon*, 1002.

[140] See V. Taylor, *Mark*, 45, 62; X. Léon-Dufour, "The Synoptic Gospels," *Introduction to the New Testament* (eds. A. Robert and A. Feuillet; New York: Desclée, 1965) 198 (especially n. 9). R. Pesch's assertion (*Naherwartungen*, 160) that periphrastic formulations with the participle are rare in Mark is incorrect; for references, see V. Taylor, *Mark*, 45, 62.

[141] See R. H. Lightfoot, *Message*, 85.

ment in vs. 24c and with the fact that the episode in 10:17-22 pointed out to them that perfect fidelity to the Law and an ardent desire to fulfil all the commandments cannot bring salvation.[142]

Mark, however, wished to introduce a further contrast, that between the disciples and the man who refused to follow Jesus. Vss. 29-31 were probably placed in their present context by the evangelist himself by means of the redactional, or redactionally retouched, vs. 28.[143] What discipleship means to Mark is indicated in 10:28-29, and in 1:16-20; 2:14. Two characteristics are mentioned: leaving behind earthly possessions and connections and following Jesus. The traditional form of 10:17-22 with its sequel of vss. 24bc, 26, 27 contained the contrast to the theme of following. To complete the picture and to draw the contrast more fully, Mark added the remark that the man who refused the call was rich and two sayings on the incompatibility of riches with entry into the kingdom. The evangelist found these two sayings in the tradition, probably joined to each other. He separated them on account of the great similarity between vss. 23b and 24c and because of his wish to establish the link between vss. 22c and 23b. He should likely be credited with constructing vss. 23a and 24a[144] and with adding *palin* and *perissōs* to vss. 24b and 26a.[145]

(3) R. Bultmann[146] considers vss. 23b and 25 as genuine sayings of Jesus. Moreover, we find rather persuasive the arguments of Walter and Légasse who consider vs. 24c as genuine. We would attribute to this saying at least a substantial authenticity. There are enough other statements, reliably stemming from Jesus, which express thoughts similar to, or identical with, those in vss. 23b and 24c.[147] An examination of the rabbinic atti-

---

[142] See Neuhäusler, *Anspruch*, 184.

[143] See R. Bultmann, *History*, 22; E. Lohmeyer, *Markus*, 216; E. Schweizer, *Mark*, 214; D. E. Nineham, *Mark*, 273; S. Légasse, *Appel*, 79; N. Walter, "Analyse," 214-15 (especially n. 39). The phrase *ērxato legein* is typically Marcan (see V. Taylor, *Mark*, 48, 63). And the verb *akoloutheō* is far more characteristic of Mark than of the other two Synoptics; this is evident from the fact that these seldom use the verb apart from the passages taken from Mark and Q.

[144] See E. Schweizer, *Mark*, 209.

[145] See the summaries of Walter's and Légasse's opinions given above, pp. 160-61. For *palin*, see J. C. Hawkins, *Horae*, 13. The phrase *apokritheis legei*, found in vs. 24b, does not always imply a previous question; see M. Zerwick (*Biblical Greek*, par. 366) and such texts as Mk 9:5; 11:14; 12:35; 14:48.

[146] *History*, 105, 117.

[147] We cannot discuss these here. For parallels to vs. 24c, see N. Walter, "Analyse," 211, and S. Légasse, *Appel*, 66-69. For parallels to 23b and 25, see E. Percy, *Botschaft*, 105-6, 92, n. 1. This reference to Percy does not mean that we agree with his opinion that "the so-called Lucan ebionitism seems to stem from Jesus." For a criticism of P. S. Minear ("The Needle's Eye: A Study in Form Criticism," *JBL* 61 [1942] 165-67), who accepts only vs. 25 as authentic, see S. Légasse (*Appel*, 81, n. 50).

tudes toward riches and the rich, and of those found in the intertestamental literature convinces us of the originality of Jesus' stand.

The Jews recognized the danger of riches becoming a hindrance to the observance of the Law, and they had had such an experience of the wealthier among their people succumbing to the temptation of a worldly Hellenism that the word "poor" had become a synonym for "pious." None the less, the rabbis strove for a balance in this matter, and their view is well expressed in Midrash Exodus Rabbah 31 on Ex 22,24: "You will find that there are riches that positively harm their possessors and other riches that stand them in good stead.[148]

In regard to vs. 24c, no one would attribute to Jesus the thought that anyone could enter the kingdom by birthright.

An argument against considering *both* logia as authentic lies in their similarity, which seems to suggest a direct dependence of one upon the other. We must note, however, that objections against the authenticity of both can be raised only on the basis of the rather singular introductory phrase "how hard it is," and not on the basis of the phrase which speaks of entering the kingdom, for the latter theme is frequent in Jesus' sayings. It may well be that there is an interdependence of the two logia with regard to the introductory words; this interdependence should likely be traced to the oral stage of tradition. Yet it does not force upon us the dilemma of accepting only one or the other logion as substantially authentic.

(B) Mark 10:23-25 on the Lips of Jesus

(1) There is a universal agreement among commentators on the meaning of and the reasons for Jesus' negative attitude toward riches. God claims man whole and entire; absolute obedience, complete devotion, and an undivided heart are required of him whom God has called. The reign of God which Jesus is announcing and bringing about in a hidden manner by his word and work manifests itself in the destruction of everything which is opposed to it. Riches, however, tend to enslave man; they make him forget the truly important realities and give him a false sense of security and independence.[149]

This view of riches is not false, but it is one-sided and too highly interiorized and spiritualized. A statement like that of C. E. B. Cranfield[150] who

---

[148] N. Perrin, *Teaching*, 143; for other rabbinic texts, see *Str-B* 1, 826-28.

[149] See R. Schnackenburg, *Moral Teaching*, 126-28; R. Bultmann, *Jesus and the Word* (New York: Scribner, 1958) 97-98; G. Bornkamm, *Jesus*, 142; T. W. Manson, *Teaching*, 276; J. Jeremias, *Parables*, 194-95; E. Percy, *Botschaft*, 105-6; T. A. Burkill, *Revelation*, 169, 185, n. 26; D. E. Nineham, *Mark*, 271; E. Neuhäusler, *Anspruch*, 180-81.

[150] *Mark*, 330; a similar opinion is voiced by E. Neuhäusler, *Anspruch*, 178.

says that "the command to give to the poor is here perhaps primarily an indication of the way to get rid of possessions which have become an idol" sounds almost buddhistic.[151]

To bring some balance into the picture, we should consider the attitude toward kings, the mighty, and the rich in *1 Enoch*, and the attitude toward the poor demanded by the OT. In *Enoch*, the three classes, i.e., the kings, the mighty, and the rich, are looked upon as one; they are accused of the same crimes and threatened with the same punishment.

Then shall the kings and the mighty perish
And be given into the hands of the righteous and holy.
And thenceforward none shall seek for themselves mercy from the
Lord of Spirits . . . (38:5-6).

The powerful and the rich will be punished because they do not recognize the fact that power and riches come from God, because of their idolatry of possessions, and because of their persecution of the righteous (paraphrase of 46:4-8). They will be punished because they "have not believed . . . nor glorified the name of the Lord of Spirits," but their "hope was in the sceptre of [their] kingdom and in [their] glory," and their "souls are full of unrighteous gain" (paraphrase of 63:7,10).

Woe to those who build unrighteousness and oppression
And lay deceit as a foundation . . .
Woe to those who build their houses with sin . . .
Woe to you, ye rich, for ye have trusted in your riches,
And from your riches shall ye depart,
Because ye have not remembered the Most High in the days
  of your riches.
Ye have committed blasphemy and unrighteousness . . . (94:6-9).

Woe unto you, ye sinners, for your riches make you appear
  like the righteous . . .
Woe to you who devour the finest of wheat,
And drink wine in large bowls,
And tread under foot the lowly with your might . . . .
Woe to you, ye mighty,
Who with might oppress the righteous . . . (96:4,5,8).

For ye men shall put on more adornments than a woman . . .
Therefore they shall be wanting in doctrine and wisdom . . .
None of your deeds of oppression are covered and hidden (98:2,3,6).

---

[151] See H. de Lubac's discussion of Christian and Buddhist concepts of charity toward one's neighbor in his *Aspects of Buddhism* (New York: Sheed and Ward, 1954) 15-52.

Woe to you who make deceitful and false measures . . .
Woe to you who build houses through the grievous toil of others . . .
Woe to them who work unrighteousness and help oppression . . .
For he shall cast down your glory . . . (99:12-13,15-16).

And they [i.e., the rulers, vs. 4] helped those who robbed us and devoured us and those who made us few; and they concealed their oppression, and they did not remove from us the yoke of those that devoured us and dispersed us and murdered us . . . (103:15).

Other passages in *Enoch* speak in the same vein: 53:1-7; 54:1-6; 62:1-12; 97:8-10; 103:9-15.

Note, however, the following features in *Enoch's* attitude toward the powerful and the rich: first, its strong eschatological coloring; secondly, the chief accusation concerns their oppression and exploitation of the poor and the helpless;[152] thirdly, their lack of a chance of repentance.

These protests against the oppression of the poor and helpless in *Enoch* are by no means something new in the tradition of Judaism. They are a constant theme appearing again and again in various layers of the OT. A few examples will suffice. The Book of the Covenant, the earliest of the three collections of laws in the Pentateuch, already defends those who cannot defend themselves:

You shall not molest or oppress an alien, for you were once aliens yourselves in the land of Egypt. You shall not wrong any widow or orphan. If ever you wrong them and they cry out to me, I will surely hear their cry" (Ex 20:20-22).

The prophets too excoriated social injustices in Israel:

Thus says the LORD:
For three crimes of Israel, and for four,
I will not revoke my word;
Because they sell the just man for silver,
and the poor man for a pair of sandals.
They trample the heads of the weak into the dust of the earth,
and force the lowly out of the way.
Son and father go to the same prostitute,
profaning my holy name.
Upon garments taken in pledge
they recline beside any altar;
And the wine of those who have been fined
they drink in the house of their god (Am 2:6-8).

---

[152] It should be noted that in the "Parables," commonly held to be the latest part of *1 Enoch,* other sins of the rich and the powerful, such as idolatry of riches and seduction of the poor, are more frequently mentioned than elsewhere in the book.

More than a century later Jeremiah threatens:

Woe to him who builds his house on wrong,
his terraces on injustice;
Who works his neighbor without pay,
and gives him no wages.
But your eyes and heart are set on nothing
except on your own gain,
On shedding innocent blood,
on practicing oppression and extortion (22:13,17).

The book of Deuteronomy commands:

If one of your kinsmen in any community is in need in the land which the LORD, your God, is giving you, you shall not harden your heart nor close your hand to him in his need. Instead, you shall open your hand to him and freely lend him enough to meet his need. The needy will never be lacking in the land; that is why I command you to open your hand to your poor and needy kinsman in your country (15:7-8,11).

The wise demand:

Remove not the ancient landmark,
nor invade the fields of orphans;
For their redeemer is strong;
he will defend their cause against you (Prov 23:10-11).

In his defense, Job adduces the following facts:

For I rescued the poor who cried out for help,
the orphans, and the unassisted;
The blessing of those in extremity came upon me,
and the heart of the widow I made joyful.
I was a father to the needy;
the rights of the stranger I studied . . . (29:12-13,16).

This is the natural background to Jesus' condemnation of riches and the rich. We have every reason to suppose that in his woes aimed at the rich the thought of oppression and exploitation of the poor and defenseless was present. There are, of course, differences between Jesus and the OT, as well as between his strictures and those of *Enoch*. OT commands and threats are not pronounced in the light of the approaching kingdom as are those of Jesus—although the prophetic preaching, denunciations, and threats were frequently uttered in the light of a coming divine judgment which can justly be called eschatological. Unlike *Enoch*, Jesus does not declare the salvation of the rich to be impossible—this apparently quite apart from the context within which Mk 10:23,25 is now imbedded. *Dyskolos* means "difficult," but not impossible; the camel saying is a hyperbole whose content must not be expressed abstractly in terms of mathematical precision. We also note the absence of exasperation and vengefulness in Jesus' sayings, something that is quite different from *Enoch*'s statements and threats. Rather than threaten the rich Jesus pities them.[153] We feel in his sayings the assurance of a person who is far above the temptation to envy the rich their illusory security and comfort.

With this background in mind we arrive at a more balanced view of Jesus' attitude toward riches and the rich. Commentators often point out that Jesus condemns riches insofar as they keep men away from total obedience to God. But, in order to escape the extreme consequences which these sayings seem to call for, they must have recourse to his manner of acting, his friendship with people of means, etc.[154] We should rather think that Jesus uttered these warnings because riches had already become a typical instrument of alienation from God and oppression of the poor in his own mind and that of many of his contemporaries.

(2) We now turn our attention to vs. 24c. S. Légasse[155] feels that the logion was pronounced by Jesus at a time when he was encountering a growing incomprehension on the part of the Jewish people. Understood against this background, the saying loses the semblance of excluding most people from the kingdom. This interpretation may be correct, but it is impossible to produce convincing proof in its favor. Since the primitive church, feeling its sting, placed it in the context of vss. 26-27, it did not regard it as a statement which had lost its force and validity with the loss of its original *Sitz im Leben*.

If we seek a background for this saying of Jesus, we might find it in Jewish views current at the time of Jesus concerning justification. S. Lyonnet, in his *Quaestiones in Epistulam ad Romanos I*,[156] draws on Wisdom, deuterocanonical, intertestamental, and rabbinic literature as well as on Josephus, and arrives at the following conclusions: The Jewish concept of salvation was founded, in practice at least, more on the thought of merit arising from works than on the thought of salvation as a free gift of God; Abraham's election was conceived of as a reward for constancy rather than as a gratuitous grace; Jews spoke far more frequently of the Covenant, upon which they looked as a bilateral contract, than about the promise of God freely bestowed upon Abraham; the Law thus became a necessary instrument for the performance of meritorious acts; the result of all this

---

[153] See C. E. B. Cranfield, *Mark*, 331.
[154] See, for instance, R. Koch, "Die Wertung des Besitzes im Lucasevangelium," *Bib* 38 (1957) 151-69.
[155] *Appel*, 71-73.
[156] Pp. 89-101.

was a certain self-sufficiency, a formalism, and a conviction of a privileged status in regard to salvation.

Jesus, however, proclaims a different message: man must accept the mysteries of the kingdom as one who lacks all wisdom (Mk 10:15) and the power and merit (Mk 10:24) which might serve as a wedge to pry open the gates of the kingdom. It is as difficult to become like one who is unwise with regard to the most important knowledge which man may attain, as it is to become like one who is powerless with regard to the most important achievement open to him.

(C) Mark's Understanding of Vss. 23-25

(1) If our reconstruction of the literary composition of vss. 23-25 is correct, some conclusions about Mark's intentions can be drawn. The evangelist is setting up a contrast between the disciples and the rich man. The contrast consists of two elements: the disciples have responded to Jesus' call, while the rich man has refused it; the disciples have given up their earthly possessions and connections, while he refuses to part with his riches. Mark is thus outlining the conditions for entry into the present hidden and the future glorious kingdom. What men must be ready to do is to follow Jesus and to renounce earthly entanglements.[157] The passage recalls the interpretation of the Sower, vs. 19 in particular, where "the anxieties over life's demands, and the desire for wealth, and cravings of other sorts come to choke off the word." These conditions are the human side of the coin; the divine element is also indicated. First, the call itself, which is portrayed not only in 10:21, but in the story which precedes that of the rich man: "Let the children come to me and do not hinder them." "Then he embraced them and blessed them." The initiative comes from Jesus in 1:17,20; 2:14, as well as in 10:21;[158] in 10:16 he does more than he is asked to do.[159] Secondly, the divine aid which is necessary to enable man to respond to the call is referred to in 10:27. Thirdly, the reward, i.e., the kingdom and a foretaste of it in the life of the community, is indicated in 10:15,23,24,29-30. Of the two elements, divine and human, it seems that

---

[157] See B. M. F. van Iersel, "La vocation de Lévi (Mc., II, 13-17, Mt., IX, 9-13, Lc., V, 27-32: Traditions et rédactions," *De Jésus aux Évangiles* (ed. I. de la Potterie) 221-22. Although his outline of the structure of vocation-stories in the gospels (pp. 215-16) is weakened by the fact that only Marcan vocation-pericopes contain all the elements which he attributes to a typical vocation-story, his remarks on Mk 10:17-22 are very much to the point, provided we attribute the message which he discovers in it to Mark, and not necessarily to the pre-Marcan tradition.

[158] S. Légasse (*Appel*, 42) thinks that the phrase "Jesus looked at him with love" is redactional.

[159] See V. Taylor's reference to Bengel in *Mark*, 424.

the latter is stressed more than the former: Mark is well aware of the divine call which is being proclaimed to the world, but he is just as conscious of the obstacles which prevent the proper response among men. It is these obstacles which he is trying to remove.

(2) These obstacles, however, do not lose their sting at the threshold of the community. If we read these verses, as we should, within the larger context of Mk 8:27—10:52, we may at least suspect that another thought was present in the evangelist's mind when he wrote them. It has frequently been remarked that in this section of the Gospel Jesus devotes himself primarily to the instruction of the disciples. In the first chapter we pointed out the echoes, and more than echoes, of the community's problems in the section. Mark is evidently struggling against abuses within a community which is still a long way from fully understanding the meaning of its Master's death on a cross and that he "came not to be served but to serve" (10:45). Unwillingness to serve and a desire to dominate are all too well known within the community: each Christian must still be exhorted to "take up his cross and follow" Jesus. He must continually strive to be subject to others, to resist the temptation to dominate. He must also resist the attraction of riches, the typical means of domination and exploitation of others. The primitive church knew of the difficulties involved when men were required to break family ties and relinquish their possessions for the sake of Jesus and the good news. It also knew that this renunciation would never be complete; the old ways would never be thoroughly rejected (cf. Mk 10:42-43). We would disagree with Légasse who suspects Mark of being too weak to resist the pull of the tradition in formulating his thought on the problem of riches.[160] Taken in isolation, vss. 23 and 25 could be interpreted as making material poverty a condition for entering the kingdom; but in conjunction with vs. 24, and in the light of the leading themes of this section of the Gospel, they serve to complete the image of "the way" which Christians must go in their following of Jesus. They are not to be taken as abstract rules, but as an expression of the radical demands made on man by the kingdom. Nothing less than a total break with the past will do, a total renunciation, and a complete subjection to Jesus' call and to the service of the community.

(3) Mark 9:47

If your eye is your downfall, tear it out! Better for you to enter the kindom of God with one eye than to be thrown with both eyes into Gehenna.

---

[160] *Appel*, 94.

(A) THE CONTEXT OF THE SAYING

The text of Mk 9:47 is hardly disputed. The only difference among critical editions concerns the article *tēn* before *geennan*, which Westcott-Hort, following B, omits. We feel, however, with all other critical editions that the codex B is not sufficient evidence against the preponderant testimony in favor of the article in other MSS.

The logion under discussion appears in a collection of sayings which are linked to one another by catchwords. The extent of this collection is not certain. That it contains vss. 42-50 is evident: vss. 42-47[161] are joined by the catchword *skandalizein*; vs. 48 is a comment on "Gehenna," introducing a new catchword, "fire," and thus forming a verbal link with vs. 49. The reference to salt in vs. 49 supplies the connection with vs. 50, which is clearly the last saying of the series. But where does the series begin? Some authors regard 9:33 as the beginning of the entire collection. Thus R. Bultmann, discussing Mk 9:33-50, says:

Clearly Mark's source was already a sort of catechism which, by providing the introduction in vv. 33f., Mark has turned into a scene in the life of Jesus. In v. 36 he has also given a most unsuitable introduction to v. 37 for which he has borrowed the motif from 10,13-16. In the source vv. 38-40 were inserted (by a later hand?) *ad vocem epi (en) tō onomati.* V. 42 in the source followed vv. 37,41 after the cue *paidia* or *mikroi*, and then vv. 43-48 . . . . Then vv. 49 and 50 are arranged . . . (50b) could well serve as the end of the catechism.[162]

R. Schnackenburg[163] would attribute a great deal more to Mark's redaction than Bultmann. He doubts whether vs. 36 is a simple imitation of 10:13,16 and turns down the suggestion that vs. 41 resumes the thought of vs. 37;[164] nor is he convinced that vss. 37,41,42 ever formed a unit prior to Mark. It is the evangelist who is responsible for the linking of vs. 40 with vs. 39, of vs. 41 with vs. 40, and of vs. 49 with vs. 48. Mark collected a number of logia at the end of ch. 9 before launching into a discussion of community problems in what is now ch. 10. Schnackenburg does not believe that we have to do with a catechism of the community and he warns against imposing an order or a unity of thought upon this series of logia.[165]

---

[161] Vss. 44, 46 are omitted in the best MSS and by the critical editions.
[162] *History*, 149-50. W. Grundmann (*Markus*, 194) seems to accept Bultmann's opinion.
[163] "Mk 9," 184-206.
[164] "Mk 9," 187. The suggestion is made by R. Bultmann (*History*, 142).
[165] R. Schnackenburg's opinion is followed by W. Pesch ("Die sogenannte Gemeindeordnung Mt 18," *BZ* 7 [1963] 229-30), and by H. Zimmermann ("'Mit Feuer gesalzen werden': Eine Studie zu Mk 9,49," *TQ* 139 [1959] 32-34).

We have already mentioned L. Vaganay[166] who thinks that he can detect an Aramaic document behind Mk 9:33-50. This Aramaic document would have been composed of a historical introduction, reproduced in Mk 9:33-34, which contained the first catchword ("great"). Thereafter followed seven short paragraphs, each one containing a thought on the theme of the previous catchword and offering a new catchword to be dwelt upon by the following paragraph. This presumed Aramaic document had, apart from minor verbal differences, additions and omissions which Matthew reproduced more faithfully than Mark, the same sequence of material, and the same content as Mk 9:33-50. From the paragraph no. 4 of the source, however, Mark, for unknown reasons, omitted two logia; these have been preserved in Mt 18:10,14 and had their place in the source between the Marcan vss. 41 and 42.

F. Neirynck[167] strongly doubts the plausibility of Vaganay's suggestion. He feels that recourse to the Aramaic source can hardly be justified since the entire composition can be explained on the basis of Mark's Greek text. He does not think that the sayings of Mt 18:10,14 were ever in the source. His own opinion is as follows.[168] Vss. 33 and 50b form an "inclusion," the first containing a reference to a dispute among the disciples and the second exhorting them to harmony. The strongly parallel vss. 43, 45, 47 possibly formed an original unit which was, prior to Mark, extended in both directions so that the source used by him already contained vss. 42-50a. Vss. 33-41, whose juxtaposition is not due to catchword bonds, have been added by Mark. Vss. 33, 34, 36, 50b are redactional compositions. J. Schniewind, like Neirynck, thinks that the beginning of the collection should be placed at 9:33.[169]

Others maintain that the first logion of the pre-Marcan collection was vs. 37. For J. Sundwall,[170] the original series consisted of vss. 37, 41-50a. V. Taylor has attributed the entire series of vss. 37-50 to "a pre-Markan compiler who sought to assist catechumens in committing the sayings to memory."[171] "A marked feature of 38-48 is the poetical form revealed when the passage is translated into Aramaic."[172] C. E. B. Cranfield[173] thinks that

---

[166] "Schématisme," 203-44.
[167] "Tradition," 67-68.
[168] "Tradition," 66-69.
[169] *Markus*, 121-22.
[170] *Zusammensetzung*, 60-63; in the main, he follows K. L. Schmidt (*Rahmen*, 223-36) and R. Bultmann.
[171] *Mark*, 409.
[172] *Mark*, 412; his authority is M. Black, *Approach*, 169-70.
[173] *Mark*, 307, 312.

the pre-Marcan collection began with vs. 38; E. Schweizer[174] places the opening verse of the pre-Marcan collection at 9:41; and A. Descamps,[175] who is reluctant to express a firm view, seems to place the opening verse at 9:42.

In this array of opinions those of R. Schnackenburg and F. Neirynck seem to be the most probable. Those who believe that they find a catechism or traces of it in these verses do not come up with much support for their view in vss. 33-41. Neirynck rightly remarks that there is no connection between vss. 41 and 42 and rejects Vaganay's improbable attempt to fill the gap between them (see our remarks above). Schnackenburg's doubts, moreover, about a pre-Marcan connection between vss. 37, 41, 42 are well justified—the logia are too diverse to allow us to draw firm conclusions about their pre-Marcan position. It is, of course, impossible to arrive at any certainty about vss. 40 and 41; yet the particle *gar*,[176] and the phrase *en onomati* in vs. 41 which seems to be reaching back to vs. 37, may indicate Mark's redactional work. It is difficult to decide whether Mark found vss. 49, 50a in the source or added them himself; the particle *gar* argues in favor of Marcan redaction. Some authors regard vs. 48 and the last phrase of vs. 43 as Mark's additions.[177] But the suggestion that vss. 33 and 50b are meant to be an "inclusion" may well be correct.

(B) THE MESSAGE OF MARK 9:33-50

R. Schnackenburg's warning against attempts to force this series of episodes and logia into the straight-jacket of a logical sequence must be heeded. A catchword arrangement, following upon a somewhat loose succession of narratives, is not designed to produce a neatly organized thought-pattern. Yet it seems that one theme is present in a number of the statements, viz., that of the radicality of the demands imposed upon a disciple of Jesus. Vss. 33-37 insist that the disciple must be "last of all and servant of all," subjecting himself even to those who are most clearly subject to others. The imagery of vss. 42-48 is frightening in its unmitigated vehemence: only the final condemnation can be more drastic than drowning with a millstone hung round one's neck;[178] and mutilation is preferable to damnation, the eternity of which is brought out by a telling OT text. The similarity between 9:33-37 and 10:35-45 has already been noted.

[174] *Mark*, 196-97.
[175] "Discours," 158.
[176] See R. Schnackenburg, "Mk 9," 196.
[177] E. Lohmeyer, *Markus*, 196; V. Taylor, *Mark*, 412.
[178] See G. Stählin, *TWNT* 7, 351.

Mark was undoubtedly not only aware of the similarity between this passage and that which follows the first prediction of the passion in 8:34-37[179] but also intended the two passages to correspond to each other to the degree that his tradition allowed. The demands of 8:34-37 are analogous to those of 9:43-48: one must sacrifice everything for the sake of the kingdom. Radical obedience and total self-renunciation are required.[180] D. E. Nineham comments on Mk 9:33-35:

> The section is, in fact, a commentary on the second prediction of the Passion, showing once again that the freely accepted suffering which awaits Jesus is not an accidental, isolated occurrence, but exemplifies a law of the kingdom which applies equally to all who would enter upon its life."[181]

These remarks are valid for vss. 42-48 as well.

Vss. 38-41 and 49-50 do not express the same thoughts as vss. 33-37 and 42-48. We must not force them into the same cast. Yet in vss. 38-41 Mark seems to warn against the arrogance and exclusiveness of Jesus' followers, which is another form of their desire to dominate instead of serving. It is difficult to decide what meaning is to be given to the "fire" of vs. 49, either in its pre-Marcan stage or in the present context.[182] Whatever the meaning of salt,[183] there is no doubt that vs. 50 is calling for an inner harmony among the disciples of Jesus; the note of self-renunciation may be present.

(C) Mark 9:47

This verse plays a part in the context of the absolute demands of the kingdom; nor does it carry a message distinct from this immediate context. The evangelist inserted it into his Gospel along with vss. 42-47 which

---

[179] See E. Schweizer, *Mark*, 200.

[180] See E. Schweizer, *Mark*, 197-200; O. Cullmann, "Das Gleichnis vom Salz: Zur frühesten Kommentierung eines Herrenwortes durch den Evangelisten," *Vorträge und Aufsätze 1925-1962* (Tübingen: Mohr, 1966) 198; E. Neuhäusler, *Anspruch*, 172; C. E. B. Cranfield, *Mark*, 314.

[181] *Mark*, 251.

[182] For various opinions, see V. Taylor, *Mark*, 413-14. H. Zimmermann ("Feuer," 39) thinks that Jesus was referring to the eschatological judgment which is already taking place. O. Cullmann ("Salz") believes that it describes suffering and persecution. E. Schweizer (*Mark*, 199) thinks in terms of persecution, or tribulations at the end, or the Holy Spirit.

[183] O. Cullmann ("Salz," 199): readiness for sacrifice. W. Nauck ("Salt as a Metaphor in Instructions for Discipleship," *ST* 6 [1953] 165-78) argues from rabbinic parallels that salt represents the wisdom to be given to men in eschatological times. J. B. Souček ("Salz der Erde und Licht der Welt: Zur Exegese von Matth. 5,13-16," *TZ* 19 [1963] 170) suggests that it is an image of the eschatological judgment.

probably had been compiled by an earlier hand. It is very likely, moreover, that vss. 43,45,47 formed a unity before vs. 42 came to be attached to them, since the identity of their content and the structure indicate this. A comparison with Mt 5:29-30 shows that vs. 45 is secondary, having been composed in imitation of the hand-logion.[184] Quite possibly vss. 43 and 47 formed a unity from the very beginning. The fact that Mt 5:29-30 reproduces them in inverse order is no argument against this possibility, since Matthew himself may have reversed them because of the preceding verse which speaks of looking lustfully at a woman. Should we consider them as authentic sayings of Jesus? A number of very close rabbinic parallels[185] would lead us to look on them as Jewish material taken over and slightly reworked by Christians. These parallels, along with the word *geenna*, show that Palestine, or at least a Jewish environment, was their original home. The structure of the second part of each saying (*kalon . . . ē*) could be an indication of their Semitic origin.[186] In itself, however, it does not afford ultimate certainty since this construction is not completely foreign to the Greek language.[187]

The three verses contain the demand for radical obedience to God and the thought that it is worth making the greatest sacrifices for the sake of the kingdom. "In contrast with 42, the theme is no longer that of causing others to stumble but of ensnaring oneself."[188] The verb *skandalizein* in these logia has the meaning of "to entice into sin." This is particularly evident in Mt 5:29-30, where they are linked with the antithesis to the sixth commandment, but it is clear enough in the Second Gospel as well.[189] The noun *geenna* is a Greek transliteration of *gy hnm*, the name given to a valley in the southern vicinity of Jerusalem. Because of child sacrifices which had been offered there during the period of monarchy (2 Kgs 23:10; Jer 7:31; 19:5-6) it was desecrated by Josiah and subsequently used for burning garbage. From the second pre-Christian century on it came to be employed in apocalyptic writings as a symbolic name for the place of future punishment (*1 Enoch* 90:26-27; 27:1-3; 54:1-5; 56:3-4; *2 Esdras* 7:36).[190] J. Jeremias[191] points out that the NT distinguishes between

---

[184] See R. Bultmann, *History*, 86.
[185] See *Str-B* 1, 302-3.
[186] See M. Black, *Approach*, 117.
[187] See K. Beyer, *Semitische Syntax im Neuen Testament: Band I: Satzlehre, Teil 1* (Göttingen: Vandenhoeck & Ruprecht, 1968) 80; Blass-Debrunner-Funk, *Grammar*, § 245 (3).
[188] V. Taylor, *Mark*, 411.
[189] See G. Stählin, *TWNT* 7, 351-52.
[190] For further intertestamental and rabbinic material, see *Str-B* 4/2, 1029-1118.
[191] *TDNT* 1, 148-49, 658.

hadēs and *geenna*. While the former is thought to be a place of provisional sojourn for the impious, the latter is considered as the place of their final punishment. The reference to Gehenna shows that the kingdom in vs. 47 is thought of as the future one. Many commentators remark that in the logia the terms kingdom and life are used interchangeably.[192] This is quite understandable, for the true life, which is here obviously thought of as future, will be given to men only when the kingdom arrives.

(4) Mark 12:34

Jesus approved the insight of this answer and told him, "You are not far from the kingdom of God." And no one had the courage to ask him any more questions.

(A) REDACTION AND TRADITION IN MARK 12:28-34

It is commonly agreed that in the pericope 12:28-34 only vss. 28a and 34b are to be attributed to Mark's redaction.[193] The multiplication of particles which we notice in vs. 28a "is a feature of Mark's style,"[194] and the purpose of vs. 34b "is to round off the series of pronouncement stories and mark a pause before the final incident in the section, in which Jesus takes the initiative."[195] It is thus most probable that the saying of Jesus which concerns us, "You are not far from the kingdom of God," was already part of the story in the pre-Marcan tradition.[196] The strongest argument in favor of the view which considers it as traditional is the depiction of a scribe in a favorable light. In the rest of the Gospel the scribes are sworn enemies of Jesus. Vs. 28a cannot be urged against this, for Mark obviously composed or reworked it with the story in mind.

Mark's version of the story is not a controversy, but a *Schulgespräch*.[197] The background of the Marcan version should, according to G. Bornkamm,

---

[192] T. W. Manson, *Teaching*, 276; J. K. Howard, "Parousia," 55; M.-J. Lagrange, *Marc*, 237.

[193] See R. Bultmann, *History*, 22, 51; C. E. B. Cranfield, *Mark*, 377; G. Bornkamm, "Das Doppelgebot der Liebe," *Geschichte und Glaube: Erster Teil: Gesammelte Aufsätze III* (München: Kaiser, 1968) 42-43.

[194] V. Taylor, *Mark*, 485; see also W. Grundmann, *Markus*, 250.

[195] D. E. Nineham, *Mark*, 328.

[196] E. Neuhäusler (*Anspruch*, 118) states that it has been frequently observed that this saying could be redactional. Unfortunately, he gives no references, and we have not been able to trace a single such observation.

[197] See R. Bultmann, *History*, 21-23; W. Grundmann, *Markus*, 250; G. Bornkamm, "Doppelgebot," 37.

be sought in a hellenistic, rather than a Palestinian, environment. He bases his opinion mainly on the scribe's reply in vss. 32-33. The reply is not structured according to the OT texts from which the commandments spring, Dt 6:5 and Lev 19:18; it rather affirms in the first member the unicity of God and in the second it combines the love of God and neighbor. This type of division reflects the apologetic and missionary efforts of diaspora Judaism and non-Palestinian Christianity, both of which were obliged to stress the fact that there is but one God.[198] The position of *dianoia* in vs. 30 as well as the terms *synesis* in vs. 33 and *nounechōs* in vs. 34 (a *hapax legomenon* in the NT) manifest an intellectualist tendency which would be at home in a hellenistic environment. The criticism of pagan sacrificial cult and the spiritualizing re-interpretation of the Jewish cult was likewise widespread among diaspora Jews.[199] Finally, the kingdom in vs. 34 is not conceived of eschatologically but has the character of a notional reality.[200]

The framework within which this pericope is found also has its home in hellenistic Judaism. It has been investigated by D. Daube[201] who suggests that in Mk 12:13-37 there is a series of episodes arranged according to the Midrash of the Four Sons, a tradition of Alexandrian Judaism. At Passover, four sons are presented asking questions of their father, a wise son, a wicked son, a perfect son, and the son who does not know how to ask (and for whom the father must formulate the question).[202] This passage would correspond to the question of the perfect son who is ready to accept good advice. Daube does not decide whether it is Mark or a pre-Marcan tradition which is responsible for this arrangement. To us it seems probable that it came into existence before Mark. Vss. 34b and 35a form a sharp

---

[198] See "Doppelgebot," 40, n. 14, for references to the literature of hellenistic Judaism.

[199] For references, see "Doppelgebot," 41, n. 17. Neuhäusler (*Anspruch*, 117, n. 13) objects to Bornkamm's suggestion by remarking that the profession of monotheism in vss. 29 and 32 is not inserted for its own sake but as an introduction to the commandment. His objection, however, misses the point of Bornkamm's argument which concerns the form, not the intention, of the OT quotation. Moreover, Neuhäusler fails to consider other indications of the hellenistic background which Bornkamm adduces.

[200] "Doppelgebot," 42.

[201] *Judaism*, 158-63. D. Daube has found fairly wide agreement. See J. Jeremias, *Jesus' Promise to the Nations* (SBT 24; London: SCM, 1958) 52-53; R. Schnackenburg, *Moral Teaching*, 91; W. Grundmann, *Markus*, 242, 250. R. Bultmann (*History*, 405) criticizes Daube, remarking on the last passage of the series, vss. 35-37 (in which, according to Daube [*Judaism*, 163] "the Alexandrians put questions of *haggada*, about apparent contradictions between passages from Scripture"), thus: "But there is no quotation of two passages of Scripture!" In reply to Bultmann, one could point to "the scribes claim" of vs. 35.

[202] *Judaism*, 163-66.

redactional incision in the flow of the compilation;[203] though the passage dealing with the greatest commandment is not a controversy in itself, Mark shows us by his redactional insertion in vs. 34b that he associates it with the controversies preceding it—undoubtedly because of the presence of the scribe. In vs. 35a, however, Jesus, having silenced his enemies, turns to the people and begins to exercize his characteristic activity of teaching. The incision shows that Mark was either no longer aware of or had no intention to preserve the unity of the traditional compilation. He does, however, preserve the traditional sequence of pericopes and, in vss. 28-34, also the traditional form of the story, despite the fact that it is at variance with his view of the scribes.

Our acceptance of Bornkamm's and Daube's interpretations does not exclude the possibility that the tradition is basically authentic.

(B) THE MEANING OF MARK 12:34a

(1) The main subject of scholarly discussion in connection with this saying is whether it refers to the present or to the future kingdom. For some exegetes "it is more probable that the *Basileia* is 'within reach,' "[204] thus already present. The presence of the kingdom is effected by the presence of Jesus in whose word and work the end has come in a hidden manner.[205] There is a hint of the messianic secret in the saying.[206] Others interpret it as referring to the future kingdom, for nothing is said about the time of its arrival. Its message concerns only the behavior necessary in those who wish to enter it.[207] It is to be looked upon as akin to the toroth of entry.[208]

There is little doubt that "taken at their face value they [i.e., the words

---

[203] See E. Schweizer, *Mark*, 254-5. F. Hahn (*Titles*, 126, n. 240) shows that teaching in the Temple is a Marcan redactional motif.

[204] V. Taylor, *Mark*, 489-90. The same opinion is held by G. E. Ladd (*Kingdom*, 193); J. Schniewind (*Markus*, 162); C. E. B. Cranfield (*Mark*, 380); E. Best (*Temptation*, 67).

[205] J. Jeremias (*Unknown Sayings of Jesus* [London: S.P.C.K., 1964] 64-73) seems to base his argument in favor of the authenticity of the *Gospel of Thomas* 82 (Jesus said: Whoever is near to me is near to the fire, and whoever is far from me is far from the kingdom) partly on this exegesis of this saying.

[206] J. Schniewind, *Markus*, 162; C. E. B. Cranfield, *Mark*, 380; E. Neuhäusler, *Anspruch*, 118, n. 18 (with reservations). We should note that admitting overtones of the messianic secret in the logion does not necessarily involve an admission that the kingdom is present. See E. Neuhäusler, ibid.

[207] W. G. Kümmel, *Promise*, 125-26; J. Schmid, *Mark*, 227; C. G. Montefiore, *Gospels* 1, 287.

[208] R. Schnackenburg, *God's Rule*, 142; H. Conzelmann, *An Outline of the Theology of the New Testament* (NT Library; London: SCM, 1969) 112.

of the logion] seem to represent the kingdom as something already present."[209] The adverb *makran*[210] seems to indicate that the kingdom is pictured as a domain, but it is evident that this spacial imagery is to be understood metaphorically.[211]

G. Bornkamm's observation, however, that the kingdom in this passage is above all a notional reality makes the discussion of its present or future character more or less academic. The question of time plays no role in the saying. The principal message of the story concerns the commandment of love; the words of Jesus in vs. 34 are a confirmation of the scribe's repetition of, and addition to, Jesus' own statement in vss. 29-31. Bornkamm quite correctly considers the saying as a litotes and interprets its meaning as: You are on the right path.[212]

The question about what the scribe still lacked in order to enter the kingdom may possibly be asked of the story in its Marcan context; but separated from this context, the story can hardly give an answer to it. It was not composed, or remembered, in order to describe the attitude of the questioner. The repetition of the commandment is typically Semitic; along with this repetition, the remark in vs. 34 that "he answered wisely" serves the purpose of stressing the importance of the commandment. It is also difficult to believe that the story, apart from its Marcan context, contained hints of the messianic secret, unless we are ready to find these hints in every authoritative statement of Jesus.[213] The center of the story is the commandment; everything else is secondary.

Hence this logion should not be invoked as an argument in favor of the thesis that the kingdom is already present. But neither should it be forced to voice a futuristic view, for it prescinds from these questions. Its purpose is to state the intimate connection between the love of God and one's neighbor and the entry into the kingdom. It should be looked upon as related to the toroth of entry, but only as related to them, since it is by no means a typical example of such toroth.

(2) What we have considered as unlikely with regard to the logion apart from the Marcan context may be true of it within this context. First of all,

---

[209] D. E. Nineham, *Mark*, 328. He refuses to decide between the present and future references.

[210] See H. Preisker, *TDNT* 4, 372.

[211] See M.-J. Lagrange, *Marc*, 303. For *makran*, see H. Preisker, *TDNT* 4, 373-74. It should be noted, however, that Preisker sees metaphors where there are none; his conclusion that *makrothen* and *makran* used metaphorically in the NT express the "numinosum and fascinosum of faith" is wide of the mark.

[212] "Doppelgebot," 42.

[213] See E. Haenchen, *Weg*, 414.

it is at least possible that the logion was understood by Mark as having a present reference. The eschatological power of Jesus has silenced his enemies one by one, and even a scribe must agree with him.[214] The enemies realize their powerlessness against Jesus' word, for "no one had the courage to ask him any more questions." Conscious of this power, Jesus proclaims the conditions of entry into the kingdom. "Of outstanding importance is the authority with which the statement is made. The speaker is the Lord and not only the Teacher."[215] "Here there speaks Someone who knows who is near the kingdom of God and who is far away from it."[216] The eschatological judgment is taking place; those who are near Jesus, those who follow him are already enjoying a share in the kingdom (cf. 10:28-30).[217] It may thus be permissible to ask what the scribe still lacks; if Mark thought of this question, his answer was that the scribe does not follow Jesus.[218]

(5) Concluding Remarks

Mark clearly gives us no tract on ethical demands of the kingdom. Yet three of the passages which bear on this subject appear in the section devoted to the instruction of the disciples. This section is dominated by the thought of the approaching death of Jesus, of the ultimate surrender and utter renunciation which he must undergo in his task of bringing about the present and future kingdom. Another thought which dominates the section is the demand that the disciple must be like his master. This is particularly evident in the redactionally introduced and redactionally closed address of Jesus after his first prediction of the passion in 8:34—9:1. It is also clearly present in the pericopes which immediately follow the second and third predictions. The disciples must be ready to follow Jesus in his total renunciation, to serve, and to die as he did.

The passages which we have considered have one thing in common: the demand for complete surrender. Mk 10:14-15 calls for a total subjection to the kingdom. Mk 10:23-25 requires that he who would follow Jesus must give up anything that may stand in the way of his turning to God and of his responding to God's eschatological action or that may serve as a means of "lording it over" his "brothers and sisters and mothers and children" (10:42,30). Mk 9:47 restates what has already been said in 8:35, that "whoever loses his life for my sake and the gospel's will preserve it."

---

[214] See W. Grundmann, *Markus*, 250.
[215] V. Taylor, *Mark*, 490.
[216] E. Lohmeyer, *Markus*, 260 (my translation).
[217] See E. Schweizer, *Mark*, 253.
[218] See D. E. Nineham, *Mark*, 328.

182   THE ETHICAL DEMANDS OF THE KINGDOM

Mk 12:34 is outside the section in which the other passages occur. Yet we find in it the same radical call for total devotion and whole-hearted service.

The coming of the kingdom which Mark announces at the beginning of the Gospel introduces a new, up-to-now unknown order of things, an order in which God effectively exercizes his lordship over all creation. When it comes to man's response to this new order, God is satisfied with no less than everything man has to offer.

Chapter IV
THE THEME OF THE KINGDOM IN THE LITURGY OF THE COMMUNITY

Chapter III has indicated the radical ethical consequences arising from the presence of the kingdom. Of those to whom this gift has been given a complete submission to God is demanded and a selfless service to the community. Chapter IV studies the influence which the theme of the kingdom exercises on the community's celebration of the Eucharist. We shall discuss the original context, form, and meaning of the eschatological logion in Mk 14:25, and the light it sheds on the Marcan form of the Eucharistic account.

Mark 14:25

I solemnly assure you, I will never again drink of the fruit of the vine until the day when I drink it new in the kingdom of God.

(A) The Text, Original Context, and Form of the Logion

(1) Critical editions are unanimous in their wording of the logion: Westcott-Hort, Nestle, Merk, the United Bible Societies edition, Tasker, von Soden, Vogels, and Tischendorf all give the same text. Instead of *piō* we find *prosthō(men) p(i)ein* in the Codices Bezae and Koridethi, in one minuscule, and some ancient versions. This reading is admittedly old and may reflect a form of the logion in the Aramaic stage of its transmission.[1] It would be unreasonable, however, to accept the testimony of only two major MSS which, in fact, do not fully agree with each other, against the preponderant evidence of all the other major MSS.

(2) Many contemporary interpreters are of the opinion that the logion as it is found in Mark is only a fragment of a larger unit of tradition. This unit is said to be preserved in Lk 22:15-18.[2] A thorough study of Lk

---

[1] See J. Jeremias, *The Eucharistic Words of Jesus* (NT Library; London: SCM, 1966) 182.

[2] See R. Bultmann, *History*, 265-66, 278-79; F.-J. Leenhardt, *Le sacrement de la sainte cène* (Serie théologique de l'actualité protestante; Neuchâtel-Paris: Delachaux & Niestlé, 1948) 41; L. Goppelt, *TDNT* 6, 153-54; F. W. Beare, *Records*, 225; E. J. Kilmartin, *The Eucharist in the Primitive Church* (Englewood Cliffs, N.J.: Prentice Hall, 1965) 38; N. Clark, *An Approach to the Theology of the Sacraments* (SBT 17; London: SCM, 1956) 40; E. Schweizer, *The Lord's Supper according to the New Testament* (FBBS 18; Philadelphia: Fortress, 1967) 20-21; J. Betz, *Die Eucharistie in der Zeit der griechischen Väter: Band 1/1, Die Aktualpräsenz der Person und des*

22:15-18 has been made by H. Schürmann in *Der Paschamahlbericht: Lk 22, (7-14.) 15-18*. He investigates the origin of the pericope with the painstaking precision and exactness that almost rival the method of a mathematician. It suffices to summarize briefly his results.

He begins with a few remarks on Lucan methods of composition. For the study of this pericope it is significant that Luke generally avoids creating parallelisms, and that he does not rearrange the order of Marcan material which he inserts into his Gospel.[3] This observation alone argues against the opinion according to which Lk 22:15-18 is merely a transposition and expansion of Mk 14:25. It is confirmed by a minute discussion of each word and phrase of the pericope. In vs. 15 pre-Lucan tradition has been retouched, to a degree, by the evangelist. The same can be said of vs. 16. It would be quite unfounded to regard vs. 17 as a Lucan redaction of Mk 14:22-23; for it is more likely, according to Schürmann, that Mk 14:22-23 has undergone the influence of the source used by Luke in composing 22:15-18. Vs. 18 is likewise independent of Mk 14:25. The conclusion to which this literary analysis leads: the source of Lk 22:15-18 was very likely an independent unit of tradition transmitted as such in the period before the canonical gospels were written.[4]

It is difficult to imagine that Mk 14:25 was transmitted without some introductory word referring to the cup and apart from the context of a meal. In Mark this introduction and context are supplied by vss. 22-24; however, a comparison with the Pauline and Lucan traditions of the words of institution shows clearly that the present Marcan combination of vss. 22-24 with vs. 25 is not original.[5]

J. Jeremias, in his work on the Last Supper,[6] arrives at a conclusion which is, for all practical purposes, identical with that of Schürmann. The reasons which lead to his conclusion are also basically the same.

Some interpreters, however, do not agree with this opinion.[7] They rather

Heilswerkes Jesu im Abendmahl nach der vorephesinischen griechischen Patristik (Freiburg: Herder, 1955) 18-19.

[3] J. Jeremias (*Eucharistic Words*, 98, 161-62) agrees with H. Schürmann on this point; see also N. Clark, *Approach*, 40; A. J. B. Higgins, *The Lord's Supper in the New Testament* (*SBT* 6; London: SCM, 1964) 32, 37, 41-42.

[4] *Paschamahl*, 52.

[5] *Paschamahl*, 43-44.

[6] *Eucharistic Words*, 161-62, 191-92.

[7] G. Bornkamm, "Lord's Supper and Church in Paul," *Early Christian Experience* (New York: Harper & Row, 1969) 157, n. 20; T. A. Burkill, *Revelation*, 276, n. 32; H. Vogels, "Mk 14,25 und Parallelen," *Vom Wort des Lebens: Festschrift für Max Meinertz* (*NTAbh*, 1. Ergänzungsband; Münster: Aschendorff, 1951) 94-95; J. M. Creed, *Luke*, 265-66; S. Dockx, "Le récit du repas pascal Marc 14,17-26." *Bib* 46

regard Lk 22:15-18 as a Lucan redactional elaboration of Mk 14:25. Many of them simply assert this, but a reasoned dissent is presented by P. Benoit, W. G. Kümmel and F. Hahn.[8] Benoit and Kümmel both dispute the assertion that Luke does not transpose the order of material which he finds in his sources. Benoit points out that Luke seems to have changed the order in his temptation narrative,[9] and Kümmel recalls the changes introduced by Luke into the Marcan material in 8:21 (Mk 3:35) and 6:17-19 (Mk 3:7-10). Yet these arguments prove only the possibility that Luke could have displaced or redactionally elaborated Mk 14:25, but not the probability. They might serve as support if the opinion were established on other grounds. Since the examples brought forward are an exception to the rule, they cannot be urged as real proof.

Kümmel further remarks that while Luke generally excises parallelisms found in his sources, he creates them occasionally. As an illustration, he refers to Lk 6:27-28 (par Mt 5:44), 10:8-12 (cf. Mt 10:13-15), and 10:16 (par Mt 10:40). But one may ask how certain it is that Luke is creating the parallelisms in the verses mentioned. Lk 6:27-28 contains no characteristically Lucan expressions, and it is difficult to determine whether Luke's or Matthew's version is closer to the source.[10] Lk 10:16 contains a logion, the many variants of which should warn us not to draw hasty conclusions about its source. Similarly, Lk 10:8-12 is much too complicated to permit any firm conclusions about its source.[11] Kümmel's claim that Lk 22:15-18 says no more than Mk 14:25 presupposes a view of the Lucan pericope which does not do justice to the probable, or at least possible, complexity of its formation[12] and the use made of it in the primitive church. His remark that the formation of a single saying out of a double one is completely inexplicable hardly takes sufficient cognizance of the winding road that many a unit of tradition had to travel in the course of its oral or written transmission.

Benoit further disputes the validity of Schürmann's assumption that a

---

(1965) 445. P. Lebeau (*Le vin nouveau du royaume: Etude exégétique et patristique sur la parole eschatologique de Jésus à la cène* [Museum Lessianum, section biblique 5; Paris: Desclée de Brouwer, 1966] 74-75) refuses to decide between the positions of P. Benoit and H. Schürmann.

[8] P. Benoit, "Le récit de la Cène dans Luc XXII 15-20," *Exégèse et théologie* 1 (Paris: Cerf, 1961) 163-203; "Les études de H. Schürmann sur Luc XXII," *Exégèse et théologie* 1, 204-9; W. G. Kümmel, *Promise*, 31; F. Hahn, "Die alttestamentlichen Motive in der urchristlichen Abendmahlsüberlieferung," *EvT* 27 (1967) 337-74.

[9] "Les études de H. Schürmann," 206.

[10] See W. Grundmann, *Lukas*, 148; J. C. Hawkins, *Horae*, 16-23, 27-29.

[11] See W. Grundmann, *Lukas*, 207.

[12] See H. Schürmann, *Paschamahl*, 46-52.

word occurring frequently in Luke, but found predominantly in his special source, cannot be ascribed, in some cases at least, to Luke's redactional activity. But the weight of proof is on the one who claims that the presence of such a word in a given Lucan text is due to his redaction. Simply to allege that it could be Lucan is not enough.

Kümmel and Benoit pay insufficient attention to two facts: Mk 14:25 is not inextricably bound up with the words of eucharistic institution,[13] and the logion could hardly have circulated all by itself without an introduction.

F. Hahn's argument moves on a different level from that of Kümmel and Benoit. He does not argue that Lk 22:15-18 is a redactional creation of the evangelist. Rather, he attempts to trace the development of the Lucan source from the logion preserved in Mk 14:25. He thinks that the source of Lk 22:15-18 has been created for the purpose of celebrating the Christian Passover by a Christian community stemming from hellenistic Judaism. He comes to this conclusion because of the impossibility of retranslating *epithymia epethymēsa* into Aramaic. Generally considered to be an authentic saying of Jesus, Mk 14:25 is, in fact, the oldest part of the Last Supper tradition, only partly dovetailing with the rest of the tradition. A confirmation of its primitive character is found in 1 Cor 11:26, where the phrase, "until he comes," still echoes it.

This opinion of Hahn seems to be merely an application of another, expressed in an earlier work,[14] according to which the most ancient form of the Palestinian Christian Eucharist focused exclusively on eschatological fulfilment. Only gradually did it begin to look backwards to the death of Jesus and its saving significance. P. Vielhauer[15] has taken issue with this reconstruction: Hahn's assertion that the reference to Jesus' death, and with it the motifs of atonement, covenant, and commemoration, were introduced only gradually is as unproven as it is unprovable. The oldest eucharistic formula which we possess, 1 Cor 11:23-25, would then consist of nothing but secondary elements. Vielhauer points out that the method of subtraction

---

[13] For the inconcinnity and lack of harmony between vss. 24 and 25 of Mk 14, see J. Jeremias, *Eucharistic Words*, 191; F. J. Leenhardt, *Cène*, 41; L. Goppelt, *TDNT* 6, 153; S. Dockx, "Récit," 447-48. F.-J. Leenhardt asks: Jésus aurait-il désigné le vin comme fruit de la vigne, après l'avoir assimilé à son sang? For a note of caution, see N. Turner, "The Style of St. Mark's Eucharistic Words," *JTS* ns 8 (1957) 108-9. Turner's arguments, however, fail to convince one of the probability of Mark's personal formulation of the account; parallel passages make it most likely that he was reproducing a liturgical formula. See also J. Jeremias (*Eucharistic Words*, 97, n. 5), where the phrases which occur nowhere else in the Second Gospel except in the account of the Last Supper are enumerated.

[14] *Titles*, 96-97; see "Motive," 340, 366-71.

[15] "Weg," 159-60.

employed by Hahn fails to arrive at the most ancient form of tradition, for the form reported by Justin seems to indicate that the way followed by the tradition was not from a simple form to a more complex one, but vice versa.

It should be added, moreover, that the inconcinnities and lack of harmony between vss. 24 and 25 in Mark's account of the Last Supper of which Hahn is himself aware, strongly suggest that the verses stem from different sources. They do not, as Hahn believes, belong to the same source (i.e., Mk 14:22-25) within which various elements had diverged in the course of time. Again, Hahn does not explain how vs. 25 was able to resist the gradual cultic schematization which other verses had undergone. Then too his contention that *epithymia epethymēsa* indicates a hellenistic environment does not take account of the probability that in this phrase we have to do with Luke's redactional work.[16] The fact that Paul refers to eschatological expectation in connection with the eucharistic tradition proves nothing about the logion in Mk 14:25. Finally, Hahn's reconstruction of the growth of the source of Lk 22:15-18 does not account for vs. 17 and the improbability of its stemming from the tradition of Mk 14:22-23.

(3) The more common opinion holds the Lucan *context* of the logion to be the more original. But it is even more commonly held that Mark gives us the more original *form* of it. The reason: the Marcan form shows stronger traces of its Aramaic substratum.[17] Jeremias has subjected the logion to a minute linguistic analysis[18] and discovers six Semitisms: (a) *amēn* which is rendered as *de* by Matthew and as *gar* by Luke;[19] (b) *ouketi ou mē*—this accumulation of negatives is barbarous Greek;[20] (c) *ek*, followed by the genitive of the thing drunk, is possible in Hebrew and Aramaic but not in Greek—on this point, however, Jeremias overstates his case;[21] (d) *to genēma tēs ampelou* was, at the time of Jesus, a firmly established formula in Jewish liturgical practices;[22] (e) *ekeinos*, used as an unemphatic demonstrative pronoun, is also a Semitism; (f) *en tē basileia tou theou* under-

---

[16] See H. Schürmann, *Paschamahl*, 5-7; W. Grundmann, *Lukas*, 393.

[17] See F.-J. Leenhardt, *Cène*, 41, n. 1; R. Bultmann, *History*, 266, n. 2; P. Benoit "Cène Luc," 188; P. Lebeau, *Vin nouveau*, 74-75.

[18] *Eucharistic Words*, 182-84.

[19] See also V. Taylor, *Mark*, 547; P. Lebeau, *Vin nouveau*, 70.

[20] See also P. Benoit, "Cène Luc," 188; P. Lebeau, *Vin nouveau*, 70; Blass-Debrunner-Funk, *Grammar*, § 431 (3).

[21] V. Taylor (*Mark*, 547) characterizes *ek* with the gen. as a possible Semitism only. Bauer-Arndt-Gingrich (*Lexicon*, 235) do not characterize it as Semitic; Blass-Debrunner-Funk (*Grammar*, § 164, 169) regard *ek* with the gen. serving as a substitute for partitive genitive as the normal tendency of the hellenistic Greek.

[22] See G. Dalman, *Jesus-Jeshua: Studies in the Gospels* (London: S.P.C.K., 1929) 150.

stood as a temporal, as opposed to a spatial, datum is likewise Semitic. Schürmann has strengthened the force of Jeremias' observations, to some degree at least, by suggesting that the following phrases in Lk 22:18 should be attributed to the hand of Luke:[23] *apo tou nyn, apo,* and *heōs hou*; these would very likely replace *ouketi, ek,* and *heōs tēs hēmeras ekeinēs* in his source.

However, some reservations must be made about certain points of Jeremias' argument. One could ask whether the phrase, "in the kingdom of God," could not have been imagined spatially from the very beginning; the properly temporal element in the latter part of the logion is the phrase "until the day." A word of caution is further expressed by Schürmann[24] about the originality of *amēn*: in Matthew we observe the tendency of the early tradition to add this word to various sayings of Jesus. In this logion it could have been added either by Mark or by the tradition that he was using. Disagreement with Jeremias has been further voiced by N. Turner on a number of counts.[25] He contends that the wording of the logion could largely be attributed to Mark himself. Strong negatives like *ouketi ou mē*[26] are not foreign to the Second Evangelist;[27] the unemphatic *ekeinos* is also found in Mk 4:11; 13:24; 14:21; *ek* in constructions similar to 14:25 occurs in 9:17; 12:44; 14:18,69,70. However, Turner's remarks are insufficient to lead us to concur with his opinion that the logion could, in its entirety, be Mark's own composition, particularly in view of the improbability that Lk 22:15-18 depends on Mk 14:25. Such words and phrases as *ouketi ou mē, genēma, ampelos, heōs . . . hotan* are not found anywhere else in Mark—this does not argue in favor of Marcan composition.[28] Yet these remarks warn us to keep open the possibility of Mark's redactional interventions in the traditional logion.

Hence, though the presence of Semitisms cannot, of itself, prove the greater antiquity of one form of a given logion, it is nonetheless likely, in view of the probable redactional interventions by Luke, that Mk 14:25 can claim to be more primitive than its Lucan counterpart.

The word which disturbs the predominantly Semitic characterization of the Marcan logion is *kainon*. G. Dalman[29] has pointed out the difficulty of

[23] *Paschamahl*, 35-38, 46.
[24] *Paschamahl*, 15.
[25] "Style," 108-11.
[26] Found only in Mk 14:25; Lk 22:16(?); and Apoc 18:14; see also H. Schürmann, *Paschamahl*, 18.
[27] He refers to 5:3; 7:12; 9:8; 12:34; 15:5; in Luke only in 4:2; 8:17; 10:19; 23:53.
[28] See J. Jeremias, *Eucharistic Words*, 97, n. 5; 184, n. 4.
[29] *Jesus*, 182.

reproducing the predicative *kainon* in Aramaic. Commentators agree with him that *pinō kainon* is a Greek construction.[30]

(B) The Message of the Logion

(1) There is no doubt that this Marcan logion voices a strong eschatological expectation. The term kingdom of God alone is sufficient to indicate that, for it connotes primarily an eschatological reality which is to arrive at the end of the aeon.

The image of the banquet in the kingdom likewise possesses a strong eschatological flavor. The roots of this image reach back to the OT sacrificial meals which the worshipper regarded as a means of union with God; the one who offered the sacrifice, his family, and his friends thought of themselves as guests at God's table:[31] "After gazing on God, they could still eat and drink" (Ex 24:11).[32] It is easy to understand how the future age which was to come after the definitive salvation of Israel came to be conceived as a banquet:

On this mountain the LORD of hosts
will provide for all peoples
A feast of rich food and choice wines,
juicy, rich food and pure choice wines (Isa 25:6);

Silence in the presence of the Lord GOD!
for near is the day of the LORD,
Yes, the LORD has prepared a slaughter feast,
he has consecrated his guests (Zeph 1:7).

The intertestamental literature also looks forward frequently to a banquet in the age to come:

And the righteous and elect shall be saved on that day
. . .
And with the Son of Man shall they eat
And lie down and rise up for ever and ever (*1 Enoch* 62:13-14).[33]

The elect will eat from the miraculous tree of life in the restored paradise;[34] or they will feed on the flesh of the slain pre-creation monsters, or on the manna which will be given again.[35] That Qumran was preparing for a

---

[30] J. Jeremias, *Eucharistic Words*, 184; H. Schürmann, *Paschamahl*, 21, n. 103; F.-J. Leenhardt, *Cène*, 41, n. 1; N. Clark, *Approach*, 40, n. 4.
[31] See E. J. Kilmartin, *Eucharist*, 8.
[32] See also Gen 31:54; Ex 18:12; Dt 12:7; 14:23,26; 15:20; 1 Sam 9:12-14; Hos 8:13; Am 2:8.
[33] See E. Lohmeyer, *Markus*, 304; P. Lebeau, *Vin nouveau*, 29-32; *Str-B* 4/2, 1154.
[34] *1 Enoch* 25:5; *Test. Levi* 18:11; *2 Esdras* 8:52.
[35] *2 Apoc. Bar.* 29:4,8; *Str-B* 4/2, 1156.

banquet in the presence of the Messiah of Israel is shown by 1QSa 2:11b-22.[36] And in the rabbinic literature the hope for the future aeon depicted as a solemn meal was very much alive.[37]

In the NT the image of a banquet is used to describe the kingdom more frequently than any other.[38] Being aware that the powers of this future kingdom are already active in him, Jesus eats with toll-collectors and sinners.[39] His meals with sinners and with his disciples are both present and eschatological, an image of the future and an anticipation of it, a parable and an event simultaneously.[40] At these meals he promises the kingdom to those who take part and gathers them already into the kingdom. They are acted parables, parables without a *tertium comparationis,* because in a hidden manner they already contain that which they promise.[41]

Besides the term "kingdom of God" and the image of banquet another word clearly indicates the eschatological character of the logion, viz., the adjective "new." Whatever its origin, it interprets correctly the message of the logion and strengthens it. Its roots are also found in the OT:

See, the earlier things have come to pass,
new ones I now foretell;
before they spring into being,
I announce them to you (Isa 42:9).

Remember not the events of the past,
the things of long ago consider not;
see, I am doing something new!
Now it springs forth, do you not perceive it? (Isa 43:18-19).

"Lo, I am about to create new heavens and a new earth;
the things of the past shall not be remembered
or come to mind" (Isa 65:17).

Newness is the quality of those realities which belong to the final fulfilment of God's saving work.

The intertestamental literature is likewise quite familiar with this notion.[42] And we encounter it often in the NT: Jesus' is a new teaching (Mk 1:27),

[36] See E. J. Kilmartin, *Eucharist,* 9; P. Lebeau, *Vin nouveau,* 56-62.
[37] *Str-B* 4/2, 1154-65.
[38] Mt 8:11/Lk 13:29; Mt 22:2-8/Lk 14:15-24; Mt 22:9-13; 25:10; Lk 22:29; Apoc 3:20; 19:9.
[39] Mk 2:15-17/Mt 9:11-13; Lk 7:33-34; 15:2.
[40] See E. Lohmeyer, *Kultus und Evangelium* (Göttingen: Vandenhoeck & Ruprecht, 1942) 91; F. Hahn, "Motive," 345-46.
[41] See E. Lohmeyer, *Markus,* 60.
[42] See *1 Enoch* 45:4; 72:1; *2 Apoc. Bar.* 44:12; Targum Micah 7:14 speaks of "the world that is about to be renewed"; see also *Str-B* 3, 796, 840-47.

which cannot coexist with the old (Mk 2:21-22); a new world is being announced (Mt 19:28) whose newness exerts its influence even now (Gal 6:15).[43]

The logion is thus a strong expression of eschatological hope. With confident certainty it announces the coming of the kingdom and the speaker's share in its joys.

(2) We have seen the inevitably universal agreement on the question of the eschatological character of this saying of Jesus. The same cannot be said of the question of its literary message. There are, fundamentally, two opinions: one considers the logion to be an act of renunciation of wine; according to the other, Jesus announces that he will no longer drink at a solemn meal with his disciples and thus, in a hidden manner, predicts that his death is near.

The first opinion is that of Joachim Jeremias; he is the most articulate exponent of the view that the logion is an act of renunciation of wine. The numerous exegetes who accept his thesis[44] can hardly be said to have contributed much to the strength of his argumentation. A brief summary of Jeremias' position and of his arguments follows.[45]

The use of the verb *epithymein* with an infinitive, as used in Luke's special source, leads Jeremias to the assertion that *epethymēsa . . . phagein* of Luke 22:15 expresses an unfulfilled wish. The reason for this unfulfilled wish is given in 22:16,18, the verses which reflect Jesus' solemn decision not to eat of the lamb nor drink wine until the coming of the kingdom. One could almost say, according to Jeremias, that the form of Lk 22:16,18/Mk 14:25 is that of an oath.[46] The feature of the logion which justifies for him this conclusion is (*ouketi*) *ou mē*, a strong negative, made even stronger by *amēn*,[47] one of the formulae of oath taking. Jeremias thus understands Jesus

---

[43] See also Apoc 21:1-5; 2 Pet 3:13; 2 Cor 5:17.

[44] See M. Thurian, *L'eucharistie: Mémorial du Seigneur, sacrifice d'action de grâce et d'intercession* (Collection Communauté de Taizé; Neuchâtel-Paris: Delachaux et Niestlé, 1963) 212-17. Note, however, that P. Lebeau (*Vin nouveau*, 76, n. 2) informs us that Thurian arrived at his conclusions independently of Jeremias. See also Lebeau, *Vin nouveau*, 75-85; W. G. Kümmel, *Promise*, 31; C. E. B. Cranfield, *Mark*, 428; E. J. Kilmartin, *Eucharist*, 39, 49; T. A. Burkill, *Revelation*, 235, n. 29; R. H. Fuller, *Mission*, 76-77; S. Dockx, "Récit," 450-51," F. Hahn ("Motive," 340, 354-55) agrees with Jeremias with regard to the Lucan context; in Mark and on the lips of Jesus, however, the logion did not express renunciation, according to Hahn. For the origin and a brief history of the opinion, see P. Lebeau, *Vin nouveau*, 75-76.

[45] *Eucharistic Words*, 207-18.

[46] In the second edition of his work Jeremias spoke of a vow of abstinence; in the third and fourth editions he has reduced this vow to a renunciation (Verzichterklärung); see *Eucharistic Words*, 207, n. 6.

[47] Implied in Luke's *gar*, preserved by Mark.

in Lk 22:17 to command the disciples that they alone should divide the contents of the cup among themselves since he has renounced all wine until the coming of the kingdom. Most likely Jesus not only abstained from eating the paschal lamb and drinking the wine, but even fasted through the entire meal. The reason for his act of renunciation is threefold: he wishes, first, to show the irrevocability of his decision to pave the way of the kingdom by his death; secondly, he wishes to manifest his total dedication to God; and thirdly, he wishes to give greater force to his prayer for a speedy arrival of the kingdom. Jeremias adds another reason: Jesus' fast is also a prayer for unbelieving Jews, deducing this from the Quartodeciman practice of fasting on the eve of the Passover for the sake of unconverted Jews. The only sufficient ground for this practice, he feels, is the influence of the first century Palestinian Christian communities on the Quartodecimans.

The substance of Jeremias' thesis has found many followers. There have been, however, disagreements either in details or as to its total content. For instance, the exegesis which interprets Lk 22:15 as an expression of unfulfilled wish is, in its present context, impermissible, since 22:7-14 clearly presupposes that Jesus has the intention of eating the passover.[48] Again, can *epithymein* with an infinitive of itself express an unfulfilled wish? It is more likely that it is the context which gives this meaning to the passages which Jeremias quotes to substantiate his contention.[49] In three of the four occurrences of this verb with an infinitive which Jeremias quotes (Lk 15:16; 17:22; Mt 13:17) it is stated that the wish remains unfulfilled. This would hardly be necessary if the verb itself were sufficient to bring it out. Further, the force of Jeremias' argument depends to a degree on the identity of the Last Supper with the passover meal; but this identification is by no means universally accepted. It is, likewise, at least questionable whether the *labete touto kai diamerisate* of 22:17 should be looked upon as an expression implying that Jesus himself does not drink; it may be, instead, a liturgical rule which was not part of the pericope from the very beginning.[50] Another objection to Jeremias' thesis is the Jewish custom which required that the host partake of "the cup of blessing" at the end of a solemn meal before passing it on to his guests.[51] It is,

---

[48] H. Schürmann, *Paschamahl*, 11; N. Clark, *Approach*, 40, n. 2.

[49] This observation I owe to my fellow-student Dr. G. Lohfink.

[50] H. Schürmann, *Paschamahl*, 65, 70-73. See, however, E. Schweizer's questioning (in his review of Schürmann's *Paschamahl* in *TLZ* 80 [1955] 157) of Schürmann's view that the original form, the double prophecy of death, was changed into a eucharistically colored narrative. In reply to Schweizer, one could point to the eucharistically colored narratives of the Feeding of the Five and Four Thousands.

[51] See G. Dalman, *Jesus*, 153; H. Schürmann, *Paschamahl*, 63, 66, n. 292.

moreover, by no means certain that *amēn* suggests an oath; in Jesus' logia it takes the place of the prophetic phrase, "thus says the Lord," and it expresses not an oath or a promise so much as a strong assertion of the certainty of what is being affirmed.[52] It is doubtful whether Jesus would have wished to break Jewish law by a refusal to eat the Passover.[53] Can we, further, be certain that *gar* refers to Jesus' renunciation of food and drink? One could well imagine that the particle expresses a different type of connection between the act of handing the cup to the disciples and the subsequent statement.[54] These and other reasons have led some authors to reject Jeremias' thesis altogether.

Other commentators, while accepting the opinion that the literary form of the logion is that of renunciation, do not follow Jeremias in his view that Jesus abstained from the meal in the course of which the logion was pronounced. Those who are not convinced of the identity of the Last Supper with the paschal meal ask whether the renunciation applies to the present meal or to the Passover on the following day.[55] Authors who are uncertain about the greater originality of the Marcan or the Lucan disposition of verses are likewise uncertain whether Jesus partook of the meal or not.[56] Against Jeremias' contention that it is unlikely that Jesus would eat and drink of the symbols of his own body and blood some have pointed out that Christian antiquity had no such scruples.[57] Others feel that Jesus' emphatic statement that he will drink no longer presupposes that he does drink now.[58] Others, again, shy away from the notion that Jesus was somehow attempting to force the coming of the kingdom by his renunciation.[59] Still others contest Jeremias' suggestion that one of the reasons for Jesus' fast was the conversion of unbelieving Jews. For the Quartodeciman practice need not be traced to traditions of Palestinian churches; it could have had its roots in the deep affinities which bound Christian and Jewish communities in Asia Minor. Finally, there is no firm evidence that

---

[52] H. Schürmann, *Paschamahl*, 66, n. 292; P. Benoit in his review of Jeremias' *Eucharistic Words* (*Exégèse et théologie*, 1, 242).

[53] H. Schürmann, *Paschamahl*, 66, n. 292.

[54] H. Schürmann, *Paschamahl*, 64, n. 287.

[55] E. Schweizer, *Lord's Supper*, 28. It should be noted that Schweizer does not accept the thesis of renunciation.

[56] See G. S. Sloyan, "'Primitive' and 'Pauline' Concepts of the Eucharist," *CBQ* 23 (1961) 8; P. Lebeau, *Vin nouveau*, 80-81.

[57] H. Vogels, "Mk 14, 25," 96-98; V. Taylor, *Mark*, 547; H. Schürmann, *Paschamahl*, 63, n. 286.

[58] E. Klostermann, *Markus*, 148; see also A. J. B. Higgins, *Supper*, 47.

[59] R. H. Fuller, *Mission*, 76-77.

Palestinian Christians of the first century fasted on the eve of the Passover.[60]

We should mention two other arguments adduced in favor of the view that the logion is an act of renunciation, arguments which do not stem from Jeremias. M. Thurian[61] has compared Num 6:3-5, where the Nazirite vow is described, with this logion. There is undoubted similarity between them: the fruit of the vine is mentioned, the duration of the binding force of the vow is stated along with a reference to fulfilment. The verb *plēroun*, however, is found only in Lk 22:16, and it is likely due to the redactional work of Luke.[62] Moreover, the subject of the verb is not the same in the two texts: in Num 6:5 the subject is the vow, in Lk 22:16 it is the Passover and the expectations linked with it. P. Lebeau, wishing to add greater depth to M. Thurian's view, links the Rechabite refusal to drink wine with the eschatological significance which wine acquired in the course of time.[63] There is little doubt about the messianic and eschatological symbolism of wine; but the Rechabite refusal to drink it probably owed nothing to this symbolism. It is more likely that their refusal was grounded in their desire to uphold the traditions of the desert. This is clearly suggested in Jer 35:6-7, where their refusal to drink wine, to engage in agriculture, and to live in houses is mentioned in the same breath.[64] Even if it could be shown that Rechabite abstinence was eschatologically motivated, it still remains to be proven that their influence persisted, in one way or another, down to Jesus' day.

Though there may be valid arguments against Jeremias' thesis, these are not of such nature as to rob it of all probability. Whether one agrees with it or not, one must admit that it is defensible, at least in its substance.

Should the objections raised against Jeremias' thesis prove too strong, there remains the interpretation whose most articulate proponent at present is H. Schürmann.[65] Jesus, according to him, is taking leave of his disciples;

---

[60] See P. Lebeau, *Vin nouveau*, 79-81.

[61] *Eucharistie*, 216.

[62] See H. Schürmann, *Paschamahl*, 20-22.

[63] *Vin nouveau*, 33-52, 81-83.

[64] See also A. van den Born and L. Hartman, *EDB*, 1992.

[65] *Paschamahl*, 65-68, 70-71; "Die Anfänge christlicher Osterfeier II," *TQ* 131 (1951) 420. Among those who hold the same opinion are R. Schnackenburg, *God's Rule*, 193; M.-J. Lagrange, *Marc*, 356; J. Dupont, "'Ceci est mon corps,' 'Ceci est mon sang,'" *NRT* 80 (1958) 1041; J. Betz, *Eucharistie Aktual*, 142; E. Schweizer, "Das Herrenmahl im Neuen Testament: Ein Forschungsbericht," *Neotestamentica*, 356; R. A. C. Leaney, *Luke*, 267-68. Hahn ("Motive," 340) attributes this meaning to the logion on the lips of Jesus and in Mark. G. Dalman (*Jesus*, 155) feels that the primary reference is to Jesus' death, but he does not exclude overtones of renunciation. A. Suhl (*Funktion*, 24, n. 72) rejects the idea of renunciation and feels that the logion serves merely to announce the coming of the kingdom.

he is telling them that the manner of life with which they have become familiar in his company is coming to an end; their common meals will no longer be taking place. He is thus implicitly predicting his imminent death. This indirect method is quite in keeping with his way of referring to his death in the synoptic gospels: in Mk 2:20; 9:12; 10:38; 14:3-8, etc. we find similar obliquely worded statements which imply, rather than clearly express, his consciousness of what is awaiting him. To P. Lebeau's objection that these predictions of death are voiced in a much less emphatic language[66] one could reply that everyone of them exhibits a different form. The emphatic language of Mk 14:25 can easily be understood within the context in which it is found: it is pronounced at the last solemn meal of Jesus with his disciples, at a moment when death is near.

It is somewhat difficult to choose between the two opinions. The thesis of Jeremias has in its favor the matter and, to a degree, the external form of the logion. Yet the objections enumerated above weaken considerably the force of his argumentation. His main argument, based on the presence of a strong negative combined with *amēn*, is not convincing. Schürmann's observation that the renunciation of the paschal meal on the part of Jesus hardly implies what Jeremias says it implies, viz., a free acceptance of death, is also a telling objection. Further, Schürmann's interpretation makes the logion fit more naturally into the context of the synoptic gospels with their frequent references to common meals, and particularly into that of the passion narrative. Jesus is taking leave of his disciples; the association with him to which they have grown accustomed is nearing its end. The manner of his departure will appear to be catastrophic. But they should not lose heart, for the meal which they are sharing now will find its counterpart in the kingdom of God.

(3) Some exegetes see in the logion only the prediction of the coming kingdom and Jesus' confidence that he will share in its joys.[67] Such a view, however, seems to restrict its message unduly, by concentrating on the directly stated message alone and refusing to hear the overtones of its historical and literary context. The logion finds its natural place in the context of a meal, or rather in a series of meals taken in common by Jesus and his disciples. It marks the end of this series. It is, therefore, significant that Jesus speaks of himself alone: he will no longer take part in such meals. This unique formulation justifies the conclusion that he expects the disciples to continue gathering at meals, that he expects, in other words, the community to continue even in his absence.[68] This conclusion is valid

---

[66] *Vin nouveau*, 77-78.
[67] A. Suhl, *Funktion*, 24, n. 72; E. Klostermann, *Markus*, 174. R. Bultmann (*History*, 266) is of the opinion that Jesus expects the kingdom to come in the immediate future.
[68] See M. Dibelius, *Tradition*, 208, n. 1; G. Bornkamm, "Lord's Supper," 158, n. 38;

whether we regard the literary form of the logion to be that of a quasi-nazirite renunciation[69] or an implicit prediction of death. A period of waiting is implied, a pause about which nothing further is said between this meal and the future banquet in the kingdom.[70] Short as this interval may have been thought to be originally, it is nonetheless implied. Should this conclusion be correct, a number of important inferences would obviously be drawn for the benefit of a NT ecclesiology.

(4) We have kept the question of authenticity of the logion until the end of this discussion because it is easier to deal with it once we have some understanding of its principal message and its overtones.

Many exegetes who express an opinion on the subject accept the logion as substantially authentic.[71] There is, first of all, strong evidence of an Aramaic substratum. This, in itself, is no conclusive proof of its authenticity, for the Aramaic church was, if anything, even more creative than its Greek counterpart; but it gives us at least the assurance of the relative antiquity of the logion. But there are other reasons which speak in favor of its authenticity. As for its vocabulary, the phrase, "the fruit of the vine," is a firmly established Jewish formula in the prayer of the host at every meal at which wine is served.[72] As for its entire tenor, it is free of features which could be construed as creations of the primitive community: Jesus

W. G. Kümmel, *Promise*, 119-21; L. Goppelt, *TDNT* 6, 142; H. Schürmann, *Paschamahl*, 65, n. 291; J. Betz, *Die Eucharistie in der Zeit der griechischen Väter: Band II/1, Die Realpräsenz des Leibes und Blutes Jesu im Abendmahl nach dem Neuen Testament* (Freiburg: Herder, 1964) 24; T. A. Burkill, *Revelation*, 275, n. 29; E. Lohmeyer, *Markus*, 304.

[69] See P. Lebeau, *Vin nouveau*, 107; W. G. Kümmel, *Promise*, 37, 77, 82.

[70] See M. Thurian, *Eucharistie*, 217; W. G. Kümmel, "Jesus und die Anfänge der Kirche," *Heilsgeschehen und Geschichte*, 300; G. Dalman, *Jesus*, 181; E. Lohmeyer, *Markus*, 304-5, 309; G. Bornkamm, *Jesus*, 160. E. Grässer (*Parousieverzögerung*, 54) feels that the main point of the saying lies in the promise and that the thought of the interim is not stressed. It may not be stressed, but it is implied; even Grässer allows the thought of the interim to appear in his own formulation of the message of the saying.

[71] J. Behm, *TDNT* 3, 731-32; L. Goppelt, *TDNT* 6, 153-54; A. J. B. Higgins, *Supper*, 41; W. G. Kümmel, "Anfänge," 300; H. Vogels, "Mk 14,25," 94; M.-J. Lagrange, *Marc*, 357; V. Taylor, *Mark*, 547; C. E. B. Cranfield, *Mark*, 427-28; G. Bornkamm, *Jesus*, 160; E. Klostermann, *Markus*, 147; J. Wellhausen, *Marcus*, 115; H. Schürmann, *Paschamahl*, 68, n. 295; F. Hahn, "Motive," 340. R. Bultmann (*History*, 266) and W. Marxsen (*The Lord's Supper as a Christological Problem* [FBBS 25; Philadelphia: Fortress, 1970] 22) seem to be in doubt. N. Turner ("Style," 108-11), as already mentioned, feels Mk 14:25 could be a Marcan composition.

[72] See L. Goppelt, *TDNT* 6, 153-54; J. Jeremias, *Eucharistic Words*, 183; *Str-B* 4/1, 62.

says nothing about his *parousia* or his unique resurrection.⁷³ Though it is difficult to deny that the logion voices the consciousness of a unique destiny,⁷⁴ it does not present Jesus as the lord of the banquet in the kingdom, but, as it were, as one of the guests. The logion, moreover, contains no messianic title; this would hardly be absent if the church were responsible for its original formulation.⁷⁵ Again, the expectation of the end which it expresses would strike no strange note in Jesus' time and environment, and the implicit prediction of death agrees with the manner in which Jesus refers to it in the synoptic tradition. This manner, presumably, had some basis in historical fact.

To argue that Jesus could not have foreseen his death is to substitute philosophical prejudice for exegetical method; there is no need, moreover, to posit a supernatural source for this knowledge. The view, further, that Jesus expected an immediate arrival of the kingdom has been disputed strongly enough in many recent works, so that it cannot be urged as a self-evident objection to the notion that the logion in its original form implied a continuation of the community. We feel, therefore, that the thesis of its substantial authenticity is amply warranted.

(C) The Logion in the Gospel of Mark

This eschatological logion was not an integral part of the tradition which transmitted the words of institution, even though all NT reports of the Last Supper, in one way or another, contain the thought of eschatological fulfilment. We have no reason to think, however, that it was Mark himself who combined the words of institution and the eschatological logion; for a typically Marcan introductory formula which would warrant such a surmise is lacking. It is most likely, therefore, that Mark found the sequence of verses which he preserves in 14:22-25 in the liturgical practice of the church which he served. Apart from vs. 23b,⁷⁶ it is difficult to point to a feature of the Last Supper narrative which could probably be designated as a Marcan redactional addition. We can, however, learn about Mark's concept of the Last Supper from the pericope; for unless we can show that his thought ran counter to that of the church from whose tradition he

---

⁷³ See J. Wellhausen, *Marcus*, 115.

⁷⁴ See H. Schürmann, *Der Abendmahlsbericht Lucas 22,7-38 als Gottesdienstordnung, Gemeindeordnung, Lebensordnung* (Schriften zur Pädagogik und Katechetik, 9; Paderborn: Schöningh, 1963) 23. Even the most radical critics do not deny to Jesus this awareness of his unique function in relation to the coming kingdom.

⁷⁵ See J. Wellhausen, *Marcus*, 115; C. G. Montefiore (*Gospels* 1, 335) questions some of Wellhausen's more extreme deductions from this verse.

⁷⁶ See H. Schürmann, *Paschamahl*, 133, n. 145.

culled the material for his Gospel, we should presume that he not only agreed with it but wished to strengthen it by incorporating it into his work.

To elucidate the function which the eschatological logion plays in Mark's concept of the Last Supper we shall first sum up the common features found in the NT traditions of the Last Supper; secondly, we shall try to see which feature is emphasized by the fact that the logion is placed at the end of the account; and thirdly, we shall seek to gain some understanding of Mark's concept of the interim period, the time which separates Jesus' departure from his disciples and their meeting again at the banquet in the kingdom.

(1) E. Schweizer[77] has summarized the common features under three headings: (a) the Last Supper and, consequently, the Christian Eucharist look forward to the coming fulfilment of the kingdom, in which the eschatological banquet serves as an expression of the full union between God and men; (b) the Eucharist is an expression of the present covenant bond between God and his people; (c) the Eucharist looks backward to Jesus' death which is the root of the community that celebrates it.

The Eucharist is, first of all, the fellowship of the interim. There is in it an inevitable tension, a tendency toward something which is not yet. It is unthinkable without a promised fulfilment in the future since it is a celebration of a victory which will become visible only when the final redemptive act takes place at the end. The community celebrating the Eucharist is still in danger, still persecuted, still living by faith and not by sight; its inner and outward existence is a paradox which will be unravelled only by the arrival of the kingdom. The community still waits for the return of its absent Lord.[78]

Yet the Eucharist is no mere waiting and commemoration of an absence. It is an anticipation and a guarantee of that toward which it tends. It is a prophecy of the future which, like every prophecy, is the first step in bringing about what it foretells. The community celebrating it is already a messianic community, not merely waiting to become such at the end, but experiencing in a hidden manner the joys of the final salvation. Its Lord's absence is not absolute; he is invisibly presiding at its common meals. The community is already sharing in his destiny by drinking of the cup which he hands to it at each Eucharist.[79]

---

[77] *Mark*, 302; also "Forschungsbericht," 359-62.

[78] See also J. M. R. Tillard, "L'Eucharistie, sacrement de l'espérance ecclésiale," *NRT* 83 (1961) 570; W. Nagel, "Der historische Jesus im Abendmahl der Kirche," *Der historische Jesus und der kerygmatische Christus* (eds. H. Ristow and K. Matthiae), 549; E. Lohmeyer, *Markus*, 305; F.-J. Leenhardt, *Cène*, 42; H. Conzelmann, *Outline*, 56-57.

[79] See also F.-J. Leenhardt, *Cène*, 29, 43-45; W. G. Kümmel, *Promise*, 119-21; E. J.

An essential part of the Lord's destiny, however, was his death. Through the Eucharist the community shares in the power of his death; it could not, in fact, exist but for his death and its participation in it. The death of Jesus "for the many" is its foundation stone.[80]

(2) Which one of these themes is stressed in Mark's account of the Last Supper? An attempt to answer this question leads to an examination of its last two verses.

It is commonly known that the last element in a Semitic narrative or other piece of literature is the one to which the greatest attention is given. It should, consequently, be looked upon as the criterion of the relative importance of other elements in the same pericope. This can be clearly seen in the pronouncement stories of the gospels, in which the logion, the ultimate reason for their preservation or creation, stands at the end. In miracle stories the choral response of the crowd at the end usually gives the interpretation of the miraculous event as God's saving intervention in the world. Thus the eschatological logion presumably expresses for Mark the most important aspect of the Eucharist.

It voices the firm hope that the Eucharist will eventually yield to something better and greater than itself. It states the belief that the fulfilment which is not yet gives the Eucharist its meaning: the common meals of the community are but a faint shadow of the banquet in the kingdom. They anticipate this banquet, but the contrast is much more evident than the analogy. The primary significance of the Eucharist consists in its tendency toward the kingdom with its fulfilment of the present observance. Mark is, of course, aware of the anticipatory character of the Eucharist; he indicates it by speaking of the "blood of the covenant." But it seems that so much of the old still clings to his Eucharist that he predicates "newness" as such only of the realities of the future kingdom.

A comparison of the Marcan version of the logion with that of Luke points in the same direction. Mark's reference to the kingdom is much fuller and more picturesquely and personally expressed than Luke's. "Until it is fulfilled in the kingdom of God" or "until the coming of the kingdom of God" is pale besides Mark's stately "until the day when I drink it new in the kingdom of God."

This emphasis on the end of time, and a broader and more imaginative

---

Kilmartin, *Eucharist*, 51-52; J. Betz, *Eucharistie Real*, 42; W. Nagel, "Historische," 547; V. Taylor, *Mark*, 547; H. Schürmann, *Paschamahl*, 67; G. Walther, *Jesus, das Passalamm des Neuen Bundes: Der Zentralgedanke des Herrenmahles* (Gütersloh: Bertelsmann, 1950) 31; E. Lohmeyer, *Kultus*, 59-60; R. H. Fuller, *Mission*, 71; A J. B. Higgins, *Supper*, 11.

[80] See also R. Schnackenburg, *God's Rule*, 193-94, 250-52; F.-J. Leenhardt, *Cène*, 143-45; E. Lohmeyer, *Kultus*, 48-49.

presentation of it, is in keeping with Mark's methods and tendencies. We need only think of his endings of the parables of the Sower and Mustard Seed in comparison with their Lucan and Matthean parallels (Mk 4:8/Lk 8:8a/Mt 13:8; Mk 4:32/Lk 13:19b/Mt 13:32b). Hence the position of the logion as well as its form show us the theme which Mark considers to be the most important in the Eucharist: that of its tending toward fulfilment, that of promising what is not yet.

If we turn to vs. 24, we see that Mark's tradition above all the others presents most clearly and single-mindedly the death of Jesus as the source of the community.[81] A comparison between his formula and those of Paul and Luke reveals interesting differences. Luke and Paul interpret the cup primarily as the new covenant, while Mark sees the cup principally as the blood which is to be shed. Luke and Paul look upon the cup as the *new* covenant, foretold in Jer 30:31-33. Since the covenant is a present reality, they are apparently concerned above all with the present effects of the Eucharist. The fact that Luke places the eschatological logion before the words of institution, and that Paul and Luke insist that the Eucharist be carried out "in memory" of Jesus, points in the same direction. Mark's formula, on the contrary, finds its OT frame of reference in Ex 24:3-8,[82] where the sprinkling of the blood of the sacrificial animals constitutes the covenant. It seems to direct our gaze more intently toward the sacrificial death of Jesus. If we, further, compare Mark's formula with that of Matthew, which depends on it but adds "for the forgiveness of sins," we see that Mark does not distract us with other thoughts apart from that of Jesus' death in which "the many" find their covenant with God.

The thought of death is enhanced by the close association which exists between vss. 24 and 25. As far as Mark is concerned, the two verses are one saying of Jesus. Almost in the same breath Jesus speaks of his death, of the interruption of the close association with him which the disciples had been enjoying up to then, and of that other banquet at the end of time. We can thus see in Mark's account of the Last Supper a strong contrast between the present death and the future glory of Jesus,[83] between the present suffering and the future joy, between the present cup which is the means of sharing in Jesus' death and the future cup which will be a

---

[81] See E. Schweizer, *Mark*, 303-5; F. Hahn, "Motive," 371. For blood as a reference to Jesus' violent death, see A. J. B. Higgins (*Supper*, 50, 52) and H. Schürmann (*Abendmahlsbericht*, 34-37) among others.

[82] See E. Lohmeyer, *Markus*, 307-8.

[83] See J. Dupont, "Ceci," 1040; E. J. Kilmartin, *Eucharist*, 51; F.-J. Leenhardt, *Cène*, 40-41.

sharing in his triumph[84]—a contrast, briefly, between the Eucharist and the banquet in the kingdom.

Mark is no stranger to such contrasts. His parable of the Mustard Seed depicts the smallness of the seed and the great size of the full-grown shrub more fully than its Matthean and Lucan parallels. Mark emphasizes the power of the eschatological kingdom working through Jesus, but lays even greater stress on Jesus' hiddenness. He juxtaposes an eschatological discourse which voices the awareness of the near kingdom more forcefully than its synoptic parallels, and a passion narrative in which the suffering of Jesus comes to the forefront more starkly than in any other gospel.

Hence the feature of the Eucharist of which Mark is most strongly aware is its interim character: only at the end will the present cup, the cup of Jesus' suffering and death, be transformed into a cup of triumph and joy.

(3) In one way or another we have already touched upon the Marcan view of the interim. In the discussion of the parables and of the demands that the kingdom places upon the followers of Jesus the condition of the community to which the parables were addressed and of which the demands were made was referred to. A few further remarks on the subject are appropriate.

The limits of the period within which the community lives are clearly delineated by the sharp contrast between the apocalyptic discourse and the Passion Narrative, and in that between Mk 14:24 and 25. What does Mark think of this period? One passage which might be examined in this connection is 2:20, a verse which refers to Jesus' death. It is difficult, however, to determine what weight Mark assigns to it. The main point of 2:18-22 is found in vss. 21-22, whether these were added by Mark, as some think,[85] or were already contained in the tradition which he is following; the reference to Jesus' death gives the impression of being an aside.[86] In order to discover Mark's thought about this interim we must consider passages which are either redactional or redactionally inserted or redactionally introduced in a manner which clearly indicates that he is drawing his readers' attention to their content.

The interim is characterized by the proclamation of the good news (13:10), an activity in which Mark himself is engaged (1:1). The seed

---

[84] We may safely assume that the phrase "with you" of Mt 26:29 is a correct explication of Mark's thought.

[85] R. Bultmann, *History*, 19; F. W. Beare, *Records*, 79.

[86] See R. Schnackenburg, *Markus*, 69.

of God's word is being sown (4:14)[87] in the world and is irresistibly progressing toward its goal. Those who hear the good news must be converted (1:15), must take up their cross and follow Jesus (8:34-38), must proclaim the good news in their turn, and be ready to suffer and be persecuted on account of their proclamation as Jesus suffered on account of his.[88] Their life, however, is not mere sorrow; for what they have lost by having accepted the good news they are being rewarded even now in the brotherhood of the community; they possess the firm hope of the joys in the future kingdom (10:28-30). But they must be watchful (13:37):[89] the obstacles are many (4:14-20), and the possibility of refusing to see the light (4:24-25) remains a permanent danger threatening not only those who are not yet in the community but members of the community itself. The temptation to be ashamed of Jesus and his words "in this faithless and corrupt age" (8:38) is ever present.

[87] The redactional vs. 13 shows that Mark makes his own the interpretation of the Sower.
[88] See R. Pesch, *Naherwartungen*, 137.
[89] Everyone accepts this verse as redactional; see R. Pesch, *Naherwartungen*, 202.

Chapter V
THE FUTURE KINGDOM

Mark's Eucharist expresses a strong awareness of the fact that the kingdom has not yet arrived in power. In this chapter we turn to Mark's thought on the future manifestation of the kingdom. The logion to be considered is that of Mk 9:1: its text, possible redactional additions by Mark, its authenticity, its message on the lips of Jesus, as well as its function within the Gospel. This function cannot be understood fully without a discussion of the eschatological discourse in Mark 13. A brief treatment of Mk 15:43 will be also in order.

(1) Mark 9:1

He also said to them: "I assure you, among those standing here there are some who will not taste death until they see the kingdom of God established in power."

(A) THE TEXT AND THE POSSIBLE REDACTIONAL ELEMENTS OF THE LOGION

(1) Critical editions are fairly unanimous in the reading of the text of Mk 9:1. Tischendorf, Westcott-Hort, Nestle, Merk, Tasker, and the United Bible Societies edition read *eisin tines hōde tōn hestēkotōn*. Exceptions to this consensus are Vogels and von Soden who read instead, *eisin tines tōn hōde hestēkotōn*. The reading preferred by the majority is found only in the Codex Vaticanus. W. G. Kümmel[1] has pointed out that a number of critical editions mistakenly attribute the same reading to the Codex Bezae; but *hōde* is missing in D altogether.[2] The reason for the preference given to the reading of B seems to be that it is the *lectio difficilior*.[3] It is questionable, however, whether the word order of the vast majority of MSS is to be disregarded in favor of the reading of a single MS on the sole basis of such a principle of textual criticism. On the other hand, the consensus of the majority of textual critics gets some support from witnesses such as P[45], which reads *tōn hestēkotōn hōde*, and D, which reads *tōn hestēkotōn met' emou*. These readings seem to indicate a certain "abnor-

---
[1] *Promise*, 28, n. 31; see also C. E. B. Cranfield, *Mark*, 288-89.
[2] See A. Huck and H. Lietzmann, *Synopse der drei ersten Evangelien* (Tübingen: Mohr, 1950) 99.
[3] See M.-J. Lagrange, *Marc*, 214; V. Taylor, *Mark*, 384; C. E. B. Cranfield, *Mark*, 288-89.

mality"[4] in the text-tradition copied by their scribes. It is possible that the text which they felt had to be improved was the one attested in the Codex Vaticanus. The evidence available does not permit a firm decision in this case.

(2) It is generally agreed that Mark is responsible for the place which the logion has in his Gospel.[5] This is indicated by the characteristic Marcan introductory formula *kai elegen autois*. The logion did not form part of the context within which it is now found before the Second Gospel came to be written. But what are possibly the redactional changes made in it by Mark? Before answering that, however, we must consider the opinion of N. Perrin with regard to its origin.

In an early work Perrin considered the logion to be an authentic saying of Jesus.[6] In his more recently published book, *Rediscovering the Teaching of Jesus*, he presents another opinion: the logion is a redactional construction of Mark.[7]

9,1 has some distinctively Markan characteristics: the concept of "seeing" the parousia, and the use of "power" and "glory" in this connexion. . . . 9,1 serves a distinct function as the climactic promise bringing to an end the pericope 8,27—9,1. As such it is a promise antithetical to the threat in 8,38.[8]

The logion "is doubly derivative. In form it is built upon 13,30 . . . and its second part is a deliberate echo of 8,38." Mk 13,30 "is not to be regarded as a Markan construction, because Mark never uses *mechri* for 'until' but *heōs*."[9]

A number of objections can be raised against this opinion. First, it does not seem that, apart from 8:21, logia inserted by means of the characteristic Marcan introductory formula are Marcan constructions. Thus no one has yet suggested that the logion in 2:27 is to be attributed to Mark; again 4:11-12 stems from pre-Marcan tradition; the fact that the logia of 4:21-25 occur in different contexts in the synoptic gospels shows them to have been independent and pre-Marcan; 6:10 contains a logion which is part of the missionary charge to the disciples, a tradition found also in Q, and thus clearly pre-Marcan. The unit of tradition introduced at 7:9 was part of the

---

[4] See F. Blass and A. Debrunner, *Grammar*, § 474, 5c.
[5] See R. Bultmann, *History*, 121; E. Klostermann, *Markus*, 79; F. W. Beare, *Records*, 140; V. Taylor, *Mark*, 218; G. Bornkamm, "Die Verzögerung der Parousie," *Geschichte und Glaube: Erster Teil: Gesammelte Aufsätze III* (München: Kaiser, 1968) 46.
[6] *Kingdom*, 138, 188.
[7] *Teaching*, 16-20, 199-201.
[8] Ibid., 199.
[9] Ibid., 200.

polemical equipment of the early church,[10] and thus not likely constructed by Mark. Perrin's remark that the concept of seeing the parousia is distinctively Marcan may be correct,[11] but the use of "power" and "glory" in this connection can hardly be considered as such. Late Judaism was expecting an eschatological manifestation of divine power,[12] and Mark is by no means the only one in the NT who is looking forward to Jesus' return "with great power and glory."[13] Since 8:38 as well as ch. 13 speak of the coming of the Son of Man, one would expect Mark, if he were composing on his own, to produce a Son of Man saying instead of a kingdom saying. Moreover, if we compare 13:30 with 9:1, it is evident that the wording (*tauta panta*) of 13:30 demands a context, while 9:1 can easily be imagined as a floating logion in no need of a context to deliver its message. If there is dependence between the two verses, we would rather assume, with R. Pesch and J. Lambrecht,[14] that 13:30 depends on 9:1. The fact that the Greek word for "until" in 13:30 is a *hapax legomenon* in the Second Gospel can hardly be decisive.

Thus it is not likely that 9:1 is a redactional composition of Mark. But we must examine the possibility of redactional additions or changes. Let us begin with the last phrase of the logion, *en dynamei*. Apart from Mk 9:1, we meet no text, either in the OT, intertestamental or Qumran literature, or in the NT, which describes the kingdom of God as coming "with power." W. G. Kümmel sees no reason to ascribe the phrase to the evangelist, for it "might simply denote the powerful manifestation to which at present nothing comparable corresponds."[15] But the very singularity of the phrase causes suspicion: if Jesus had spoken of the kingdom in this manner one would expect some other trace of it in the rest of the NT. A. Vögtle[16] attributes the phrase to the creativeness of the church; wishing to distinguish between the arrival of the kingdom in the work of the earthly Jesus and that to be brought about by him who is now constituted "Son of God in power according to the spirit of holiness, by his resurrection from the dead" (Rom 1:4), the church added the phrase to the logion. He points also to Mk 13:26 for a formulation similar to that of this logion. Yet if we are to ascribe the

---

[10] See R. Bultmann, *History*, 49.
[11] See R. Pesch, *Naherwartungen*, 168-70.
[12] See W. Grundmann, *TDNT* 2, 294-96.
[13] See W. Grundmann, *TDNT* 2, 305.
[14] *Naherwartungen*, 181-88; *Redaktion*, 202-11.
[15] *Promise*, 26.
[16] "Exegetische Erwägungen über das Wissen und Selbstbewusstsein Jesu," *Gott in Welt: Festgabe für K. Rahner* (Freiburg: Herder, 1964) 1, 646-47; see also C. H. Dodd, *Parables*, 44, n. 24.

formula of "the kingdom coming in power" or the "kingdom of God being established in power" to the creativity of the church, we should be able to find some other evidence of it in the rest of NT.

It has been suggested[17] that the phrase "with power" is due to Mark. That this is not entirely improbable is shown by the following considerations. First, the fact, seen more and more clearly in recent years,[18] of strong editorial activity of Mark in 8:27—9:1. The evangelist leads his reader from the confession of Jesus' messiahship to the thought of his death, then insists on the necessity of the disciple's following Jesus in his suffering, and finally shows that the goal of "coming after Jesus" is to share in the glory of the coming Son of Man and in the joys of the future kingdom. Although their interpretation of the connection differs, some interpreters see a link between 8:38—9:1 and 13:26.[19] The traditional expectation preserved in 13:26 looks forward to the coming of the Son of Man "with great power and glory." Since Mark is speaking of the same future reality in 8:38—9:1, it is not unreasonable to suppose that he would wish to reproduce, insofar as the logia would permit, the vocabulary of the traditional theme of 13:26. 8:38 already speaks of "glory"; Mark fills out the picture in 9:1 by adding a reference to "power." The difference in prepositions used in 13:26 and 9:1 is due to that used with the term "glory" in 8:38.

A further reason may have induced Mark to add the phrase, viz., a comparison between the present, hidden condition of the kingdom and the future fully manifested one.

The Son of man was humiliated, his secret was hidden, his end was to be the complete and incomprehensible poverty and powerlessness of the cross. Similarly, the Kingdom was obscure and unobservable; it was . . . a tiny seed sown in the earth, so small and insignificant that men did not notice it. But the time would come for the Son of man to appear on the clouds, with great power and glory, and all men would see him.[20]

We would conclude then that it is at least possible to consider *en dynamei* as a redactional addition of Mark.

[17] E. Klostermann, *Markus*, 85.
[18] See E. Haenchen, "Leidensnachfolge," *Weg*, 293-300: E. Schweizer, *Mark*, 165-66; G. Strecker, "Voraussagen," 35; W. Grundmann, *Markus*, 174. A more detailed discussion follows below.
[19] G. H. Boobyer, "St. Mark and the Transfiguration Story," *JTS* 41 (1940) 125-26; C. K. Barrett, *Holy Spirit*, 73; J. Lambrecht, *Redaktion*, 182-84; R. Pesch, *Naherwartungen*, 171-72.
[20] C. K. Barrett, *Holy Spirit*, 73. Whether Barrett is correct in not restricting these ideas to the Second Gospel need not be discussed here; his remarks certainly apply to Mark.

Many commentators are of the opinion that the phrase *eisin tines hōde tōn hestēkotōn* indicates the influence of the delay of the parousia.[21] Statements which we find in Mk 13:30 and Mt 10:23 are not so carefully qualified concerning those who will live to see the end. They do not seem to have been affected by the experience of the primitive church which saw its first generation die one by one. The logion here still clings to the hope that the promise which it contains will come true, but "this generation" or "those standing here" is, due to the Lord's procrastination, modified into "some standing here."

W. Marxsen[22] disagrees with this opinion; he feels that Mk 9:1 contains the same message as 13:30, for the phrase "this generation" was certainly not meant to imply that everyone living at the moment of the prediction would still be alive at the time of fulfilment. Marxsen is undoubtedly right with regard to the meaning of "this generation." Yet the words "some standing here" do betray a reflection on an element of the logion which would, at first, be hardly considered. This reflection was in all probability forced upon Christians by the gradual dying out of all those who had heard and seen Jesus in the flesh. The phrase seems intended to prescind from the age of those "standing here" in order to extend the validity of Jesus' promise to as late a date as possible.[23] It is generally assumed that the tradition of the community is responsible for its formulation. It may be, however, that this too is the work of the evangelist. If the reading of the Codex Vaticanus be accepted as original, the awkwardness of its word sequence would be an argument in favor of a Marcan redaction. Coupled with this is the fact that Mark seems to be more fond of the construction of the indefinite *tis* with a following partitive genitive than the other two Synoptists.[24] There may have been another influence at work also, viz., the words, "this faithless and corrupt age," of 8:38.[25] If Mark's source contained, as in 13:30, the words "this generation," Mark would have been compelled to change them for the simple reason that 9:1 is intended to be a promise of final salvation to those who have "taken up their cross and followed" Jesus. Since it is the evangelist who combined 8:38 and 9:1, and since the "generation" of 8:38 is quite different from that in the presumed

---

[21] See G. Bornkamm, "Verzögerung," 48; H. Conzelmann, *RGG*³ 2, 671; R. Bultmann, *History*, 121; W. Grundmann, *Markus*, 177; E. Percy, *Botschaft*, 177; A. Vögtle, "Erwägungen," 645-46.

[22] *Mark*, 205, n. 193.

[23] See A. Vögtle, "Erwägungen," 646.

[24] See above, pp. 33-34.

[25] See A. Vögtle, "Erwägungen," 646.

source of 9:1, we are justified in attributing the formulation of this phrase to his editorial activity.

These reasons in favor of Marcan redactional formulation can at best produce a degree of probability, which should be considered in our discussion of Mk 9:1. The phrase, "some standing here," may have replaced, as has been suggested by Vögtle, the words "this generation," but it could just as well have replaced *hōde hoi hestēkotes*.

Should *idōsin* be attributed to Mark? It may be correct to say with Perrin that the concept of seeing the parousia is distinctively Marcan. Yet the evidence is much too limited to give us any certainty on the subject; the only passages which may offer some evidence are 13:26 and 14:62. In both cases the reference to seeing could be due to the influence of Dan 7:13 or a pre-Marcan tradition. In favor of Marcan redaction may be the fact that the kingdom forms the object of the verb *horaō* in only two other NT passages, i.e., Lk 9:27, which is parallel to Mk 9:1, and Jn 3:3. However, Jn 3:3b derives in all probability from an oral tradition firmly set before the Fourth Gospel was written,[26] and can be looked upon as independent evidence in favor of treating the phrase "seeing the kingdom" in Mk 9:1 as a traditional datum.

Other words and phrases of the logion most likely stem from Mark's source. It is a commonplace of NT interpretation that there is no Jewish usage comparable to the manner in which *amēn* is employed in NT.[27] Since the phrase, "Truly, I say to you," is common to all strands of the synoptic tradition, we would be hard put to prove that it is redactional in Mk 9:1. The Jews did not speak of the kingdom as coming,[28] but in the NT *erchesthai* is predicated of the kingdom frequently enough (Mt 6:10 par; Lk 17:20; 22:18; Mk 9:1; 11:10). Only in Mk 9:1 is the perfect tense of the verb used. Vögtle thinks that this formulation should be attributed to the community since it would be singular on the lips of Jesus.[29] T. A. Burkill[30] remarks that the evangelist wished to emphasize the difference between the state of the kingdom coming and that of the kingdom come. This need not imply that Mark is responsible for the formulation of the perfect participle. In case, however, that *idōsin* is redactional, another form

---

[26] See R. Schnackenburg, *The Gospel according to St. John*, Vol. I (Herder's Theological Commentary on the New Testament; Freiburg/Montreal: Herder/Palm, 1968) 367.

[27] See J. Jeremias, "Characteristics," 112-15.

[28] See *Str-B* 1, 418 and 4/2, 968-76 (the only exception is Targum Micah 4:8); G. Dalman, *Words*, 107.

[29] "Erwägungen," 647.

[30] *Revelation*, 167.

of the verb must be presumed to have existed in the source. The phrase *geusōntai thanatou* can hardly be a creation of Mark. Apart from this verse, he never employs the verb. The phrase is also rare in the NT, occurring elsewhere only in Jn 8:52 and Heb 2:9. Its counterpart is, for that matter, rarely attested in intertestamental literature; the only examples which can with some certainty be dated as early as the first century A.D. occur in *2 Esdras* 6:26 and Targum Yerushalmi I Deut 32:1.[31] I could find no instance of it in Qumran literature. Hence the assertion of N. Perrin[32] that it is a "stock phrase from apocalyptic" is not supported by available evidence.

It is, then, at least possible that the phrase "in power" has been added by Mark, and that "some standing here" is a redactional qualification of a more sweeping statement in his source. It may be that the reference to "seeing" was inserted by Mark; however, we consider it unlikely. The rest of the logion is traditional.

(B) THE MESSAGE AND THE AUTHENTICITY OF THE LOGION

(1) There is little doubt about the message of the logion before its insertion into the present context. Had it not been for the unfulfilled prediction it contains, there would likely have been a more limited number of interpretations given to it in the past. Many of the Fathers explained it as referring to the transfiguration of Jesus,[33] "other interpreters have found the fulfilment of the prophecy in the Fall of Jerusalem, the gift of the Holy Spirit, or the spread of Christianity throughout the Roman Empire."[34] V. Taylor has suggested that the logion "voices the belief of Jesus at a time when He still looked for the speedy inbreaking of the Divine Rule of God," and that it speaks of "a visible manifestation of the Rule of God displayed in the life of an Elect Community ... but what this means cannot be described in detail because the hope was not fulfilled in the manner in which it presented itself to Him."[35] We feel, however, with R. H. Fuller, that "all these interpretations overlook the plain sense of the words."[36] The contextual connection with the Transfiguration has been made by

---

[31] See *Str-B* 1, 751-52.
[32] *Teaching*, 201.
[33] For a number of references, see F. J. Schierse, "Historische Kritik und theologische Exegese: Erläutert an Mk 9,1 par.," *Scholastik* 29 (1954) 528-30.
[34] V. Taylor, *Mark*, 386; for a similar enumeration with references to modern authors, see C. E. B. Cranfield, *Mark*, 285-89.
[35] *Mark*, 386.
[36] *Mission*, 27. The same judgment is voiced by A. Vögtle ("Erwägungen," 612); G. Bornkamm ("Verzögerung," 47-49); E. Grässer (*Parousieverzögerung*, 132, n. 4).

Mark; it cannot be invoked in the interpretation of the saying in its detached condition. As for other suggestions, T. W. Manson says:

For the Fall of Jerusalem as a fulfilment of the prophecy there is simply nothing to be said. . . . Against the identification of the coming of the Kingdom with the outpouring of the Spirit and the astonishing progress of Christianity in the first century is to be set the fact that the people who lived through these great events did not make the identification.[37]

Should the phrase "in power" be Mark's redactional addition, it would simply serve to indicate how he understood the saying. Moreover, Taylor's opinion rests on the supposition that the development of Jesus' thought in the course of his ministry can be reconstructed. According to C. H. Dodd, the saying should be translated either "until they see the Kingdom of God as something that has already come" or "until they see that the Kingdom of God has come."[38] This interpretation has often been **refuted**,[39] and Dodd himself has modified his view of the entire problem of realized eschatology.[40]

The logion of Mk 9:1 thus speaks of the definitive arrival of the kingdom at the end of time. Its future reference prevents us from understanding it in the sense of the Q saying preserved in Lk 11:20 and Mt 12:28, for this future coming was not thought of as gradual.

(2) Many commentators consider the logion in Mk 9:1 to be a product of the community,[41] and for various reasons. For G. Bornkamm and E. Grässer the decisive fact seems to be that the logion reflects the delay of the parousia; they feel that in Mk 9:1 we meet an early Christian prophetic pronouncement given in the name of the exalted Lord and intended to reassure a community troubled by the non-fulfilment of its fondest hopes. E. Schweizer, referring to Lk 17:20-21, points out that Jesus refused to indulge in apocalyptic calculations.[42] H. Conzelmann and E. Grässer have given comprehensive presentations of the entire problem of Jesus' eschatology.[43] Conzelmann's position is well summarized by J. M. Robinson:

---

[37] *Teaching*, 281; see also J. Schniewind, *Markus*, 115; R. Schnackenburg, *God's Rule*, 206.

[38] *Parables*, 43, n. 23.

[39] See J. Y. Campbell, "'The Kingdom of God Has Come,'" 93. C. H. Dodd's reply to Campbell, "'The Kingdom of God Has Come,'" 141-42. Further criticism of Dodd is offered by J. M. Creed, "'The Kingdom of God Has Come'"; T. A. Burkill, *Revelation*, 165-67; W. G. Kümmel, *Promise*, 25-27; K. W. Clark, "Realized," 373.

[40] See N. Perrin, *Kingdom*, 67.

[41] R. Bultmann, *History*, 121, 400; G. Bornkamm, "Verzögerung," 48; E. Grässer, *Parousieverzögerung*, 133-36; E. Schweizer, *Mark*, 178-79; E. Percy, *Botschaft*, 177.

[42] See also R. Schütz, *RGG*³ 1, 468.

[43] H. Conzelmann, "Gegenwart," 277-96; *RGG*³ 2, 666-68; E. Grässer, *Parousieverzögerung*, particularly 3-75.

Christology replaces chronology as the basic meaning of Jesus' message: the kingdom which Jesus proclaims is future, but the 'interim' is of no positive significance to him. Rather Jesus confronts man with an unmediated and consequently determinative encounter with the kingdom. This is the common significance of various themes which when taken literally could be contradictory: the nearness of the kingdom, the suddenness of its coming, and Jesus himself as the last sign. None of this is meant by Jesus temporally, but only existentially. Although the nearness is presented temporally, its 'meaning lies in qualifying the human situation in view of the coming of the kingdom'. Predictions of coming reward and punishment, like the present beatitudes and woes, represent the alternatives of salvation and lostness involved in one's present situation.[44]

Conzelman himself says: "The future is not conceived as such *formally* (as time yet to elapse), it is rather salvation or damnation."[45]

Grässer, on the other hand, is convinced that Jesus fully shared the contemporary apocalyptic expectations and believed that the kingdom was coming in the near future. This expectation dominated him from the beginning to the end of his ministry. The kingdom is for him a purely future eschatological reality; if it is said to be already present, then only in the sense that the present is an overture to the future. Down to the moment of his death, which came to him as a surprise, Jesus was convinced that the parousia of the Son of Man was about to take place. All the texts of the NT which indicate or imply that he envisioned a period of time elapsing between his death and the parousia are creations of the community which had to face the problem of the Lord's delay. Jesus thus did not count on a development after his death; his ethical demands should not be claimed to prove any such expectation on his part. In short, the delay of the parousia was not Jesus' problem; it is a problem which arose and received various answers in the community after his death.

These arguments and views are not entirely convincing. We have pointed out the possibility that Mark is responsible for the qualifying phrase, "some standing here." If this is to be attributed, not to Mark, but, as Bornkamm and Grässer suppose, to the source used by Mark, one may wonder whether a prophet speaking in the name of the exalted Lord and addressing a community disappointed in its hopes would produce such a qualification. The comparison with 1 Cor 15:51 and 1 Thes 4:15-17, which the two authors draw in support of their interpretation, is not valid. In 1 Thessalonians Paul is attempting to relieve the anxieties of the community about the fate of those who have already died (cf. 4:13); in 1 Corinthians he is discussing the qualities of the risen body. The logion of Mk 9:1, however, seems to be addressed to people who are beginning to question whether *they* will live

---

[44] *New Quest*, 18.
[45] "Die Zukunft ist ja nicht *formal* verstanden (als noch ausstehende Zeit), sie ist Heil oder Verlorenheit" (*RGG*³ 2, 667).

to see the arrival of the kingdom. The reassurance that only some of those present will live to see it is a poor reassurance indeed. According to Bornkamm and Grässer, this word of consolation comes from the exalted Lord who is addressing the community in the present, not out of an already distant past. But it seems to us that it is precisely the weakening of the promise, i.e., "*some* standing here," which indicates that the logion was looked upon as a saying of the earthly Jesus. The community which preserved it and weakened its broad sweep knew that many of those to whom the promise was spoken had already died, but it refused to abandon the hope awakened by it.

Regarding Conzelmann's and Schweizer's opinion, the following may be said: though it is rather commonly agreed today that Jesus refused to indulge in apocalyptic calculations, yet is this a sufficient reason to doubt the substantial authenticity of this logion? It does not attempt to determine "the day or the hour" of the arrival of the kingdom; it simply puts into words the temporal component—not of primary importance, but nonetheless real—of Jesus' call for watchfulness. The note of urgency in Jesus' call to his own people is clearly heard: the disciples must waste no time on their missionary travels, for it may soon be too late because the kingdom is near.[46] We can hardly attribute *all* the calls for watchfulness, and the expectation of a speedy arrival of the kingdom to the creativity of the community alone. Here T. W. Manson's remark is very apt: "There would be no point in telling men to be on the alert for something which might not happen until centuries after they had died."[47] It is true that the aspect of the kingdom which was uppermost in Jesus' mind was the definitive salvation which it was to bring and the decision for which its approach called, but its nearness was not for that reason conceived by him in a futureless manner. It is very much open to doubt whether Jesus was making the distinctions which later philosophical frameworks make possible, and which the problems of a later day are forcing upon us. The fact that the futurity of the kingdom is not the primary object of attention makes it no less real. Jesus was addressing the present in the light of a salvation which was imminent; it is difficult to imagine that this imminence, instead of being a concrete temporal reality, was to him a mere abstraction employed to disclose to men the urgency of the present moment.

Grässer's approach to the problem of Jesus' expectation of the end has

[46] See J. Jeremias, *Parables*, 53-55, 160-69; A. Vögtle, "Erwägungen," 615-20; R. Schnackenburg, *God's Rule*, 197-200; B. Rigaux, "La seconde venue de Jésus," *La venue du Messie: Messianisme et Eschatologie* (*RechBib* 6; Bruges: Desclée de Brouwer, 1962) 178-99.

[47] *Teaching*, 278.

deservedly met with some strong criticism.⁴⁸ There are objections to points of detail, but the main objections are raised against his method of procedure. The postulate with which he opens his discussion already contains the results at which he arrives in the end: Jesus shared the apocalyptic expectations of his day, consequently anything which can be understood as implying a delay of the parousia should be looked upon as a formulation of the community. One, and by no means the chief, aspect of NT eschatology is thus raised into a criterion by which the authenticity of the entire body of material is judged. Opinions which would attribute more to the historical Jesus than Grässer's simple schema would allow are all too frequently, and uncritically, cast aside as uncritical. Though he will often admit that a certain saying could be Jesus' own, he opts for its inauthenticity because as creation of the community it fits better within the system which he is constructing. The NT evidence, instead of being studied in all its variety and complexity, is measured by the stern rod of his postulate, and it will not permit him, among other things, to do justice to the present aspects of Jesus' proclamation of the kingdom.

Two other interpreters should be mentioned in this connection, T. F. Glasson and J. A. T. Robinson. They look upon Mk 9:1 as an authentic saying misunderstood by the community.

Glasson believes that Jesus' ethical teaching, which was meant to supersede the Old Law, his teaching on the church as the new Israel, and his calling of the Gentiles encourage "the view that He looked forward to a long period in which the Kingdom of God would spread through the world until at last the old prophecies would come to pass and the knowledge of the Lord would cover the earth."⁴⁹ The words of Jesus in Mk 9:1 were fulfilled at the Pentecost when "the Kingdom of God came with power and the conflagration began."⁵⁰ Remarking that Matthew (16:28, par Mk 9:1) inserted the thought of the parousia into a saying of Jesus which originally did not mention it,⁵¹ he maintains that Mk 9:1 itself was probably misunderstood by the community.⁵² Jesus himself did not look forward to a second advent.⁵³ But we may note with N. Perrin:

---

⁴⁸ See O. Cullmann, "Parousieverzögerung und Urchristentum," *Vorträge und Aufsätze 1925-1962*, 443-44; J. Gnilka, " 'Parousieverzögerung' und Naherwartung in den synoptischen Evangelien und in der Apostelgeschichte," *Catholica* 13 (1959) 277-90; N. Perrin, *Kingdom*, 146-47; H.-W. Bartsch, "Zum Problem der Parousieverzögerung bei den Synoptikern," *EvT* 19 (1959) 116-18.

⁴⁹ *The Second Advent* (London: Epworth, 1947) 148; see the entire ch. 15.

⁵⁰ *Advent*, 112.

⁵¹ Ibid., 72.

⁵² Ibid., 196.

⁵³ Ibid., 105.

Actually, far from proving that the teaching of Jesus is incompatible with the assumption that he expected a speedy end of the world, these three aspects of his teaching tend to prove the exact opposite! They do, in fact, show us how thoroughly eschatological his teaching was. The New Law is the eschatological Torah, which replaces the old Torah because the End-time has come; the New Israel is the eschatological community, the Israel of the End-time; and although the purpose of Jesus embraced the Gentiles, he expected that they would be brought into the Kingdom by an eschatological act of God.[54]

The fact that references to the coming of the Son of Man had been introduced into some contexts in the course of transmission is no indication that Jesus' words on the coming of the kingdom should be interpreted as speaking of the history of the church.

While J. A. T. Robinson admits that "Jesus' belief in the final consummation of God's purpose is never in question,"[55] he insists that we must distinguish between the future coming and "two closely related and often inseparable ideas, both of which are integrally involved in the conception of the *Parousia*. These are the themes, on the one hand, of *vindication*— of victory out of defeat—and, on the other, of *visitation*—of a coming among men in power and judgement."[56] Neither vindication nor visitation need refer to the end-time; the vindication of Jesus takes place in his resurrection and its results, and his visitation in his ministry and its consequences.[57] It is the church which is responsible for linking Jesus' expectation of vindication and his statements on visitation with the thought of the parousia at the end. On Mk 9:1 Robinson remarks: "There is no actual mention here of a *Parousia*. . . . But in the Gospel tradition [it is] closely associated with this expectation."[58] We may again quote Perrin's criticism:

He makes a sharp distinction between the expectation of Jesus concerning a final consummation and his expectation concerning vindication and visitation. But this distinction is a false one; because the expectation of Jesus concerning his vindication is part of his whole expectation concerning the final consummation, and the visitation that took place in his ministry does not exclude the expectation of a future visitation.[59]

In concluding this discussion, we find it practically impossible to doubt that the phrase "some standing here" reflects the church's experience of the delay of the parousia, and that its present formulation should be at-

---

[54] *Kingdom*, 137-38. In support of the last statement he refers to J. Jeremias, *Promise*, 55-73.
[55] *Jesus and His Coming* (London: SCM, 1957) 38.
[56] *Coming*, 39 (Robinson's italics).
[57] Ibid., 40-82.
[58] Ibid., 84.
[59] *Kingdom*, 141-42.

tributed either to pre-Marcan tradition or to Mark himself. This phrase alone, however, should not be taken as evidence that the entire logion is a creation of the community. The arguments against its basic authenticity are not convincing. And much less convincing is the view which considers the logion to be authentic, but contends that the early church misinterpreted it by, implicitly or explicitly, applying it to the coming of the kingdom at the end of time.

(3) We would thus agree with those exegetes who consider this logion to be substantially an authentic utterance of Jesus.[60] This opinion, however, entails a number of consequences, one of them being that Jesus erred; and it raises the question whether his eschatology has retained any validity after his own expectations and those which he inspired have turned out to be false. To discuss all or even some of the solutions which have been proposed is quite impossible here. A useful presentation of opinions held by a number of Protestant scholars has been made by W. G. Kümmel,[61] and some Catholic solutions have been presented by A. Vögtle.[62] We limit ourselves to a few of the solutions proposed in recent years.

One of the most oustanding contributions to the discussion of this problem has undoubtedly been that of Kümmel himself. According to him, there are two themes clearly discernible in Jesus' proclamation: "the threatening approach of the Kingdom of God within his generation,"[63] and the fact that "the *eschaton* showed itself effective in his own person."[64] Jesus proclaims that the arrival of the kingdom is imminent; yet the significance of this proclamation "cannot lie in the *fact* that the end of the world is near."[65] It serves rather to emphasize, on the one hand, that God's redemptive action which has begun with Jesus is unfailingly directed toward its future consummation and, on the other hand, to confront men with this eschatological act of God, demanding of them a decision.[66] "Thus the fundamental presupposition for the future eschatological judgement was

---

[60] T. W. Manson, *Teaching*, 278-83; J. Schniewind, *Markus*, 115; R. H. Fuller, *Mission*, 27-28; W. G. Kümmel, *Promise*, 25-28; E. Trocmé, "Marc 9,1: prédiction ou réprimande?" *SE* 2 (*TU* 87, 1964) 263; A. Nisin, *Histoire de Jésus* (Paris: Seuil, 1961) 255; R. Schnackenburg, *God's Rule*, 205; A. Vögtle, "Erwägungen," 639-46; B. Rigaux, "Venue," 184; R. E. Brown, *Jesus God and Man: Modern Biblical Reflections* (London: G. Chapman, 1968) 74-75.

[61] *Promise*, 141-53.
[62] "Erwägungen," 611-15, 635-37.
[63] *Promise*, 87.
[64] Ibid., 105.
[65] Ibid., 104 (Kümmel's italics).
[66] Ibid., 152-55; see also "Die Eschatologie der Evangelien," *Heilsgeschehen und Geschichte*, 51, 57.

created already in the present, in which Jesus was the determining factor."[67] Essential to Jesus' message is thus the proclamation that the kingdom is already active in him, and that its definitive arrival will take place in the future, while "the *imminent* expectation, being a necessarily contemporary form of expression, can ... be detached from Jesus' message."[68]

Some recent Catholic authors, while agreeing with many of Kümmel's conclusions, are disturbed by his suggestion that Jesus erred. But to their objection one may respond with R. E. Brown,[69] "Is it totally inconceivable that, since Jesus did not know when the parousia would occur, he tended to think and say that it would occur soon? Would not the inability to correct contemporary views on this question be the logical effect of ignorance?" There is a more pertinent objection to Kümmel's solution of the problem. Is it legitimate to make the distinction which he makes between the permanently valid and time-conditioned elements in Jesus' eschatological proclamation? Is this distinction part of the message itself, or is it forced upon it externally by the simple fact of non-fulfilment of Jesus' predictions? Can we not find some other explanation which will interpret the sum total of Jesus' statements on the end and its imminence in the light, not of our problems so much, but of the nature of Jesus' message as such?

One recent attempt to give an explanation of Jesus' expectation of the end and, at the same time, to avoid the conclusion that Jesus erred is that of B. Rigaux.[70] We must distinguish, according to him, Jesus' teaching from his expectations. These expectations are couched in apocalyptic language. But Jesus' teaching is contained primarily in Mk 13:32, where he affirms that knowledge of when the end is to arrive is reserved to the Father alone. Such logia as Mk 9:1 and 13:30 must be judged in the light of this teaching. Although these logia seem to be absolute affirmations, they must not be taken as teaching—their apparent absoluteness is due to the apocalyptic literary form in which they are cast—but are to be understood as voicing a desire, a hope, an exhortation to vigilance. Therefore we cannot speak of truth and falsehood in their regard.

A. Vögtle[71] has reservations about this attempt at solving the problem.

---

[67] *Promise,* 105.

[68] Ibid., 152 (Kümmel's italics). A position very similar to Kümmel's is that of T. W. Manson: "the belief in the nearness of the Day of the Lord is not one of the unique features in the eschatology of Jesus but a belief which, like the belief in demons or the Davidic authorship of the Psalter, was the common property of his generation" (*Teaching,* 283).

[69] *Jesus,* 78.

[70] "Venue," 190-99.

[71] "Erwägungen," 636-38.

He remarks, first, that Mk 9:1 and 13:30 cannot be considered as more apocalyptic than 13:32, for ignorance of the moment when the end is to come is a typical trait of apocalyptic eschatology. Secondly, Mark certainly seems to consider the logia of 9:1 and 13:30 as teaching—the introductory formula, "I assure you," shows this clearly; he gives no indication that he looks upon 13:30 as less true than 13:32. Thirdly, if we take the three sayings as they stand, it is more likely that 13:32 should be judged in the light of 9:1 and 13:30 than vice versa, i.e., the end is to come during the lifetime of Jesus' generation, but the precise moment within this period is unknown. One way of removing the difficulty might consist in the assumption that the statements were made at different points in the growth of Jesus' self-understanding,[72] but the state of evidence offers no possibility of discovering the course of such a growth.

These objections to Rigaux's position are valid. "Yet Vögtle himself manages to explain away by exegesis all reference to the parousia in the promises of what will happen in the lifetime of Jesus' hearers."[73] To Vögtle's solution we now turn our attention. He begins by pointing out that recent Catholic exegesis is no longer afraid of admitting a true lack of knowledge on the part of Jesus concerning the time when the end is to come,[74] but that it is still unwilling to admit that he was in error. When he comes to the treatment of Mk 9:1 and 13:30,[75] he remarks that these logia, should they be understood to refer to the end on the lips of Jesus, indicate that he did not expect its arrival in the immediate future. The imminence of the kingdom is not the central motif of his warnings and exhortations to watchfulness. The central motif of his eschatology should thus not be sought in these logia but in the sayings preserved in Mk 13:32 and Lk 17:20, where Jesus asserts his lack of knowledge about the "day or the hour" and points out the uselessness of calculations in this regard. In the light of this teaching of Jesus it becomes doubtful whether Mk 13:30 and 9:1 originally referred to the end. Mk 13:30 could have been an answer to the question about the time of the destruction of Jerusalem, and 9:1 can be looked upon as a community reformulation of 13:30. The structure of the two sayings is very similar, and reformulation by the community becomes likely if we keep the following in mind. It is probable that the primitive community combined the destruction of Jerusalem with the coming of the kingdom; the destruction thus became an aspect of the final

---

[72] A suggestion half-heartedly made by A. Nisin (*Histoire*, 255, n. 3) among others.
[73] R. E. Brown, *Jesus*, 78.
[74] "Erwägungen," 610; for references, see n. 5 on the same page.
[75] Ibid., 639-51.

eschatological tribulation. However, the community longed for, not the tribulation, but its end, i.e., the coming of the kingdom and with it definitive salvation. Jesus' threatening prediction of the destruction of Jerusalem within the lifetime of his generation thus became a promise of definitive salvation in the community.

Serious objections can be raised against such a theory. Mk 13:30 might well be a redactional formulation of the evangelist formed on the pattern of 9:1.[76] Even if we accept it as traditional, we can hardly say to what the phrase, "all these things," originally referred. If we recognize the secondary character of the composition that we meet in Mark 13—as Vögtle does also—we cannot say what the phrase has replaced. It would seem more correct to look on 9:1 as the more original logion since it does not need a context to deliver its message, whereas it is clear that 13:30, as now formulated, demands a context. One wonders, moreover, whether it is correct to pit Mk 13:32 and Lk 17:20 against Mk 9:1 and 13:30. These logia find their unity in Jesus' call for watchfulness: the kingdom is near, it is coming within the lifetime of this generation, but no one knows the moment of its arrival; men must therefore be continually on the alert. The phrase, "some standing here" (9:1), on the lips of Jesus would tend to decrease the need of watchfulness, but precisely this phrase betrays the concern about the delay of the parousia. It should be attributed either to the evangelist or to the community. The temporal proximity of the end is not the chief motive for watchfulness, yet it would be difficult not to overhear the overtones of it in Jesus' preaching and acting. Vögtle sees that.[77] Must we engage all exegetical ingenuity to purge certain logia of a thought which so many other sayings and acts of Jesus manifestly imply?

Like many scholars who delve into the problem of Jesus' expectation of the end today, R. Schnackenburg is convinced that Jesus did not engage in such apocalyptic calculations and speculations as we find in Jewish apocalyptic literature. His proclamation undoubtedly possessed eschatological traits. It would be poor apologetics to deny, or to play down, the fact that he was looking forward to the arrival of the kingdom in the near future. His exhortations to watchfulness make sense only if the possibility exists that the end is soon coming. Jesus' proclamation is not mere religious preaching presenting to men God, his acts, his mercy, and his demands, into which the primitive church inserted apocalyptic features. Rather, he proclaims the kingdom that is coming, and he is firmly convinced that the powers of this kingdom are already active in his words and works. This is

---

[76] See R. Pesch, *Naherwartungen*, 181-87; J. Lambrecht, *Redaktion*, 202-11.
[77] "Erwägungen," 615-20.

shown by his parables whose eschatological message cannot be questioned, and confirmed by what we know of the longings of his contemporaries. Though it would be false to limit his proclamation to the theme of the nearness of the end or to consider it as its only motive force, this nearness is nonetheless an integral part of it. Parousia is seen by Jesus above all within the framework of humiliation and exaltation; rejection by men will be overturned by God's acceptance.[78]

On Mk 9:1 specifically, Schnackenburg thinks it unlikely that it referred to the destruction of Jerusalem, the resurrection or Pentecost, for there is no evidence in the gospels that Jesus expected a gradual coming of the kingdom or that he associated this coming with any of these events. It is equally unlikely that the logion on the lips of Jesus spoke of his Transfiguration. To discover its meaning, we would have to know the situation in which it was spoken, or at least its context in the tradition; its pre-Marcan isolatedness, however, does not permit us to discover either of these with any certainty. This scepticism about our ability to discover the meaning of the logion, voiced in his earlier work, *God's Rule and Kingdom*, is no longer present in a later article, "Kirche und Parousie," where he understands it as an indication that Jesus expected the end in the near future.[79]

If we admit that Jesus expected the kingdom's definitive arrival in the near future, how do we escape the conclusion that he was in error? Schnackenburg does it by characterizing Jesus' proclamation as prophetic.[80] Jesus' principal intention, like that of OT prophets, is to address men in the present and to confront them with the need of decision at the moment of proclamation. This type of preaching has the effect of drawing the future into temporal proximity, creating a "prophetic perspective" in which certitude of the future act of God entails a certain indifference to the lapse of time dividing the future from the present.

But the term "prophetic perspective" needs elaboration. Matthew (1:23) may see the fulfilment of the Emmanuel prophecy in Jesus, but this is no reason to think that Isaiah had such a distant future in mind. The context of the Isaian prophecy demands an event which is about to happen.[81] The

---

[78] See *God's Rule*, 195-99, 209-10; "Kirche und Parousie," *Gott in Welt: Festgabe für K. Rahner*, 568-71.

[79] See *God's Rule*, 206-7; "Parousie," 568-69.

[80] See *God's Rule*, 199-201, 210-14; "Parousie," 569; also *LTK*² 7, 778. The same opinion is voiced by J. Gnilka, "Naherwartung," 288-90. R. H. Fuller (*Mission*, 28) also seems to accept the thesis of prophetic perspective.

[81] See O. Kaiser, *Der Prophet Jesaja, Kap. 1-12* (*ATD* 17; Göttingen: Vandenhoeck & Ruprecht, 1963) 80.

same is true of the messianic prophecies in Isa 9:2-6 and 11:1-8. Impossible as these predictions may appear to be, the immediately preceding verses, 8:23b—9:1, make it evident that in 9:2-6 the prophet is consoling his people at a time of a concrete catastrophe in its history, viz., the annexation of territories of the northern kingdom to a province of the Assyrian empire. He foretells to suffering Israelites the destruction of their oppressors, promising them a king whom he describes by a series of current royal titles. In 11:1-8 the prophet looks forward to a new David who will replace the totally corrupt Davidic dynasty. This same expectation we find in his contemporary Micah (5:1).[82] Second Isaiah likewise foretells what God is about to do; the "new Exodus" with all its wonders is no distant event in a remote future, but is already beginning in the exploits of Yahweh's Anointed, Cyrus.[83] How imminent the messianic expectations of the prophets could be is shown by such passages as Hag 2:20-23 and Zech 6:13, which designate Zerubbabel as the future king who will reign over Israel after God has "shaken the heavens and the earth and overthrown the thrones of kingdoms" (Hag 2:21-22).[84]

These examples indicate that the prophetic perspective must not be imagined to imply that the prophets were somehow aware that their words would be fulfilled only in the distant future. Witness to it should rather be sought in the fact that their words and predictions were preserved, cherished, and eventually reinterpreted despite their non-fulfilment. Another element which undoubtedly contributed to the rise of this notion is the fulfilment of OT expectations in Jesus and the application of OT passages to him by NT writers and by much of Christian exegesis. In speaking about prophetic perspective, we must be careful not to attribute the hindsight of the Christian church to the prophets themselves.

But the "prophetic perspective" view of Jesus' expectation of the end

---

[82] See G. von Rad, *Theology* 2, 169-75; O. Kaiser, *Jesaja*, 99-103, 125-30; A. Weiser, *Die Propheten: Hosea, Joel, Amos, Obadja, Micha* (*ATD* 24; Göttingen: Vandenhoeck & Ruprecht, 1967) 228, 273-75.

[83] See C. Westermann, *Isaiah 40-66* (OT Library; Philadelphia: Westminster, 1969) 39, 152-62; "Sprache und Struktur der Prophetie Deuterojesajas," *Forschung am Alten Testament* (Theologische Bücherei 24; München: Kaiser, 1964) 139-40, 144, 151; G. von Rad, *Theology* 2, 243-50; J. Begrich, *Studien zu Deuterojesaja* (Herausgegeben von W. Zimmerli; Theologische Bücherei 20; München: Kaiser, 1963) 101-14; W. Zimmerli, "Der 'neue Exodus' in der Verkündigung der beiden grossen Exilspropheten," *Gottes Offenbarung: Gesammelte Aufsätze zum Alten Testament* (Theologische Bücherei 19; München: Kaiser, 1963) 199-200.

[84] See K. Elliger, *Die Propheten: Nahum, Habakuk, Zephanja, Haggai, Sacharja, Maleachi* (*ATD* 25; Göttingen: Vandenhoeck & Ruprecht, 1967) 97-98, 129-30; G. von Rad, *Theology* 2, 283-85.

would seem to offer the best solution to the problem. It does justice to the complicated evidence which the NT offers on the subject. It does not attribute twentieth-century mental constructs to Jesus; it refuses to take the escape route of community-formulation when a saying does not fit into the picture of Jesus' expectations which we have formed. Nor does it have recourse to ingenious exegesis when faced with difficult texts. And it does not introduce distinctions when they are not warranted by the NT evidence. This solution is founded on the character of Jesus' preaching itself, a proclamation addressing the present in view of the definitive salvation which is approaching. Though future, the final saving act of God exerts its influence in the present, demanding of men that they decide now. A comparison of Jesus' message with that of the prophets is thus justified.

C. Westermann has pointed out that one of the essential traits of prophecy consists in the fact that it addresses the concrete historical situation in which it is being proclaimed. Second Isaiah proclaims nothing but salvation; yet it would be false to imagine that there is a break in the continuity between him and his predecessors who were announcing judgment and punishment. This judgment has taken place with the fall of Jerusalem; a new situation has arisen which demands a new word from God. Ezekiel, who announced judgment up to the moment of the fall of Jerusalem, began proclaiming salvation after that catastrophe had taken place. God remains the same, but his word is not always the same. The prophetic word can thus never assume the character of teaching which remains the same in every situation. It remains a living word, proclaiming a new message when a new hour strikes. Essentially, prophecy is not prediction.[85] Though the non-fulfilment of some of the prophetic words must have been a most painful trial for the prophet himself and for his followers, it is significant that these non-fulfilled words have been preserved. Thus they must have been looked upon as holding out an expectation beyond the present hope or hopelessness, and as containing a validity which survives all failure.[86] Hope in a God who will unfailingly accomplish his purpose— the hope which the words of OT prophets and of Jesus awakened and kept

---

[85] *Isaiah,* 9-10. W. Zimmerli, in his comparison of Ezekiel's and Second Isaiah's treatment of the theme "new Exodus" ("Exodus," 201-4), points out that Ezekiel's view of God as judge is no longer present in Second Isaiah. For a discussion of God's word of condemnation and his word of salvation in Ezekiel, see G. von Rad, *Theology* 2, 225-28, 233-37; W. Zimmerli, "Das Gotteswort des Ezechiel," *Gottes Offenbarung,* 133-47.

[86] See G. von Rad, *Theology* 2, 165-69; K. Elliger, *Propheten,* 130; J. Begrich, *Deuterojesaja,* 115-16. Note, however, C. Westermann's critical remark on Begrich's division of Second Isaiah's prophecies into epochs in "Sprache," 99.

alive—is so fundamental to the faith of Israel and of the Christian church that it can afford to preserve, cherish, and nourish itself on the disappointed hopes of the past, for the simple reason that the hopes of the past are also the hopes of the present. To discuss the presence or absence of error within such a framework seems to be more or less a matter of semantics.

We thus conclude that Mk 9:1 reproduces substantially an authentic word of Jesus in which he expresses his expectation of the approaching kingdom. It is a prophetic saying putting into words a hope which is implicitly contained in other words and actions.

(C) Mark 13

We must now examine the function of the logion in the Second Gospel. To do that, however, we must try to understand Mark's own expectation of the end. This is most eloquently expressed in ch. 13 of his Gospel.

(1) The composite character of Mark 13 has long been recognized. Until the advent of redaction criticism the efforts of exegetes were devoted to the discovery of the road travelled by the discourse, or its antecedents, in the tradition, without paying much attention to the work and intentions of the evangelist.[87] Lately the interest has centered on Mark's redaction of the discourse.[88] That Mark is not simply recording a tradition, but has taken a hand at the work of bringing the discourse to its present shape is admitted by everyone.[89] The extent of it is, and will undoubtedly remain, a matter of dispute. Opinions range from the very reserved judgment of F. Hahn[90] to that of J. Lambrecht, who attributes to redaction more than anyone yet,[91] and more than most would be ready to admit.[92] For our purpose we hope to discover the central intention of the evangelist in the composition of the discourse, and in the light of this intention to determine his thought about the end.

The last part of the discourse tells us most clearly what aim the evangelist is pursuing. The last verse, undoubtedly redactional,[93] bursting, as it does, the framework of esoteric instruction imposed upon the discourse by

---

[87] For a history of these attempts, see G. R. Beasley-Murray, *Future*, 1-171.

[88] For a presentation and discussion of various attempts on this level, see R. Pesch, *Naherwartungen*, 27-47.

[89] See W. Marxsen, *Mark*, 161.

[90] *Mission*, 68-73. As Marcan additions he considers vss. 1-4, 7c, probably vss. 33-36, a little less probably vs. 32.

[91] *Redaktion*, passim, particularly 256-60.

[92] See R. Pesch, *Naherwartungen*, 43-44; J. H. Elliott's review of J. Lambrecht's *Redaktion*, *CBQ* 30 (1968) 267-69.

[93] See R. Pesch, *Naherwartungen*, 202.

Mark in vs. 3, insists that the watchfulness urged upon the disciples is also the attitude demanded of all Christians. The key word, and the only message of this verse, *grēgoreite*, leads us upward into the verses which precede it and of which it is such a forceful conclusion. We find the verb again as the opening word of the preceding unit, vss. 35-36, which gives the reason for the demand, i.e., ignorance of the moment of the Lord's return, and the danger that his sudden arrival might find them asleep. The particle *oun* of vs. 35 shows that the unit is an application to the disciples of the parable in vs. 34, the last word of which is the verb with which vs. 35 opens. Like the doorkeeper, the disciples must be on the watch. Making a further step backwards, we see that vs. 33, beginning with the typical Marcan *blepete*, enunciates the same message as vss. 35-36: the imperative need for watchfulness because of the ignorance of the decisive moment. Thus we have in vss. 33-36 an antiphonally framed parable whose antiphon and ending contain an unmistakably clear demand for watchfulness. Vs. 37 states that this message applies to all Christians. Even without inquiring into the provenance of each word and image in this final section of the discourse, it is impossible not to detect strong traces of redaction in its arrangement and formulation.[94]

Vss. 33 and 35-36, however, are not merely a comment on the thought of the parable which they enclose; they spell out the message of vs. 32. This verse states an objective fact: no one but the Father knows "of that day or that hour." Vss. 33, 35-36 draw the consequence for the attitude of the disciples, "be on the watch, stay awake." The verb "to know" is the catchword linking these verses with vs. 32.[95] In its turn, vs. 32 forces us to probe still further backwards—the particle *de*, while distinguishing, connects; the verse is obviously meant as a qualification of the preceding verses.[96] The vss. 28-31 speak of the nearness of "all these things." What "all these things" are will be discussed later; at the moment it is the links weaving the entire section into one unit that are of interest.

However we judge the origin of the unit itself and the provenance of its various elements,[97] two features of its function within the discourse are

---

[94] See W. Marxsen, *Mark*, 162; W. Grundmann, *Markus*, 260; R. Pesch, *Naherwartungen*, 195-202; J. Lambrecht, *Redaktion*, 228-56. See also L. Hartman, *Prophecy Interpreted: The Formation of Some Jewish Apocalyptic Texts and of the Eschatological Discourse Mark 13 Par.* (Coniectanea Biblica, NT Series 1; Lund: Gleerup, 1966) 174-76, where he points out the difference between the parenesis in vss. 33-37 and in the rest of the discourse.

[95] See J. Sundwall, *Zusammensetzung*, 78.

[96] See V. Taylor, *Mark*, 523; D. E. Nineham, *Mark*, 360; W. Grundmann, *Markus*, 271; G. R. Beasley-Murray, *Future*, 261.

[97] For F. Hahn's opinion, see above note 90. W. Marxsen (*Mark*, 162) thinks that

easily discernible. The unit serves as an introduction to the final parenesis in vss. 33-37, and it answers the question posed in vs. 4. The fact that "he is near, even at the door," and that "all these things" will take place before this generation is to die, as well as the fact of the ignorance of the moment when "that hour" is to strike, serve as foundation for the watchfulness demanded of the disciples.

It is becoming more and more clearly recognized that it is this exhortation to watchfulness which reveals Mark's true purpose and fundamental intention in the composition of the discourse.[98] L. Hartman[99] has shown that the parenetic tendency was present in early Christian eschatological and apocalyptic teaching before Mark's time. Mark is thus not innovating but continuing a tradition; by constructing and adding vss. 33-37, and probably vss. 28-32 as well, he shows what the entire apocalyptic discourse means to him and what it should produce in his readers: an attitude of intense wakefulness, of readiness for the return of the Lord who may come to gather his elect at any moment. The Lord is near, but the moment of his coming cannot and must not be calculated, since such calculations would only have the effect of allowing Christians to "sleep." The parenetic features are not limited to the last section alone; the characteristic Marcan *blepete* occurs at salient junctures of the discourse (vss. 5b, 9, 23, 33).[100] Numerous imperatives and pointed appeals to the listeners (vss. 9, 23, 29) are found throughout the discourse, which begins and ends with the warning to "be on guard" (vss. 5b, 37).

Besides serving as the introduction to the final parenesis of the discourse, stating the objective facts of nearness and incalculability of "the hour," vss. 28-32 also answer the question put to Jesus by the disciples in vs. 4:[101] *tauta* and *tauta panta* of vs. 4ab find their echo and their answer

---

Mark inserted vss. 28-29 and possibly 32. E. Schweizer (*Mark*, 278-80) is of the opinion that the entire unit belongs to the *Vorlage*. On the other hand, W. Grundmann (*Markus*, 260) seems to think that Mark is responsible for the present position of the verses. R. Pesch (*Naherwartungen*, 175-95) maintains that the composition of the unit is due to Mark who is partly using older traditions and partly formulating freely (vss. 28a, 29, 30). J. Lambrecht (*Redaktion*, 193-240) considers vss. 28a, 29, 30, 32 as Mark's redactional formulations. Pesch's presentation and argumentation are the most convincing.

[98] See R. Pesch, *Naherwartungen*, 77, 202; S. G. F. Brandon, "The Date of the Markan Gospel," *NTS* 7 (1960-61) 137; L. E. Keck, "Introduction," 365, n. 6; N. Walter, "Tempelzerstörung und synoptische Apokalypse," *ZNW* 57 (1966) 40; H. Conzelmann, "Eschaton," 220-21; J. Schmid, *Mark*, 233; E. Schweizer, *Mark*, 280.

[99] *Prophecy*, 145-226; see also H. Conzelmann, "Eschaton," 213-15.

[100] See R. Pesch, *Naherwartungen*, 77; J. Lambrecht, *Redaktion*, 274.

[101] See W. Marxsen, *Mark*, 187; H. Conzelmann, "Eschaton," 220; E. Schweizer, *Mark*, 280-81; J. Lambrecht, *Redaktion*, 227; R. Pesch, *Naherwartungen*, 79; G. Minette de Tillesse, *Secret*, 423-24; W. Grundmann, *Markus*, 260.

in *tauta* and *tauta panta* of vss. 29 and 30. These words also provide the discourse with its redactional clasp.

(2) The primary purpose of the discourse is thus parenetic; Jesus' entire instruction on the last things is made by the evangelist to result in the exhortation to be awake and ready for the moment when "the hour" should strike. With this in mind we may now ask, What does Mark think about the end, and what are the hopes which he is holding out to his readers? He gives the answer in vss. 28-32. But to understand his answer, we must understand his question. For that reason we must examine vs. 4 more closely.

It is commonly agreed that in vss. 1-5a we have to do with a Marcan construction. Vs. 1, certainly a Marcan composition in part and most likely in its entirety,[102] shows Jesus definitively turning away from the Temple. It also expresses a disciple's surprised exclamation, the purpose of which is to set the scene for Jesus' answer in vs. 2. The manner in which this answer echoes the disciple's exclamation argues in favor of the view that vs. 2 is also a redactional formulation.[103] Should vs. 2c be looked upon as a traditional logion, its present position is undoubtedly due to the redactional activity of the evangelist. The short scene in vss. 1-2 thus closes the previous section of the Gospel. After fruitless controversies with his enemies in the Temple Jesus departs from it and announces its destruction. The scene also opens the parting discourse given by the Master to his disciples before he dies. That the next scene, vss. 3-4, is Marcan in its entirety is hardly open to doubt. It is "not a self-contained narrative, but an introduction to 5-37";[104] its vocabulary is Marcan, as is also the theme of instruction given to the disciples in private. The disciples' question contains the redactional clasp mentioned above, *tauta* and *tauta panta*; their *pote* and *hotan* echo the oft recurring *hotan* and *tote* of the discourse. It is evident, then, that the question is formulated with the answer in mind.

What does the question ask? It obviously refers to Jesus' prediction of the destruction of the Temple in vs. 2. That reference, according to G. Minette de Tillesse,[105] exhausts the content of the question; he sees no reason whatever to distinguish between the "this" of vs. 4a and "all this" of vs. 4b. Thus vs. 30 would refer, not to the parousia of the Son of Man,

---

[102] See J. Sundwall, *Zusammensetzung*, 76; V. Taylor, *Mark*, 500. R. Pesch (*Naherwartungen*, 84-87) makes a very good case in favor of the entire verse being a free formulation of Mark.

[103] For an exhaustive discussion, see R. Pesch, *Naherwartungen*, 84-96. J. Lambrecht (*Redaktion*, 72-79) thinks that Mark is composing on his own.

[104] V. Taylor, *Mark*, 501-2; see also W. Grundmann, *Markus*, 261.

[105] *Secret*, 422-25.

but to the historical events preceding the parousia.[106] In this he has few companions. One argument against his opinion is the term *synteleisthai* which has eschatological overtones;[107] another is the discourse itself speaking, as it does, of the coming of the Son of Man. R. Pesch[108] has shown that the second part of the question by no means merely repeats the first. He compares this verse with 11:28 where the authorities ask Jesus: "By what authority are you doing these things (*tauta*), or (*ē*) who gave you the authority to do them (*tauta*)?" The question refers to Jesus' cleansing of the Temple, yet it would be false to imagine that it remains limited to that action alone. The enquiry about the origin of Jesus' power in vs. 28b concerns the entire activity of Jesus. In 13:4, similarly, the question about the destruction of the Temple broadens into a question about the sign of the end. The weight of the question lies in its second half, the chief preoccupation being the end and the sign announcing it. The destruction of the Temple is somehow related to the end. What this relationship is the discourse will reveal; vs. 4 leaves various possibilities open.

(3) Though we do not intend to discuss every redaction-critical treatment of Mark 13, it will help our exposition if we present the first of these, viz., that of W. Marxsen.[109] Marxsen observes that Mark editorially combines the destruction of the Temple with the end in vss. 1-4. From this he concludes that the destruction must, in Mark's eyes, form part of the end-event. Vs. 14 is thus the pivot round which the entire discourse turns. However the source used by Mark may have conceived the "abomination of desolation," Mark has, by editorially prefixing vs. 2 to the discourse, indicated that he understands it as referring to the destruction of the Temple. At the moment of destruction Antichrist will have ensconced himself in the holy place—from this the end of the world will result. The evangelist is thus writing before the destruction, vs. 2 is not a *vaticinium ex eventu*, but a traditional piece which Mark takes quite literally. The present of the evangelist and of the community for which he is writing is reflected in vss. 5-13. He attempts to correct a false apocalyptic expectation which sees the end in the war already raging (vss. 7c, 10, 13) by insisting that this

---

[106] For a similar recent opinion, based on different arguments, see J. Winandy, "Le logion de l'ignorance (Mc., XIII, 32; Mt., XXIV, 36)," *RB* 75 (1968) 67-72. J. Lambrecht's view will be discussed below.

[107] See G. Dalman, *Words*, 154-56; E. Lohmeyer, *Markus*, 269; G. Delling, *TWNT* 8, 65-67. Luke seems to have understood the verb as implying an eschatological event; since he wishes to separate the destruction of the Temple from the immediate eschatological context he replaces the verb; see H. Conzelmann, *The Theology of St Luke* (New York: Harper & Row, 1960) 126; W. Grundmann, *Lukas*, 379.

[108] *Naherwartungen*, 101-5; see also W. Marxsen, *Mark*, 168.

[109] *Mark*, 151-89, particularly from p. 166 on.

war is only the beginning of the final tribulation (vs. 8c). The present is not yet the end, but its beginning. Marxsen believes that he can date the redaction of the Second Gospel very closely: the Roman-Jewish war has already begun, but Jerusalem has not yet fallen; vss. 5-13 place us in the period between the years 66 and 70. The phrase "let the reader take note" of vs. 14b addresses the readers of the Gospel most directly. When they see devastation and depopulation following in the wake of invading armies, they should understand that Antichrist, spoken of in vs. 14a, is already casting his shadow over the world. They must flee from Jerusalem and Judea to the mountains, i.e., to Galilee, where the community, according to Marxsen's understanding of Mark, is to await the Lord's parousia.[110] Vss. 14-27 describe the conditions and warn of the dangers between the flight of the Jerusalem community and the end. This end is very near—when everything predicted in the discourse has taken place (vs. 29), then the parousia is just around the corner. Although its precise date is known to the Father alone (vs. 32), it will undoubtedly come within the life-span of one generation.

Marxsen's interpretation of Mark 13 has been severely criticized. H. Conzelmann and E. Schweizer[111] point out that vss. 14-27 never refer to Galilee, that the only tradition which we have of the flight of the Jerusalem community speaks of Transjordan, and that Galilee was a rather unlikely refuge since it was there that the war began. R. Pesch,[112] subjecting Marxsen's exposition to a thorough examination, also points out a number of defects and false conclusions: the structure of the discourse is incorrectly assessed. With W. Grundmann[113] he asks why Mark would bother writing a Gospel a few years before the end. We would add a further criticism: Vs. 14 speaks of the "abomination of desolation" as the signal for the flight; but Mark, according to Marxsen, identifies the abomination with the destruction of the Temple. If this identification be accepted, then according to the text the flight should take place only after the destruction and not, as Marxsen would have it, at the threat of the destruction mirrored in the devastation of the countryside. The phrase "let the reader take note" does not change the content of vs. 14a from fact to threat, as Marxsen seems to suggest.

To arrive at the proper understanding of the relationship between the destruction of the Temple and the end, we must note the structure of the discourse. Where does Mark place the main incisions and, consequently, at

---

[110] *Mark*, 75-95.
[111] "Eschaton," 215; *Mark*, 273.
[112] *Naherwartungen*, 28-31.
[113] *Markus*, 19.

what point in the discourse do we pass from the description of the past and the present to that of the future? There is universal agreement that vs. 28 is the opening verse of the last section of the discourse. As to the question where the chief break is to be placed in vss. 5-27 there is less agreement. For Marxsen the description of the evangelist's present ends with vs. 13. Conzelmann,[114] sharply distinguishing between the destruction and the eschatological events, also regards vss. 14-23 as referring to the future; they speak of the last epoch of history which is to come before the parousia. Recent discussions of Mark 13, however, contend that the main break is to be placed between vss. 23 and 24.[115] What are the reasons for this contention? G. Minette de Tillesse draws attention to the apocalyptic predictions in chs. 7, 8, 11 and 12 of the Book of Daniel. In 7:1-8, for instance, the four beasts are easily recognizable because the writer is "predicting" what is already history. When, however, he moves from the earth into heaven and from the past and the present into the future the picture becomes blurred. In Mark 13 the "prediction" of historical events reaches down to vs. 23; what is described in vss. 5-23 has thus already happened. These verses reflect events concomitant with the destruction of the Temple, not something which has yet to take place.

R. Pesch and J. Lambrecht have studied the literary structure of vss. 5-23 and have shown that vss. 5b-6 and 21-22, warning against false prophets, form an inclusion in the entire first section of the discourse. Moving inwards, there is another pair of units corresponding to each other, both beginning with *hotan de (akousēte-idēte)* and describing war and destruction, viz., vss. 7-8 and 14-20. The second of these units speaks of the destruction of Jerusalem and the Temple. W. Marxsen has understood correctly that vs. 14 should be read in the light of vs. 2.[116] The heart of the first section of the discourse, in structure and thought, is to be placed in vss. 9-13. This subsection opens with the characteristic *blepete*, addresses the disciples directly, and describes the experiences and tasks of the community in the period which is yet to elapse before the return of the Lord. The first section ends with the redactional vs. 23; its renewed exhortation to "be on guard" echoes the beginning of the discourse in vs. 5b, and *proeirēka* recalls the disciples' question (*eipon hēmin*) in vs. 4. The prediction of what is to happen on earth (*panta*) is complete. The community, having already seen its

---

[114] "Eschaton," 219-20.

[115] See R. Pesch, *Naherwartungen*, 78-82, 139-47, 155-57; J. Lambrecht, *Redaktion*, 267-92; S. G. F. Brandon, "Date," 130-38; G. Minette de Tillesse, *Secret*, 420-29.

[116] N. Walter ("Tempelzerstörung," 42-44) thinks that vs. 14 refers to the appearance of Antichrist at the end in an unnamed place; for criticism, see R. Pesch, *Naherwartungen*, 45-46.

fulfilment, can thus be armed against the siren call of false prophets and confirmed in its hope for the future parousia.

There is no need to go into further detail. We agree with the conclusion that Mark is writing shortly after the destruction of Jerusalem, and that he separates this destruction from the parousia by creating the break between vss. 23 and 24. Historical events are described in vss. 5-23, the parousia in vss. 24-27. Vss. 24-27 are the centerpiece of the entire discourse. The phrase "after that tribulation" of vs. 24, which may be redactional,[117] underlines once more Mark's desire to keep the destruction of Jerusalem distinct from the coming of the Son of Man; it voices the same "retarding" theme as vs. 7c.

(4) We now return to the question which has led us to the consideration of Mark 13: What is Mark's own expectation of the end? He gives his answer in vss. 28-32; but this answer cannot be understood unless we know what he is asking in vs. 4; and vs. 4, in its turn, cannot be understood unless we know how Mark sees the relationship between the end and the destruction of the Temple. We have seen that for him there is a link between the two events, but that he clearly distinguishes them, thus stripping the destruction of the eschatological dignity with which an intimate connection with the parousia would endow it.[118] This conclusion is valid even apart from the supposition of an apocalyptic propaganda leaflet whose existence seems to be almost universally accepted, but on whose content there has been little agreement.[119]

The verses which are of particular interest to us are 29, 30 and 32. Vs. 29 undoubtedly cannot refer to the parousia, for a thing cannot be a sign of its own nearness. Does vs. 30 refer to the parousia? J. Lambrecht argues against this view,[120] pointing to the parallelism between vss. 29 and 30 (*tauta-tauta panta, ginomena-genētai*). He recalls the *panta* of vs. 23, the word which refers to the devastation and tribulation described in vss. 5-22 and the similar relationship between "this" and "all this" in vs. 4. He rejects the opinion according to which "this" in vss. 29 and 4a would refer to the destruction, and "all this" of vss. 30 and 4b to the destruction and the parousia. *Tauta* does refer to the destruction, but the "all" that *panta* adds to it is the thought of devastation which will endure for a long time.

---

[117] See R. Pesch, *Naherwartungen*, 157-58.

[118] Ibid., 107-18; H. Conzelmann, "Eschaton," 214-16; *RGG*³ 2, 671; N. Walter, "Tempelzerstörung," 41.

[119] See G. R. Beasley-Murray, *Future*, 1-80. The most recent attempt to reconstruct it is that of R. Pesch, *Naherwartungen*, 207-23.

[120] *Redaktion*, 207-8; see also C. E. B. Cranfield, *Mark*, 408-9. For the similar opinions of G. Minette de Tillesse and J. Winandy, see notes 106 and 115.

Thus vs. 30, like vs. 29, speaks only of the events which lie before the parousia. R. Pesch, along with many others,[121] thinks differently. The phrase *hotan idēte tauta ginomena* of vs. 29 echoes vss. 4, 14 and 30. The difference between "this" of vs. 29 and "all this" of vs. 30, however, is the same as in vs. 4. Vs. 30 speaks not only of the destruction but also of the parousia. This opinion seems to us to be correct for a number of reasons. One is Pesch's analysis of vs. 4 which we have summarized above; it is difficult to escape the impression that vs. 4b refers to the final consummation, and it is equally difficult to miss Mark's intention in his arrangement of the same terms in the same succession in the answer as well as in the question. Another reason is the urgency of the exhortation in vss. 33-37; a simple statement of the uncertainty of the day and the hour would hardly seem to justify it. A solemn affirmation, strengthened by the declaration of the irrevocability of Jesus' words, that the end will come within a generation seems to be the only adequate explanation of this urgency. Moreover, the primary interest of the last section of the discourse is the parousia; since this is evidently true of vss. 33-37 one would presume that the same holds true of the more didactic unit which forms the basis and introduction to the final exhortation. It is not likely that four verses of this unit (28-31) would be devoted to the destruction of Jerusalem and its concomitant horrors, and only one verse (32) to the parousia.

The meaning of vs. 32 can hardly be missed: Mark asserts that Jesus himself did not know the precise moment of the parousia. The primary purpose of the assertion lies, of course, in the parenesis which follows; since Jesus did not know, Christians do not know, and any attempt to calculate that moment is an impermissible and sinful prying into secrets which are God's alone. Though Mark is convinced that there is a connection between the destruction of the Temple and the end (vss. 1-4), and though this destruction is a sign of the nearness of the end (vs. 29), he clearly separates the historical, and for him past, event from the future coming of the Son of Man. He insists that it must not be abused for the purpose of laying hands on what God has reserved for himself. Instead of being a reckoning device, the destruction of the Temple is made to serve ethical exhortation. Mark thus counteracts apocalyptic propaganda based on the illusion that with the fall of Jerusalem the end has arrived. "The end is not yet"; Christians must be awake and watchful because they do not know when the Lord will return.

The end, however, cannot be far away; vs. 30 places it within the life-span of one generation. Attempts to evade the obvious temporal implications of

---

[121] R. Pesch, *Naherwartungen*, 179-87; see also W. Marxsen, *Mark*, 187; E. Schweizer, *Mark*, 280-82; W. Grundmann, *Markus*, 270-71; J. Schmid, *Mark*, 243-44; V. Taylor, *Mark*, 521; R. Schnackenburg, *God's Rule*, 208; L. Hartman, *Prophecy*, 225-26.

"this generation"[122] are most unconvincing. The term most likely contains the reference to sinfulness as it does elsewhere in the NT; but neither the saying itself nor its context permits the moral connotation to be taken as its only or its primary meaning.

(D) The Function of the Logion within the Gospel

(1) As already mentioned many of the Church Fathers saw the fulfilment of Mk 9:1 in the immediately following transfiguration. It was this thought which induced the men responsible for the chapter division to link the verse with ch. 9 instead of ch. 8.[123] Yet there can be no doubt that the verse belongs to what precedes, not to what follows. Decisive in this respect is the manner in which Mark employs his characteristic *kai elegen autois*. After a minute study of Mark's use of the verb *legein*, M. Zerwick[124] comes to the conclusion that the phrase is employed to indicate the continuation of an address already begun; to introduce an address, other forms of the verb are used. Other considerations confirm this conclusion: the subject matter of 8:38 is, like that of 9:1, eschatological; there is great similarity between the phrases *elthē en tē doxē* of 8:38 and *elēlythuian en dynamei* of 9:1.[125]

Mk 9:1 is thus evidently meant to close what precedes. It is a comment considered necessary to round off the thought of the unit. Before we can ask about the significance which the logion has for the evangelist we must thus try to discover the message of the unit of which it forms the ending and, in a certain sense, the climax. That vs. 34 opens this unit has never been disputed; vs. 34a is typically Marcan,[126] introducing a series of logia which, isolated and scattered in the synoptic tradition, had been gathered only secondarily either in the pre-Marcan stage of tradition[127] or by Mark.[128] The redactional character of vs. 34a, and the difference between the form and the content of the verses on each side of it, strongly suggest that it was Mark who inserted vss. 34-38 into the present context.[129] To see the reason for this insertion, we must go back to 8:27.

---

[122] See J. Schniewind, *Markus*, 178-79; F. Mussner, *Christ and the End of the World: A Biblical Study in Eschatology* (Contemporary Catechetics Series; South Bend, Ind.: University of Notre Dame, 1965) 57; M. Meinertz, "'Dieses Geschlecht' im Neuen Testament," *BZ* n.s. 1 (1957) 284-88.
[123] See E. P. Gould, *Mark*, 159.
[124] *Markusstil*, 67.
[125] See J. Sundwall, *Zusammensetzung*, 57.
[126] See R. Bultmann, *History*, 229-30; J. Sundwall, *Zusammensetzung*, 56.
[127] This is the opinion of R. Bultmann, *History*, 82-83; E. Lohmeyer, *Markus*, 171.
[128] The opinion of E. Haenchen, "Leidensnachfolge," 116-17.
[129] See E. Lohmeyer, *Markus*, 161; J. Schniewind, *Markus*, 112; E. Haenchen, "Leidensnachfolge," 114.

The entire passage 8:27—9:1 bears strong traces of redactional work. There is no need to discuss in detail the various opinions about the nature and extent of the sources which Mark had at his disposal in constructing 8:27-33;[130] we shall simply attempt to identify redactional passages within this unit. Vs. 30 is undoubtedly redactional, voicing, as it does, the theme of the messianic secret. Redactional also are the introductory words of vs. 31.[131] With E. Haenchen, F. Hahn and E. Schweizer we would also attribute vs. 32 to Mark's hand. In its first clause we come across familiar, if not characteristic, Marcan vocabulary; the term *parrēsia*, a *hapax legomenon* in the Synoptists, is called for by Mark's theological schema, for there is no question about the fact that the confession of Peter constitutes the decisive moment in the course of the narrative toward which the entire first half of the Gospel tends and in which it finds its climax and explanation. In vs. 32b there occurs the typical *ērxato* together with *epitiman* which we have already met in vs. 30.[132] Vs. 32 is, moreover, in perfect harmony with the schema which we encounter after each of the predictions of the passion in Mk 8:27—10:52: prediction, lack of understanding on the part of the disciples, further instruction given by Jesus. That vs. 33 has also been at least retouched by Mark is shown by the third occurrence of the verb *epitimaō* and by the phrase *tous mathētas autou*.[133] These remarks on the redactional material in 8:27-33 permit us to conclude that it was the evangelist who combined the confession of Peter with the prediction of the passion. He formulated the injunction of silence, Peter's protest, and at least reformulated the introduction to Jesus' rebuke of the disciples in vs. 33. The opinion that it was Mark who linked the confession with the prediction can be described as common among today's exegetes.[134]

By means of the redactional vs. 34a Mark attached new material to what he had already skillfully arranged in vss. 27-33. The sudden introduction of the crowd into what has been a conversation between Jesus and his disciples is his way of emphasizing the fact that the words which follow have a

---

[130] For various opinions, see R. Bultmann, *History*, 257-59, 427; F. Hahn, *Titles*, 223-28; E. Haenchen, "Leidensnachfolge," 103-20; E. Schweizer, *Mark*, 165-66; W. Grundmann, *Markus*, 167.

[131] W. Grundmann's remark (*Markus*, 167) that *ērxato didaskein* is an unusual Semitism for Mark is contradicted by facts: see V. Taylor, *Mark*, 48.

[132] Against W. Grundmann (*Markus*, 167), who seems to believe that the presence of the verb argues in favor of a source, we think that it speaks in favor of Marcan redaction.

[133] See C. H. Turner, "Usage," *JTS* 26 (1924-25) 235-37.

[134] Besides the authors referred to in note 130 (with the exception of W. Grundmann), see also G. Strecker, "Voraussagen," 32-35; J. Roloff, "Markusevangelium," 88; E. Lohmeyer, *Markus*, 161.

universal application; he is, in other words, addressing the reader more directly than is his custom.[135] For purposes of our discussion we need not decide whether the combination of the logia in vss. 34b-38 should be attributed to Mark or to the tradition. By pointing to the explicative *gar* in vss. 35, 36, 37, 38, E. Haenchen suggests that the collection was first made by Mark. It is, of course, impossible to deny that the particle may be an indication of an editorial collection of logia, but it is just as difficult to affirm it with any degree of certainty. We would think that the redactional introductory phrase in 9:1 argues against Haenchen's suggestion, for it gives the impression of adding a logion to an already existing unit. On the supposition that Mark has combined the logia one would expect this phrase to occur within vss. 34-38, as it does in 4:21-25.

(2) Within 8:27—9:1 we thus have four blocks of traditional material redactionally combined: the confession of Peter, the prediction of the passion, and sayings on cross-bearing, and on the coming of the kingdom. What is the message of this Marcan composition?

The confession of Peter is the confession of the church; whatever the historical background of vss. 27-29 may be, R. Bultmann[136] is surely right when he says: "the disciples represent the Church, and (the passage gives) expression to the specific judgement which the Church had about Jesus, in distinction from that of those outside." The fact that Jesus does not confirm or deny the confession of Peter, the prediction of the passion which follows and the rebuke hurled at Peter and other disciples: all this should not mislead us into questions about the historical Jesus' attitude toward the title of messiah, and misconceptions nurtured by the Twelve about that title at the time of confession. The injunction of silence in vs. 30 is redactional; to see it as an indication of the historical Jesus' reservations about the title would be about as valid as to conclude, from the injunctions of silence given in connection with the miracles, that the historical Jesus had doubts about their realness, or to conclude that he questioned the message of the divine voice on the Mount of Transfiguration because he commanded silence about the vision. Peter's remonstrances in vs. 32 should not lead to conjectures about the historical disciples' Jewish ideas about the messiah. The entire section 8:27—10:52 is devoted to the instruction of the disciples; Jewish authorities, so much in evidence in chs. 2, 3, 7, 8, 11, 12 and in the Passion Narrative, appear here only twice (9:14; 10:2) and play only a

---

[135] See R. Schnackenburg, "Vollkommenheit," 431; E. Haenchen, "Leidensnachfolge," 114; E. Lohmeyer, *Markus*, 171; see our discussion above of Mk 4:10 (in ch. II).

[136] *History*, 258; see also E. Haenchen, "Leidensnachfolge," 109.

marginal role. It is not the Jewish misconceptions of the role and task of the messiah which Mark is combatting, but those of Christians. The problems with which he is dealing and the abuses against which he is protesting in the name of the Messiah-Son of Man, who must suffer and die in order to rise again, are those of the community. Peter's remonstrance manifests the same lack of understanding as the discussion about greatness in 9:33-34, and the request of James and John and the anger of other disciples in 10:35-41. Mark is at pains to guard against Christian misconceptions of the confession that Jesus is the Messiah; at the very outset he insists that Jesus' messiahship does not exclude but, on the contrary, includes death.[137] In 8:31 Jesus teaches "quite openly" that the divinely decreed destiny of the Son of Man leads to glory only through total renunciation.

The purpose of 8:34—9:1 is to show what is essentially involved and demanded whenever the confession of Jesus as the Messiah takes place:[138] the disciple of Jesus must share his destiny; he must deny himself, take up his cross, be ready to lose everything, his life included, if he is to share in the glory of the Son of Man at his coming.[139] The structure and theme of this unit is the same as that of the prediction of the passion: suffering, death, glory.[140] Every Christian must walk this path; the refusal to do so brands his confession of Jesus' messiahship as an empty formula. Of such the Son of Man will be ashamed when he comes "in his Father's glory."

(3) We now turn our attention to the last two logia of the unit, 8:38 and 9:1. As for 8:38, we may take it for granted that Mark identifies Jesus with the Son of Man—in this, he is simply following the tradition of the community.[141] It is also certain that the verse refers to the parousia,[142] and that, in the form and context in which we find it in the Second Gospel, it sounds a threatening note of judgment.[143] We need not go into the problem of its genuineness and the far more complicated and disputed question of Jesus' own thought on the Son of Man. C. E. B. Cranfield states the message of the logion in this context: "Now he [i.e., Jesus] is one of whom men can be ashamed; then he will be manifest as the one who has the glory of God."[144]

[137] See E. Haenchen, "Leidensnachfolge," 110.
[138] See D. E. Nineham, *Mark*, 227; E. Lohmeyer, *Markus*, 161.
[139] See G. Strecker, "Voraussagen," 35.
[140] See E. Lohmeyer, *Markus*, 171; W. Grundmann, *Markus*, 174.
[141] See H. E. Tödt, *The Son of Man in the Synoptic Tradition* (NT Library; London: SCM, 1965) 42-43.
[142] R. Pesch, *Naherwartungen*, 169.
[143] See E. Schweizer, *Mark*, 177-78; E. Haenchen, *Weg*, 298-300.
[144] *Mark*, 285.

The saying of 9:1 in its isolated condition could hardly refer to anything but the coming of the kingdom at the end of time. If the phrase "in power" was added by Mark in imitation of 13:26, as we have suggested, the eschatological reference becomes even more evident. The perfect participle of the verb "to come" points in the same direction. "Some of those standing by will witness the triumphant consummation of the eschatological process which has its humble beginnings in the earthly life of the Messiah."[145] The phrase "some standing here" in all probability reflects the delay of the parousia; if it is Marcan, it expresses with yet greater poignancy the evangelist's and his community's awareness of the lapse of time since the words had first been spoken, and of the fact that in the meantime many of Jesus' contemporaries had died. It may also be expressing his consciousness of the difference between the community and the "faithless and corrupt age" (8:38) in the midst of which it must live and give witness to Jesus and his words.[146] The logion is clearly speaking of the future, not any future, but of that eschatological day which will put an end to the twilight in which the community must endure persecution, ostracism, and the temptation to "judge not by God's standards but by man's" (8:33). Since 9:1 is intended to be a continuation of the preceding verses, and of 8:38 in particular, the redactional linking makes it all the more evident that the second verse treats of the same reality, or better, of another facet of the same reality as the preceding one. The terms "glory" and "power" and the references to the coming of the Son of Man and the kingdom[147] show that Mark has in mind the event of which he will speak again in 13:24-27.

To the Christian reader of the Gospel Mk 9:1 thus gives the assurance of glory after humiliation and suffering. The present condition of the community will not last forever; it will, in fact, end very soon—if some of those who stood with Jesus at the time when the logion was pronounced are not to die before the arrival of the kingdom this arrival cannot be far away. This verse thus conveys the same message as 13:30. It is a consoling message, a promise of salvation to those who refuse to be ashamed of Jesus and his words in this world. It brings balance to the picture of the end which 8:38 portrays in the threatening colors of judgment and rejection. The logion, finally, conveys a sense of contrast between the present of the community and the future fulfilled kingdom. The kingdom has come with Jesus (1:15), the community participates in its mystery (4:11), and eschatological forces are already at work. All this, however, is hidden: Mark's Jesus

---

[145] T. A. Burkill, *Revelation*, 167.

[146] See A. Vögtle, "Erwägungen," 646.

[147] The two "comings" obviously coincide, at least in the eyes of Mark; see N. Q. Hamilton, "Resurrection," 419; R. Schnackenburg, *God's Rule*, 287.

is one whose greatness is manifested in humiliation, whose strength reveals itself in the total weakness of death; his Lordship is perceived in service, the Father's acceptance in men's rejection, his supreme independence in the complete obedience to God's will. The world is not convinced, and the community which must follow its Lord is tempted to be ashamed of him. Complete manifestation of the kingdom is still in the future; only the sworn word of Jesus vouches for the certainty of its coming.

(4) What remains to be discussed is the relationship between 9:1 and the scene of transfiguration. The logion in its isolated condition could hardly have referred to the transfiguration. "In fact it seems impossible that it should have done so for at least two reasons. (i) However we interpret what the disciples saw, it could scarcely, in reality, have been called *the kingdom of God come with power*. (ii) If someone says: 'So and so will happen while some of you in this very audience are still alive,' he is thinking not of something that will happen a week later but of something that will happen after a lapse of a good many years when it is reasonable to suppose that some of his hearers will still be alive, while others will not."[148] On the face of it, these remarks should be valid not only for the logion in its isolated condition but also within the context of Mark. It is most unlikely that Mark saw in the scene of transfiguration a fulfilment of the logion in 9:1; nor do we think that he saw in it a partial, proleptic fulfilment.

There are, however, a number of authors who think that this is the case. We must examine their arguments. G. H. Boobyer[149] is probably the most eloquent proponent of the view that Mk 9:1 finds its fulfilment in the transfiguration. Mark, according to him, saw the vision on the mountain as a forecast of the parousia. Besides the apocalyptic imagery of the story itself (cloud, voice, the presence of Elijah, eschatological booths), this is confirmed by the two immediately preceding verses which speak of the parousia:

Both verses . . . anticipate for Mark Christ's coming again in the clouds at the End, and express the promise that certain bystanders were in some way to witness that coming of the Kingdom before they died. What is the explanation of such an introduction to the transfiguration, unless Mark reads the transfiguration in a way resembling the use made of it in the Apocalypse of Peter and 2 Peter?[150]

We should mention that, according to Boobyer, the Apocalypse of Peter and 2 Peter understand the transfiguration as an anticipated parousia.[151]

---

[148] D. E. Nineham, *Mark*, 263.

[149] In his article, "St. Mark and the Transfiguration Story," and in his book with the same title (Edinburgh: T. & T. Clark, 1942); we refer to the article as "Transfiguration" and to the book as *Story*.

[150] "Transfiguration," 125-26; see also *Story*, 29, 58, 69, 87.

[151] "Transfiguration," 119-23.

Another proponent of the same view is U. W. Mauser:[152] "The vision of the three disciples corresponds to the promise that some will see (9, 1), and the transfiguration is told in the terminology of a theophany which reveals the powerful coming of the kingdom of God." F. J. Schierse[153] analyzes Mk 9:1 and finds in it four motifs: election, seeing, not dying, and the kingdom. The election motif is fulfilled in the fact that Peter, James, and John are allowed to see their transfigured Master. The motif of seeing is fulfilled in the three disciples' witnessing the glory of the One who will return in the future parousia. The fulfilment of the promise that they will not die before the end consists in the fact that they can witness its anticipation. Transfiguration itself, finally, is the anticipation of the kingdom. R. Pesch[154] gives a similar interpretation. S. Schulz claims that Mark rejects the primitive community's expectation and reinterprets the logion of 13:30 in this manner: "The kingdom of God was near to this generation because Jesus, God and man, sojourned with them."[155] C. E. B. Cranfield, after examining various opinions, also comes to the conclusion that "the kingdom of God come with power" "is not unfair description of what the three saw on the mount of Transfiguration. For the Transfiguration points forward to, and is as it were a foretaste of, the Resurrection, which in turn points forward to, and is a foretaste of, the Parousia; so that both the Resurrection and the Parousia may be said to have been proleptically present in the Transfiguration."[156] Others will not go so far as to affirm that the promise of 9:1 finds its fulfilment in the transfiguration. V. Taylor, for instance, is of the opinion that "Mark introduces the saying [i.e., 9:1] at this point because he sees at least a partial fulfilment in the Transfiguration."[157] W. Grundmann[158] says no more than that the verse forms a transition between the logia on cross-bearing and the scene of transfiguration.

The common feature in all these opinions seems to be the assumption that Mk 9:1 forms an introduction, or at least a transition, to the scene of transfiguration. We regard this assumption as false. Mk 9:1 should be looked upon as a conclusion of what precedes, not as an introduction to what follows. If we are to discover any relationship between this logion

---

[152] *Wilderness*, 111.
[153] "Kritik," 530-36.
[154] *Naherwartungen*, 187-88.
[155] "Nahe war die Gottesherrschaft deshalb diesem Geschlecht, weil der Gottmensch Jesus unter ihnen weilte" (*Stunde*, 98).
[156] *Mark*, 288.
[157] *Mark*, 385. A similar opinion is held by F. W. Beare (*Records*, 140-41). D. E. Nineham (*Mark*, 236) admits this opinion as possible.
[158] *Markus*, 177.

and the transfiguration, we must not separate it from the context with which Mark has linked it. The only valid procedure is to establish the relationship between the unit of which it forms the conclusion and the scene which follows it.

On the origin and growth of the transfiguration story in the pre-gospel tradition of the community there is no agreement among commentators.[159] Was it originally a resurrection story which has been read back into the earthly life of Jesus, as R. Bultmann[160] would have it? Is it a forecast of the parousia, as G. H. Boobyer[161] thinks? M. Dibelius[162] looked upon the story as an epiphany revealing Jesus as the Son of God. E. Lohmeyer[163] thought of it as a symbolic narrative in which Jesus was revealed to the nucleus of the eschatological community as their Lord, as their Master, and as the One by whom the eschatological fulfilment was to be brought about. Some authors[164] refuse to agree with Lohmeyer and others who deny all historicity to the account. The list of opinions is long and we cannot hope to resolve the discussion.[165] We feel that the redactional remarks of the evangelist will help us to see how he understood the pericope.

These remarks are easy to detect. The most evident one is vs. 9, "As they were coming down the mountain, he strictly enjoined them not to tell anyone what they had seen, before the Son of Man had risen from the dead." Another one is vs. 6, "He hardly knew what to say, for they were all overcome with awe." This verse, with its reference to the disciples' lack of understanding and their fear, is attributed to the hand of the evangelist almost as universally as vs. 9. The phrase *kat' idian* in vs. 2 may also be Marcan. Thus we meet in this pericope three themes typical of the Second Gospel: the injunction of silence, the disciples' failure to understand, and esoteric instruction. They are all subsumed under the heading of the messianic secret. These themes are particularly evident in the section of the Gospel within which this pericope is located, 8:27—10:52. Jesus'

---

[159] For brief outlines of various opinions, see V. Taylor, *Mark*, 386-88; W. Grundmann, *Markus*, 178-80.

[160] *History*, 259; for others holding the same view, see V. Taylor, *Mark*, 386-88, and R. Bultmann, *History*, 428. For a criticism of Bultmann's view, see G. H. Boobyer, *Story*, 11-16; V. Taylor, ibid.; H. P. Müller, "Die Verklärung Jesu," *ZNW* 51 (1960) 56-62.

[161] For a criticism of G. H. Boobyer, see E. Grässer, *Parousieverzögerung*, 149-51.

[162] *Tradition*, 275-77.

[163] *Markus*, 178-81; for reservations about this view, see R. Bultman, *History*, 428; V. Taylor, *Mark*, 388; W. Grundmann, *Markus*, 180.

[164] V. Taylor and W. Grundmann; also C. E. B. Cranfield, *Mark*, 292-96.

[165] E. Schweizer, *Mark*, 180: "It is no longer possible to explain the history of the tradition of this passage."

attention and instruction are devoted almost exclusively to the disciples; at its very beginning they confess him to be the Messiah, a fact which they are forbidden to divulge. Their failure to understand his words is continually stressed.

Since the injunction of silence and the failure to understand concern Jesus' present status—in other words, since the disciples fail to perceive and after having perceived are forbidden to publicize what Jesus already is—it is only reasonable to conclude that for Mark the transfiguration story spoke principally of Jesus' already present divine sonship and unique teaching authority. The messianic secret refers to the past, not to the future. This is confirmed by the manner in which Mark's version of the pericope focuses on the disciples:

> The whole event, from first to last, takes place solely for the sake of the three disciples. "He was transfigured *before them*"; "there appeared *unto them* Elijah with Moses"; "there came a cloud overshadowing *them*"; "this is my only Son; hear *ye* him"; "and suddenly, looking round about, they saw no one any more, save Jesus only *with themselves*."[166]

We must consider this passage in connection with Jesus' baptism and the centurion's confession in 15:39; that the three scenes are linked in Mark's editorial intention is generally acknowledged.[167] What Jesus alone hears at baptism is revealed to the three disciples on the mountain and is publicly proclaimed at his death on the cross. The divine voice heard by the disciples proclaims what Jesus already is, not what he is to be in the future; now he is the Son of God and, as such, the only Master to whom they must listen.

Apocalyptic coloring of the scene should not be taken as an indication that the future is primarily in view—at least as far as Mark is concerned. The baptism, transfiguration and death of Jesus are the only events in the Gospel accompanied by visible apocalyptic phenomena.[168] All three manifest the present status of Jesus, Son of God, in whom God reveals his saving presence and, particularly in the last of the three scenes, his judging presence. The transfiguration, in Mark's eyes, is not a fulfilment or anticipation of a fulfilment, but a revelation of what Jesus is, and of the ultimate meaning of what has been happening from the first moment of his ministry on; ". . . the content of the revelation, or that which is revealed, at the trans-

---

[166] R. H. Lightfoot, *Message*, 44; see also E. Lohmeyer, *Markus*, 178.

[167] See W. Bousset, *Kyrios*, 95; R. H. Lightfoot, *Message*, 57; P. Vielhauer, "Christologie," passim; E. Schweizer, *Mark*, 356-59.

[168] See E. Schweizer, *TWNT* 8, 369-71; P. Vielhauer, "Christologie," 161-62; R. G. Bratcher, "Introduction to the Gospel of Mark," *RevExp* 55 (1958) 365; E. Lohmeyer, *Markus*, 344-48; W. Grundmann, *Markus*, 315-17.

figuration, is an unchanging, secret fact—the constant presupposition of Jesus' manifold activities on earth."[169] The question, "Who is this?" repeatedly asked throughout the first part of the Gospel has been answered by Peter's confession. This confession now receives the final and irrefutable confirmation from God himself. But the divine voice does not speak immediately after the confession, it speaks only after the implications of his messiahship for Jesus, and the implications of the confession of this messiahship for the community, have been spelled out. It thus confirms the way of the Son of Man described in 8:31, the demands imposed upon his followers in 8:34-38, and the promise of 9:1. The words of Jesus are words of the Son of God; they must therefore be accepted and obeyed.

The passage of which the logion in 9:1 forms the conclusion addresses the readers of the Gospel most directly. It tells them that their way to glory is to be the same as that of their Master: through suffering and death. It has a most immediate relevance for the life of the community of Mark's day. The transfiguration, however, is a past event, proclaiming to the disciples, and now to the readers, the innermost secret of the person who speaks the words on cross-bearing and promises the speedy coming of the kingdom. The community is still waiting for the fulfilment of the promise, but the period of waiting is not a time of confusion and despair because the promise throws light on and gives meaning to the darkness of its suffering. In this promise the community can place absolute trust since it has been given by the Son of God, revealed as such in his transfiguration. Mk 9:1 looks to the future, Mk 9:2-8 tells of the past. The transfiguration is not a fulfilment, either partial or total, of the community's hope, but its confirmation.

(2) A Note on Mark 15:43

Joseph from Arimathea arrived—a distinguished member of the Sanhedrin. He was another who looked forward to the kingdom of God. He was bold enough to seek an audience with Pilate and urgently requested the body of Jesus.

Today it seems to be rather widely recognized that cultic interests of the earliest community played an important role in the formation and preservation of the story of Jesus' burial. G. Schille[170] has shown that the entire Passion Narrative echoes the community's commemoration of Jesus' death. This commemoration which took place on Good Friday formed part

---

[169] T. A. Burkill, *Revelation*, 159.
[170] "Das Leiden des Herrn: Die evangelische Passionstradition und ihr 'Sitz im Leben,'" *ZTK* 52 (1955) 193-99.

of the community's celebration of Easter. The tradition found in Mk 15:42-47 must be seen within this context if we are to understand its original function and message. We do not owe the story to the mere desire of preserving the historical remembrance of Jesus' burial,[171] or, along with the rest of the Passion Narrative, to the apologetic interests attempting to take the sting out of the shameful death of Jesus by pointing out that such was the will of God and by stressing the glorious ending and reversal of the Good Friday tragedy on Easter Sunday.[172] It may be that vss. 44-45 were inserted later in order to counter the claim that Jesus' death had been only apparent,[173] but L. Schenke[174] correctly assesses the primary purpose of the original story when he asserts that it was meant to assure the community that the site venerated as the tomb of Jesus was really such. The women serve as witnesses of the location; vs. 47, far from being adventitious,[175] reveals to us the original purpose of the story: the women "saw *where* he had been laid." The place and time indications which occur throughout the Passion Narrative, and which are to be traced to the cultic interests of the community, appear in the story of the burial also; vss. 42a and 47 thus belong to the tradition.[176]

J. Jeremias[177] has shown that the veneration of the graves of great figures of the past was by no means foreign to the Jews of Jesus' time and has suggested that the Christian community of Jerusalem had all the greater reason to venerate the place where the body of Jesus had been laid.[178] It is likely that this veneration was expressed in a cultic manner.[179] The entire Passion Narrative seems to have been ordered originally accord-

---

[171] As R. Bultmann (*History*, 274) seems to suggest.

[172] As M. Dibelius (*Tradition*, 183-89) suggests for the entire Passion Narrative.

[173] This is the opinion of R. Bultmann, *History*, 274; E. Lohmeyer, *Markus*, 350; E. Schweizer, *Mark*, 362; W. Grundmann, *Markus*, 318. V. Taylor (*Mark*, 599) thinks otherwise: "The surprise of Pilate . . . and the questioning of the centurion are easily credible . . . ."

[174] *Auferstehungsverkündigung und leeres Grab: Eine traditionsgeschichtliche Untersuchung von Mk 16,1-8* (SBS 33; Stuttgart: Katholisches Bibelwerk, 1968) 18-19.

[175] As R. Bultmann (*History*, 274) and V. Taylor (*Mark*, 599, 602) assert. For a discussion of this view, see L. Schenke, *Auferstehungsverkündigung*, 18, and J. Blinzler, *Der Prozess Jesu* (Regensburg: Pustet, 1969) 413-14.

[176] See G. Schille, "Leiden," 194-96; L. Schenke, *Auferstehungsverkündigung*, 15.

[177] *Heiligengräber in Jesu Umwelt (Mt. 23,29; Lk. 11,47): Eine Untersuchung zur Volksreligion der Zeit Jesu* (Göttingen: Vandenhoeck & Ruprecht, 1958), particularly p. 114.

[178] *Heiligengräber*, 145.

[179] See G. Schille, "Leiden," 199; W. Nauck, "Die Bedeutung des leeren Grabes für den Glauben an den Auferstandenen," *ZNW* 47 (1956) 261.

ing to the hours of liturgical gatherings.[180] A number of authors stress that the cultic background of this story does not deprive it of its basic historical reliability.[181]

Reports which were originally connected with the cultic gatherings later came to be separated from them.[182] They were thus more strongly exposed to various tendencies to which the entire oral tradition was subject. V. Taylor[183] remarks that there is reason to think that the account of Mk 15:42-47 was compiled in a Gentile environment. He enumerates seven characteristics which speak in favor of this supposition. Five of these, however, occur in vss. 44-45, verses which also seem to be secondary on other grounds, and two of them in vs. 46a. These characteristics thus speak more in favor of the secondary nature of vss. 44-45 and of a reworking of vs. 46a after these verses had been inserted than in favor of the view that the entire story stems from a Gentile environment.

The indications of time and place which the account of the burial has in common with the rest of the Passion Narrative suggest very strongly that it had been joined to this narrative long before Mark came to write his Gospel. We should probably not seek particular Marcan redactional intentions in the story apart from his desire to continue and complete the redactional *vita Jesu* which he gives us in his Gospel.

In vs. 43, Joseph is described as a rich property owner from Arimathea and, apparently at least, as a member of the Sanhedrin.[184] P. Winter's opinion[185] that "he was a member of a lower Beth Din (there were three Jewish Courts in Jerusalem) whose duty it was to ensure that the bodies of executed persons were given a decent burial before nightfall" is seriously weakened by J. Blinzler's observation[186] that there is no evidence that the burial of executed persons was a duty of members of the lower courts.

Joseph is further described as one "who looked forward to the kingdom of God." This clause could be redactional;[187] an indication of this is the characteristic Marcan periphrastic construction *ēn prosdechomenos*. What

---

[180] G. Schille, "Leiden," 198.

[181] See L. Schenke, *Auferstehungsverkündigung*, 98-103; W. Nauck, "Bedeutung," 263-65; J. Blinzler, *Prozess*, 405-11. See also the quotation of H. von Campenhausen, "Der Ablauf der Osterereignisse und das leere Grab," in W. Nauck, "Bedeutung," 265.

[182] See G. Schille, "Leiden," 198.

[183] *Mark*, 599.

[184] See J. Jeremias, *Jerusalem in the Time of Jesus: An Investigation into Economic and Social Conditions during the New Testament Period* (London: SCM, 1969) 96, 223; V. Taylor, *Mark*, 600; J. Blinzler, *Prozess*, 392, n. 39.

[185] "Marginal Notes on the Trial of Jesus II," *ZNW* 50 (1959) 244, n. 99.

[186] *Prozess*, 392, n. 39.

[187] V. Taylor, *Mark*, 599.

does this clause tell us about Joseph? Many authors feel that he is not to be considered as a follower of Jesus, but as being in sympathy with his expectation and annunciation of the kingdom.[188] Others associate him with the pious Jews (cf. Lk 2:25, 38) who were longing and praying for the fulfilment of Israel's hopes.[189] Others, however, think that in Mark's Gospel at least, if not in the pre-Marcan tradition, the clause suggests that he was expecting the kingdom to come through Jesus.[190] J. Blinzler[191] seems to be right when he suggests that Matthew's (27:57) and John's (19:38) description of Joseph as a *mathētēs* of Jesus does not go beyond what Mark says of him. Since Mark applies the term to the Twelve only,[192] he could not use it of Joseph of Arimathea. He does, however, wish to indicate that Joseph had accepted Jesus' message of the kingdom.

The clause is intended primarily to describe Joseph of Arimathea. Whether we consider it as redactional or traditional, it can hardly be said to contribute a great deal to our understanding of Mark's concept of the kingdom. It may, however, be significant that a person of high standing performs the last honors for the dead Son of God. Like the confession of the centurion under the cross, his act of piety may be meant to foreshadow what is to happen after the resurrection.[193]

---

[188] V. Taylor, ibid.; E. Haenchen, *Weg*, 542; E. Schweizer, *Mark*, 362; E. Lohmeyer, *Markus*, 350.

[189] W. Grundmann, *TDNT* 2, 58, who refers to *Str-B* 2, 124-26, 141, to support his views; cf. J. Schmid, *Mark*, 301. We fail to find in *Str-B* the reference to a definite group among the Jews which Grundmann seems to discover.

[190] J. Schniewind, *Markus*, 209; E. Klostermann, *Markus*, 169; D. E. Nineham, *Mark*, 434.

[191] *Prozess*, 329, n. 39.

[192] See R. P. Meye, *Twelve*, 98-99, 120.

[193] E. Best, *Temptation*, 65: "Here Mark may be indicating the attitude of the believer of his own time."

Chapter VI

CONCLUSIONS

In Chapter I we attempted to show that for Mark the primary message of *euangelion* concerns the kingdom of God, a kingdom yet to come which is, paradoxically, already present. The very proclamation of its future coming brings it about: Jesus, by his powerful word and deed, is putting an end to the time of waiting and introducing the new era of God's eschatological salvation. Mark's redactional "life of Jesus" is summarized in the opening statement of Jesus' ministry; it tells us of the present kingdom striving for its future completion.

The present state of the kingdom was explored in Chapter II. This state can best be described as hidden. Being a divine reality, it grows irresistibly and by its own power towards its final goal. As such, it is also inscrutable. Since it has not yet been fully manifested, it remains exposed to unbelief and misinterpretation outside the Christian community and to weakness of faith and discouragement within it. Though the community has been given its "mystery," it is not as firm as it should be in its acceptance of the only way available to enter the kingdom, viz., Jesus' word, deed, and destiny. It must be exhorted continually to understand, to struggle against hardness of heart which threatens to turn Jesus' saving words into words of condemnation. The community is asked to trust, in the teeth of outward appearances, that the kingdom will not remain hidden forever.

Chapter III considered the ethical demands of the kingdom. Since the establishment of the kingdom is the definitive saving act of God, subjection to it entails a radical change in the life of the community and that of its members. The main effect and sign of the already present kingdom consists in a radical obedience to Jesus' call and a selfless service to others. Mark is struggling against the ever present temptation to treat the gift of God as one's own possession and to become alienated from God and the community by the abuse of riches for purposes of domination. The crowning act of God's work of salvation calls for total surrender on the part of man.

Chapter IV examined the awareness voiced by the Marcan form of the eucharistic tradition of the sharp contrast between the present condition of the community and the future kingdom. The cup now being distributed is the cup of Jesus' suffering and death; only at the end of time will this cup give way to a joyful sharing in Jesus' triumph.

Chapter V was devoted to Mark's expectation of the definitive manifestation of the kingdom at the end of time. This manifestation is the goal of proclamation, cross-bearing, and constant watching which constitute the present duty of the community. Though rejecting attempts to compute the exact moment of its arrival, Mark has no doubt that the kingdom is near. To some it will bring condemnation, but for those who are not ashamed of the Son of Man it will be the moment of irrevocable triumph.

There can be no doubt of Mark's awareness of the difference between the time before and after Jesus' death and resurrection, between the period of the earthly life and ministry of Jesus and the period of the community. This awareness is expressed in the theme of the messianic secret: what was hidden, or should have remained hidden, during the time of Jesus' ministry is now being openly proclaimed. The evangelist knows that Jesus' injunctions of silence no longer bind him, that, in fact, the opposite is the case: he must proclaim Jesus' miracles, he must write about his acts of power. It is his duty to tell everyone what Jesus alone heard at the baptism, what Peter confessed at Caesarea Philippi, and what the divine voice communicated to the three chosen disciples on the Mount of Transfiguration. Everyone who cares to listen or read may now be told that Jesus is the Messiah and the Son of God. The process of revelation in the ministry and destiny of Jesus has been completed with his death and resurrection; from the moment of his death Jesus may, indeed must, be proclaimed as the Son of God. The task of Jesus is, however, not yet completed. He is still to come to gather his elect at the end of time. He is still to manifest his power in a manner which will brook no opposition, which will shatter all resistance, whose unambiguous perceptibility will dispense with further need of proclamation.

Since this final act of the Son of Man has yet to take place, the similarity between the time of Jesus and that of the community is as evident as is the difference. The community proclaims the good news which Jesus had brought and proclaimed. The power manifested in the word of Jesus is being manifested in that of the community; the seed of the Word, once having been sown, grows irresistibly toward its goal. But the opposition to the good news is as virulent in the time of the community as it was during the ministry of Jesus. As he was opposed, contradicted, plotted against, condemned, and put to death because he was carrying out the task given to him by the Father, so are the Christians being delivered up to councils, beaten in synagogues, made to stand before governors and kings, hated, and put to death for the sake of Jesus and the good news which they are proclaiming (13:9,11-13). The similarity of Jesus' destiny and that of the Christians is brought out most strongly in the redactional composition 8:27—9:1. As the Son of Man must suffer, be rejected, and be killed, so must those who confess and follow him be ready to face persecution, spoliation, ostracism, shame, and death. As he came "not to be served, but to serve" (10:45), so also does their greatest distinction lie in being last of all and servants of all (10:43; 9:35). They must become like children whose duty it is to obey and be subject to others; their material possessions must never be abused as a means of dominating, exploiting, and enslaving others (10:14-15, 23-25). Finally, the spiritual condition of Christians is not totally unlike that of the Twelve during the earthly life of their Master. The resurrection of Jesus did not free them from fear and lack of understanding. Like the Twelve, they confess Jesus to be the Messiah, they have received the teaching on his death and resurrection, like the three disciples on the Mount of Transfiguration they know that Jesus is the Son of God whom alone they must follow, but like the Twelve they are puzzled by him and his humanly incomprehensible way to the cross, they are tempted to be ashamed of him and his words, they still feel the pull of "the anxieties over life's demands, and the desire for wealth, and cravings of other sorts" (4:19). There still exists the danger that Satan will come and take away the word sown in them (4:15). They may still be led astray by false Christs and false prophets, by their signs and wonders (13:5-6, 21-22); their longing for the last saving act of the Son of Man is not free of the danger of feeding on illusions. They must be continually exhorted to watch (13:37). The temptation to lord it over others (10:42-43) is ever present; the threat of their being untrue to Jesus' destiny which is also their own is never completely dispelled. Even the feature of the Gospel which most clearly expresses the difference of times, i.e., the messianic secret, serves a purpose in the present; it is meant to combat false conceptions about Jesus and Christian life which circulate in the community.

Christians thus still live in the twilight of the morning. Jesus' ministry marked the end of the night and the arrival of the eschatological light; but the bright sun of the eschatological day has yet to rise. Jesus' messiahship and divine sonship are no longer hidden, but they are not recognized by the world. Satan has been defeated, but his power is yet to be totally annihilated. "Those outside" already stand condemned, but they can still persecute the followers of Jesus. Christians are saved but not yet freed from the tendency to "judge not by God's standards but by man's" (8:33). Christian existence thus remains a paradox: the time is fulfilled, but the fulfilment which will destroy all doubt and eliminate all weakness is still in the future.

The condition of the kingdom of God after the death and resurrection of Jesus is, in the eyes of Mark, not fundamentally different from its condition during his life on earth. It has become a present reality in Jesus' proclamation of the good news. The content of "the good news of God" is that "this is the time of fulfillment; the kingdom of God is at hand." This statement, being the foundation and the summary of everything that Mark intends to say in the rest of the book, tells us that Jesus' words and deeds are a manifestation and a bringing about of God's kingdom, of the eschatological reality promised by the prophets quoted at the very beginning of the Gospel. Having come, it cannot be done away with; there is no demonic or human power which is able to extinguish its light and to arrest its irresistible growth. The good news of its having arrived in Jesus Christ, the Son of God, must be proclaimed to all nations (13:10). This divine "must" cannot be repealed. The community proclaims what Jesus proclaimed; its proclamation has the same content and manifests the same divine act of eschatological salvation. In the word preached by the community the same power is at work as in that preached by Jesus—the Spirit which had descended upon Jesus at his baptism assists Christians who give testimony of their faith (13:9-11). The community has been given the mystery of the kingdom; it has received this greatest of God's gifts to men, has entered it, and is already participating in its joys (4:11; 10:14-15,30).

Yet the fact itself and the content of the proclamation, as well as the condition of the world and of the community in it, show clearly that the kingdom, though already present, is still coming. The very fact that it must be proclaimed is a witness to its hiddenness, for at the end there will be no need to tell of what will be obvious to all, the elect and condemned alike. The peculiar juxtaposition of the fulfilment of time and the approach of the kingdom in 1:15a indicates the paradoxical state of completion awaiting its plenitude. The kingdom still seems to be small, insignificant and powerless (4:30-32); its unfailing certitude of its achieving its goal seems as inexplicable now as it was during the ministry of Jesus (4:26-29). The enemies of Jesus have found their successors in "those outside," the unwillingness to follow in the footsteps of Jesus is as painfully evident in the community as is the lack of understanding and the fear of the Twelve. The kingdom remains a hidden reality and will remain such until the moment when it will have come with power (9:1). The community has its share in it, but is not to be identified with it. For the world does not deny the existence of the community—by persecuting it, it shows its awareness of it. What the world does deny, and will be able to keep on denying until the day when the Son of Man comes "in clouds with great power and glory," is the existence of the kingdom. The kingdom in the time of Jesus' life on earth and in the time of the community is a hidden kingdom. It is a reality, however, which cannot but become manifest at the end of time.

BIBLIOGRAPHY

Aland, K., M. Black, B. M. Metzger, A. Wikgren, *The Greek New Testament* (Stuttgart: United Bible Societies, 1966).

Albright, W. F., *From the Stone Age to Christianity: Monotheism and the Historical Process* (2nd ed.; Garden City, N.Y.: Doubleday, 1957).

Arndt, W. F. and F. W. Gingrich, *A Greek-English Lexicon of the New Testament and Other Early Christian Literature: A Translation and Adaptation of Walter Bauer's Griechisch-deutsches Wörterbuch zu den Schriften des Neuen Testaments und der übrigen urchristlichen Literatur* (Chicago: University of Chicago, 1952).

Asting, R., *Die Verkündigung des Wortes Gottes im Urchristentum* (Struttgart: Kohlhammer, 1939).

Baltensweiler, H., "Das Gleichnis von der selbstwachsenden Saat (Mc 4,26-29) und die theologische Konzeption des Markusevangelisten," *Oikonomia: Heilsgeschichte als Thema der Theologie: Festschrift für Oscar Cullmann* (ed. F. Christ; Hamburg-Bergstedt: H. Reich, 1967) 69-75.

Barrett, C. K., *The Holy Spirit and the Gospel Tradition* (London: S.P.C.K., 1966).

Bartsch, H.-W., "Zum Problem der Parousieverzögerung bei den Synoptikern," *EvT* 19 (1959) 116-31.

Beare, F. W., *The Earliest Records of Jesus: A Companion to the Synopsis of the First Three Gospels by Albert Huck* (Oxford: Blackwell, 1962).

Beasley-Murray, G. R., *Jesus and the Future: An Examination of the Criticism of the Eschatological Discourse, Mark 13, with Special Reference to the Little Apocalypse Theory* (London: Macmillan, 1954).

Begrich, J., *Studien zu Deuterojesaja: Herausgegeben von W. Zimmerli* (Theologische Bücherei 20; München: Kaiser, 1963).

Behm, J., "*Noeō*, etc.," *TDNT* 4, 948-80, 989-1022.

Behm, J., "*Klaō*, etc.," *TDNT* 3, 726-43.

Benoit, P., "Note sur une étude de J. Jeremias," *Exégèse et théologie* (Paris: Cerf, 1961), 1, 240-43.

Benoit, P., "Le récit de la Cène dans Luc XXII 15-20," *Exégèse et théologie*, 1, 163-203.

Benoit, P., "Les études de H. Schürmann sur Lc XXII," *Exégèse et théologie*, 1, 204-9.

Berkey, R. F., "*EGGIZEIN, PHTHANEIN* and Realized Eschatology," *JBL* 82 (1963) 177-87.

Best, E., *The Temptation and the Passion: the Marcan Soteriology* (SNTSMS 2; Cambridge: Cambridge University, 1965).

Best, E., "Uncomfortable Words: VII. The Camel and the Needle's Eye (Mk 10:25)," *ExpT* 83 (1970-71) 83-89.

Betz, J., *Die Eucharistie in der Zeit der griechischen Väter*, Band I/1: Die

Aktualpräsenz der Person und des Heilswerkes Jesu im Abendmahl nach der vorephesinischen griechischen Patristik (Freiburg: Herder, 1955).

Betz, J., *Die Eucharistie in der Zeit der griechischen Väter*, Band II/1: Die Realpräsenz des Leibes und Blutes Jesu im Abendmahl nach dem Neuen Testament (2d ed.; Freiburg: Herder, 1964).

Beyer, K., *Semitische Syntax im Neuen Testament*, Band I: Satzlehre, Teil 1 (2d ed.; Studien zur Umwelt des Neuen Testaments, 1; Göttingen: Vandenhoeck & Ruprecht, 1968).

Beyer, W., "*Blasphēmeō*, etc.," *TDNT* 1, 621-25.

Bieler, L., *THEIOS ANĒR: Das Bild des "göttlichen Menschen" in Spätantike und Frühchristentum* (Darmstadt: Wissenschaftliche Buchgesellschaft, 1967).

Bieneck, J., *Sohn Gottes als Christusbezeichnung der Synoptiker* (*ATANT* 21; Zürich: Zwingli, 1951).

Black, M., *An Aramaic Approach to the Gospels and Acts* (3rd ed.; Oxford: Clarendon, 1967).

Black, M., "The Kingdom of God Has Come," *ExpT* 63 (1951-52) 289-90.

Blank, J., "Marginalien zur Gleichnisauslegung," *Bibel und Leben* 6 (1965) 50-60.

Blass, F. and A. Debrunner, *A Greek Grammar of the New Testament and Other Early Christian Literature: A Translation and Revision of the Ninth-tenth German edition by R. W. Funk* (Chicago: Chicago University, 1961).

Blinzler, J., *Der Prozess Jesu* (4th ed.; Regensburg: Pustet, 1969).

Boobyer, G. H., "St. Mark and the Transfiguration Story," *JTS* 41 (1940) 119-40.

Boobyer, G. H., *St. Mark and the Transfiguration Story* (Edinburgh: Clark, 1942).

Boobyer, G. H., "The Secrecy Motif in St. Mark's Gospel," *NTS* 6 (1959-60) 225-35.

Boobyer, G. H., "The Redaction of Mark IV. 1-34," *NTS* 8 (1961-62) 59-70.

Born, A. van den and L. Hartman, "Rechabites," *EDB*, 1992-93.

Bornkamm, G., *Jesus of Nazareth* (New York: Harper, 1960).

Bornkamm, G., "Das Wort Jesu vom Bekennen," *Geschichte und Glaube: Erster Teil: Gesammelte Aufsätze III* (Müchen: Kaiser, 1968) 25-36.

Bornkamm, G., "Das Doppelgebot der Liebe," ibid., 37-45.

Bornkamm, G., "Lord's Supper and Church in Paul," *Early Christian Experience* (New York: Harper & Row, 1969) 123-60.

Bornkamm, G., "Die Verzögerung der Parousie," *Geschichte und Glaube: Erster Teil: Gesammelte Aufsätze III*, 46-55.

Bornkamm, G., "*Mystērion*," *TDNT* 4, 802-28.

Bornkamm, G., "Evangelien, Synoptische," *RGG*³ 2, 753-66.

Bornkamm, G., G. Barth and H. J. Held, *Tradition and Interpretation in Matthew* (NT Library; Philadelphia: Westminster, 1963).

Bosch, D., *Die Heidenmission in der Zukunftsschau Jesu: Eine Untersuchung zur Eschatologie der synoptischen Evangelien* (*ATANT* 36; Zürich: Zwingli, 1959).

Bousset, W., *Kyrios Christos: A History of the Belief in Christ from the Beginnings of Christianity to Irenaeus* (Nashville: Abingdon, 1970).

Brandon, S. G. F., "The Apologetical Factor in the Markan Gospel," *SE* II, 34-46.

Brandon, S. G. F., "The Date of the Markan Gospel," *NTS* 7 (1960-61) 126-141.

Bratcher, R. G., "Introduction to the Gospel of Mark," *RevExp* 55 (1958) 351-66.

Brown, R. E., "The Pre-Christian Semitic Concept of 'Mystery,'" *CBQ* 20 (1958) 417-43.

Brown, R. E., "The Semitic Background of the NT *mystêrion*," *Bib* 39 (1958) 426-48; 40 (1959) 70-87.

Brown, R. E., "Parable and Allegory Reconsidered," *NovT* 5 (1962) 36-45.

Brown, R. E., *Jesus God and Man: Modern Biblical Reflections* (London: Chapman, 1968).

Bruce, F. F., *Biblical Exegesis in the Qumran Texts* (Grand Rapids: Eerdmans, 1959).

Bultmann, R., *The History of the Synoptic Tradition* (New York: Harper & Row, 1962).

Bultmann, R., *Theology of the New Testament* (2 vols.; London: SCM, 1965).

Bultmann, R., *Jesus and the Word* (New York: Scribner, 1958).

Bultmann, R., "*Pisteuō*, etc.," *TDNT* 6, 174-82, 197-228.

Burger, C., *Jesus als Davidssohn: Eine traditionsgeschichtliche Untersuchung* (*FRLANT* 98; Göttingen: Vandenhoeck & Ruprecht, 1970).

Burkill, T. A., "The Cryptology of Parables in St. Mark's Gospel," *NovT* 1 (1956) 246-62.

Burkill, T. A., "The Hidden Son of Man in St. Mark's Gospel," *ZNW* 52 (1961) 189-213.

Burkill, T. A., *Mysterious Revelation: An Examination of the Philosophy of St. Mark's Gospel* (Ithaca, N.Y.: Cornell University, 1963).

Burkill, T. A., "Mark 3,7-12 and the Alleged Dualism in the Evangelist's Miracle Material," *JBL* 87 (1968) 409-17.

Burkitt, F. C., "W and Θ: Studies in the Western Text of St. Mark: Hosanna," *JTS* 17 (1916) 139-50.

Campbell, J. Y., "'The Kingdom of God Has Come,'" *ExpT* 48 (1936-37) 91-94,

Cerfaux, L., "'L'aveuglement d'esprit' dans l'évangile de Saint Marc," *Muséon* 59 (1946) 267-79.

Cerfaux, L., "La connaissance des secrets du royaume d'après Mt., XIII,11 et parallèles," *NTS* 2 (1955-56) 238-49.

Childs, B. S., *Memory and Tradition in Isreal* (*SBT* 37; London: SCM, 1962).

Clark, K. W., "Realized Eschatology," *JBL* 59 (1940) 367-83.

Clark, N., *An Approach to the Theology of the Sacraments* (*SBT* 17; London: SCM, 1956).

Conzelmann, H., "Gegenwart und Zukunft in der synoptischen Tradition," *ZTK* 54 (1957) 277-96.
Conzelmann, H., "Geschichte und Eschaton nach Mc 13," *ZNW* 50 (1959) 210-21.
Conzelmann, H., *The Theology of St Luke* (New York: Harper & Row, 1960).
Conzelmann, H., *An Outline of the Theology of the New Testament* (NT Library; London: SCM, 1969).
Conzelmann, H., "Eschatologie, (IV) im Urchristentum," *RGG*³ 2, 665-71.
Conzelmann, H., "*Syniēmi*, etc.," *TWNT* 7, 886-94.
Coutts, J., "The Authority of Jesus and of the Twelve in St. Mark's Gospel," *JTS* 8 (1957) 111-18.
Coutts, J., "'Those Outside' (Mark 4, 10-12)," *SE* II 155-57.
Cranfield, C. E. B., *The Gospel according to Saint Mark* (2nd ed.; Cambridge Greek Testament Commentary; Cambridge: Cambridge University, 1963).
Creed, J. M., *The Gospel according to St. Luke* (London: Macmillan, 1965).
Creed, J. M., "'The Kingdom of God Has Come,'" *ExpT* 48 (1936-37) 184-85.
Cullmann, O., *Baptism in the New Testament* (*SBT* 1; London: SCM, 1950).
Cullmann, O., "Parousieverzögerung und Urchristentum," *Vorträge und Aufsätze 1925-1962* (Tübingen: Mohr, 1966) 427-44.
Cullmann, O., "Das Gleichnis vom Salz: Zur frühesten Kommentierung eines Herrenwortes durch den Evangelisten," ibid., 192-201.

Dahl, N. A., "The Parables of Growth," *ST* 5 (1951) 132-66.
Dalman, G., *The Words of Jesus Considered in the Light of Post-Biblical Jewish Writings and the Aramaic Language* (Edinburgh: Clark, 1902).
Dalman, G., *Jesus-Jeshua: Studies in the Gospels* (London: S.P.C.K., 1929).
Daube, D., *The New Testament and Rabbinic Judaism* (London: Athlone, 1956).
Delling, G., "*Telos*, etc.," *TWNT* 8, 50-88.
Delorme, J., "Aspects doctrinaux du second Evangile," *ETL* 43 (1967) 74-99.
Descamps, A., "Du discours de Marc IX,33-50 aux paroles de Jésus," *La formation des évangiles: Problème synoptique et Formgeschichte* (RechBib 1; Bruges: Desclée de Brouwer, 1957) 152-77.
Dibelius, M., *From Tradition to Gospel* (New York: Scribner, n. d.).
Dockx, S., "Le récit du repas pascal: Marc 14,17-26," *Bib* 46 (1965) 445-53.
Dodd, C. H., *The Apostolic Preaching and its Developments* (London: Hodder and Stoughton, 1963).
Dodd, C. H., *According to the Scriptures: The Sub-structure of New Testament Theology* (London: Collins, 1965).
Dodd, C. H., *The Parables of the Kingdom* (rev. ed.; London: Collins, 1967).
Dodd, C. H., "'The Kingdom of God Has Come,'" *ExpT* 48 (1936-37) 138-42.
Duncan, G. S., *Jesus, Son of Man: Studies Contributory to a Modern Portrait* (London: Nisbet, 1948).
Dupont, J., *Les béatitudes* (Bruges: Abbaye de Saint-André, 1954).
Dupont, J., "'Ceci est mon corps,' 'Ceci est mon sang,'" *NRT* 80 (1958) 1025-41.
Dupont, J., "Le chapitre des paraboles," *NRT* 89 (1967) 800-20.

Dupont, J., "Les paraboles du sénevé et du levain," *NRT* 89 (1967) 897-913.
Dupont, J., "La parabole de la semence qui pousse toute seule (Mc 4, 26-29)," *RSR* 55 (1967) 367-92.

Ebeling, H. J., *Das Messiasgeheimnis und die Botschaft des Markus-Evangelisten* (Berlin: Töpelmann, 1939).
Elliger, K., *Die Propheten: Nahum, Habakuk, Zephanja, Haggai, Sacharja, Maleachi* (*ATD* 25; Göttingen: Vandenhoeck & Ruprecht, 1967).
Elliott, J. H., Review of J. Lambrecht, *Die Redaktion der Markus-Apokalypse*, *CBQ* 30 (1968) 267-69.

Farrer, A., *A Study in St. Mark* (Westminster: Dacre, 1951).
Fenton, J. C., "Paul and Mark," *Studies in the Gospels: In Memory of R. H. Lightfoot* (ed. D. E. Nineham; Oxford: Blackwell, 1955) 89-112.
Fitzmyer, J. A., "The Son of David Tradition and Matthew 22,41-46 and Parallels," *Concilium* 20 (1966) 75-87.
Flender, H., "Lehren und Verkündigung in den synoptischen Evangelien," *EvT* 25 (1965) 701-14.
Friedrich, G., "*Euangelizomai*, etc.," *TDNT* 2, 707-37.
Fuchs, E., *Hermeneutik* (3rd ed.; Bad Cannstatt: R. Müllerschön, 1963).
Fuchs, E., "Bemerkungen zur Gleichnisauslegung," *Zur Frage nach dem historischen Jesus: Gesammelte Aufsätze II* (Tübingen: Mohr, 1965) 136-42.
Fuchs, E., "The Quest of the Historical Jesus," *Studies of the Historical Jesus* (*SBT* 42; London: SCM, 1964) 11-31.
Fuchs, E., "Jesus' Understanding of Time," ibid., 104-66.
Fuchs, E., "The Theology of the New Testament and the historical Jesus," ibid., 167-90.
Fuller, R. H., *The Mission and Achievement of Jesus* (*SBT* 12; London: SCM, 1963).
Fuller, R. H., Review of G. Minette de Tillesse, *Le secret messianique dans l'évangile de Marc*, *CBQ* 31 (1969) 109-12.

Gardner, H., *The Business of Criticism* (Oxford: Clarendon, 1959).
Gealy, F. D., "The Composition of Mark IV," *ExpT* 48 (1936-37) 40-43.
George, A., "Le sens de la parabole des semailles (Mc., IV,3-9 et parallèles)," *SacPag*, 2, 163-69.
Gerhardsson, B., "The Parable of the Sower and its Interpretation," *NTS* 14 (1967-68) 165-93.
Glasson, T. F., *The Second Advent* (London: Epworth, 1947).
Gnilka, J., *Die Verstockung Israels: Isaias 6,9-10 in der Theologie der Synoptiker* (*StANT* 3; München: Kösel, 1961).
Gnilka, J., "'Parousieverzögerung' und Naherwartung in den synoptischen Evangelien und in der Apostelgeschichte," *Catholica* 13 (1959) 277-90.
Goppelt, L., "*Pinō*, etc.," *TDNT* 6, 135-60.
Gould, E. P., *A Critical and Exegetical Commentary on the Gospel according to St. Mark* (*ICC*; Edinburgh: Clark, 1955).

Grässer, E., *Das Problem der Parousieverzögerung in den synoptischen Evangelien und in der Apostelgeschichte* (2d ed.; *BZNW* 22; Berlin: Töpelmann, 1960).

Grant, F. C., "The Gospel according to St. Mark," *The Interpreter's Bible*, 7 (New York: Abingdon, 1951).

Grobel, K., "Idiosyncracies of the Synoptists in Their Pericope Introductions," *JBL* 59 (1940) 405-10.

Grundmann, W., *Das Evangelium nach Matthäus* (Theologischer Handkommentar zum NT, 1; Berlin: Evangelische Verlagsanstalt, 1968).

Grundmann, W., *Das Evangelium nach Markus* (Theologischer Handkommentar zum NT, 2; Berlin: Evangelische Verlagsanstalt, 1965).

Grundmann, W., *Das Evangelium nach Lukas* (Theologischer Handkommentar zum NT, 3; Berlin: Evangelische Verlagsanstalt, 1961).

Grundmann, W., "*Dynamai*, etc.," *TDNT* 2, 284-317.

Grundmann, W., "*Dechomai*, etc.," *TDNT* 2, 50-59.

Güttgemanns, E., *Offene Fragen zur Formgeschichte des Evangeliums: Eine methodologische Skizze der Grundlagenproblematik der Form- und Redaktionsgeschichte* (*BEvT* 54; München: Kaiser, 1970).

Guthrie, D., *New Testament Introduction 1: Gospels and Acts* (London: Tyndale, 1965).

Haenchen, E., *Die Botschaft des Thomas-Evangeliums* (Theologische Bibliothek Töpelmann; Berlin: Töpelmann, 1961).

Haenchen, E., *Der Weg Jesu: Eine Erklärung des Markus-Evangeliums und der kanonischen Parallelen* (2d ed.; Berlin: de Gruyter, 1968).

Haenchen, E., "Leidensnachfolge," *Die Bibel und wir: Gesammelte Aufsätze* (Tübingen: Mohr, 1968), 2, 102-34.

Hahn, F., *Mission in the New Testament* (*SBT* 47; London: SCM, 1965).

Hahn, F., *The Titles of Jesus in Christology: Their History in Early Christianity* (London: Lutterworth, 1969).

Hahn, F., "Die alttestamentlichen Motive in der urchristlichen Abendmahlsüberlieferung," *EvT* 27 (1967) 337-74.

Hamilton, N. Q., "Resurrection Tradition and the Composition of Mark," *JBL* 84 (1965) 415-21.

Harder, G., "Das Gleichnis von der selbstwachsenden Saat: Mark. 4,26-29," *Theologia Viatorum* 1 (1948-49) 51-70.

Hartman, Lars, *Prophecy Interpreted: The Formation of Some Jewish Apocalyptic Texts and of the Eschatological Discourse Mark 13 Par.* (Coniectanea biblica: NT Series, 1: Lund: Gleerup, 1966).

Hatch, E. and H. E. Redpath, *A Concordance to the Septuagint and the Other Greek Versions of the Old Testament (Including the Apocryphal Books)* (Graz: Akademische Druck- u. Verlagsanstalt, 1954).

Hauck, F., *Das Evangelium des Markus* (Theologischer Handkommentar zum NT, 2; Leipzig: Deichert, 1931).

Hauck, F., "*Parabolē*," *TDNT* 5, 744-61.

Hawkins, J. C., *Horae Synopticae: Contributions to the Study of the Synoptic Problem* (Oxford: Clarendon, 1909).

Hermaniuk, M., *La parabole évangélique: Enquête exégétique et critique* (Bruges: Desclée de Brouwer, 1947).

Higgins, A. J. B., *The Lord's Supper in the New Testament* (SBT 6; London: SCM, 1964).

Hoskyns, E. and N. Davey, *The Riddle of the New Testament* (London: Faber & Faber, 1958).

Howard, J. K., "Our Lord's Teaching Concerning His Parousia: A Study in the Gospel of Mark," *EvQ* 38 (1966) 52-58, 68-75, 150-157.

Huck, A. and H. Lietzmann, *Synopse der drei ersten Evangelien* (10th ed.; Tübingen: Mohr, 1950).

Hunter, A. M., "Interpreting the Parables," *Int* 14 (1960) 70-84, 167-85, 315-32, 440-54.

Hutton, W. R., "The Kingdom of God Has Come," *ExpT* 64 (1952-53) 89-91.

Iersel, B. M. F. van, "Fils de David et Fils de Dieu," *La venue du Messie: messianisme et eschatologie* (RechBib 6; Bruges: Desclée de Brouwer, 1962) 113-32.

Iersel, B. M. F. van, "La vocation de Lévi (Mc., II,13-17, Mt., IX,9-13, Lc., V,27-32): Traditions et rédactions," *De Jésus aux évangiles: Tradition et rédaction dans les évangiles synoptiques* (ed. I. de la Potterie; Gembloux: Duculot, 1967) 212-32.

Jeremias, J., *Heiligengräber in Jesu Umwelt (Mt. 23,29; Lk. 11,47): Eine Untersuchung zur Volksreligion der Zeit Jesu* (Göttingen: Vandenhoeck & Ruprecht, 1958).

Jeremias, J., *Jesus' Promise to the Nations* (SBT 24; London: SCM, 1958).

Jeremias, J., *Infant Baptism in the First Four Centuries* (Library of History and Doctrine; London: SCM, 1960).

Jeremias, J., *The Parables of Jesus* (NT Library; London: SCM, 1963).

Jeremias, J., *Unknown Sayings of Jesus* (London: S.P.C.K., 1964).

Jeremias, J., *The Eucharistic Words of Jesus* (NT Library; London: SCM, 1966).

Jeremias, J., *Jerusalem in the Time of Jesus: An Investigation into Economic and Social Conditions during the New Testament Period* (London: SCM, 1969).

Jeremias, J., "Characteristics of the *ipsissima vox Jesu*," *The Prayers of Jesus* (SBT 2/6; London: SCM, 1967) 108-15.

Jeremias, J., "The Lord's Prayer in the Light of Recent Research," ibid., 82-107.

Jeremias, J., "Die älteste Schicht der Menschensohn-Logien," *ZNW* 58 (1967) 159-72.

Jeremias, J., "*Hadēs*," *TDNT* 1, 146-49.

Jeremias, J., "*Geenna*," *TDNT* 1, 657-58.

Johnson, A. R., "*Māšāl*," *Wisdom in Israel and the Ancient Near East, presented* to H. H. Rowley (*VTSup* 3; eds. M. Noth and D. W. Thomas; Leiden: Brill, 1960) 162-69.

Johnson, S. E., *A Commentary on the Gospel according to St. Mark* (Black's NT Commentaries; London: Black, 1960).

Jülicher, A., *Die Gleichnisreden Jesu* (Zwei Teile in einem Band; Darmstadt: Wissenschaftliche Buchgesellschaft, 1963).

Jüngel, E., *Paulus und Jesus: Eine Untersuchung zur Präzisierung der Frage nach dem Ursprung der Christologie* (Hermeneutische Untersuchungen zur Theologie 2; Tübingen: Mohr, 1967).

Käsemann, E., "The Problem of the Historical Jesus," *Essays on New Testament Themes* (SBT 41; London: SCM, 1964) 15-47.

Käsemann, E., "Blind Alleys in the 'Jesus of History' Controversy," *New Testament Questions of Today* (NT Library; London: SCM, 1969) 23-65.

Kahlefeld, H., *Parables and Instructions in the Gospels* (Montreal: Palm, 1966).

Kaiser, O., *Der Prophet Jesaja: Kap. 1-12* (ATD 17; Göttingen: Vandenhoeck & Ruprecht, 1963).

Karnetzki, M., "Die galiläische Redaktion im Markusevangelium," *ZNW* 52 (1961) 238-72.

Keck, L. E., "Mark 3,7-12 and Mark's Christology," *JBL* 84 (1965) 341-58.

Keck, L. E., "The Introduction to Mark's Gospel," *NTS* 12 (1965-66) 352-70.

Kee, H. C., "The Terminology of Mark's Exorcism Stories," *NTS* 14 (1967-68) 232-46.

Kilmartin, E. J., *The Eucharist in the Primitive Church* (Englewood Cliffs, N.J.: Prentice-Hall, 1965).

Kilpatrick, G. D., "The Punctuation of John VII. 37-38," *JTS* 11 (1960) 340-42.

Klostermann, E., *Das Markus-Evangelium* (HNT 3; Tübingen: Mohr, 1950).

Klostermann, E. and H. Gressmann, *Das Lukasevangelium* (HNT 4; Tübingen: Mohr, 1919).

Knox, W. L., *The Sources of the Synoptic Gospels I-II* (ed. H. Chadwick; Cambridge: Cambridge University, 1953).

Koch, R., "Die Wertung des Besitzes im Lucasevangelium," *Bib* 38 (1957) 151-69.

Kuby, A., "Zur Konzeption des Markus-Evangeliums," *ZNW* 49 (1958) 52-64.

Kümmel, W. G., *Promise and Fulfilment: The Eschatological Message of Jesus* (SBT 23; London: SCM, 1961).

Kümmel, W. G., "Die Eschatologie der Evangelien," *Heilsgeschehen und Geschichte: Gesammelte Aufsätze 1933-1964* (eds. E. Grässer et al.; Marburger theologische Studien 3; Marburg: Elwert, 1965).

Kümmel, W. G., "Jesus und die Anfänge der Kirche," ibid., 289-309.

Kümmel, W. G., "Die Naherwartung in der Verkündigung Jesu," ibid., 457-70.

Kuhn, H.-W., "Das Reittier Jesu in der Einzugsgeschichte des Markusevangeliums," *ZNW* 50 (1959) 82-91.

Kuss, O., "Zum Sinngehalt des Doppelgleichnisses vom Senfkorn und Sauerteig," *Bib* 40 (1959) 641-53.

Lacan, M.-F., "Conversion et royaume dans les évangiles synoptiques," *LumVie* 9 (1960) 25-47.

Ladd, G. E., *Jesus and the Kingdom: The Eschatology of Biblical Realism* (New York: Harper & Row, 1964).

Lagrange, M.-J., *Evangile selon Saint Marc* (Paris: Gabalda, 1920).

Lagrange, M.-J., *Evangile selon Saint Luc* (Paris: Gabalda, 1921).

Lagrange, M.-J., "Le but des paraboles d'après l'évangile selon Saint Marc," *RB* 7 (1910) 5-35.

Lambrecht, J., *Die Redaktion der Markus-Apokalypse: Literarische Analyse und Strukturuntersuchung* (AnBib 28; Rom: Bibelinstitut, 1967).

Leaney, R. A. C., *A Commentary on the Gospel according to St. Luke* (Black's NT Commentaries; London: Black, 1958).

Lebeau, P., *Le vin nouveau du royaume: Etude exégétique et patristique sur la parole eschatologique de Jésus à la cène* (Museum Lessianum, section biblique 5; Bruges: Desclée de Brouwer, 1966).

Leenhardt, F.-J., *Le sacrement de la sainte cène* (Série théologique de l' "actualité protestante"; Neuchâtel: Delachaux et Niestlé, 1948).

Légasse, S., *L'appel du riche (Marc 10,17-31 et parallèles): Contribution à l'étude des fondements scripturaires de l'état religieux* (VS, collection annexe 1; Paris: Beauchesne, 1966).

Légasse, S., "Jésus a-t-il annoncé la conversion finale d'Israël? (à propos de Marc x 23-27)," *NTS* 10 (1963-64) 480-87.

Léon-Dufour, X., "The Synoptic Gospels," *Introduction to the New Testament* (eds. A. Robert and A. Feuillet; New York: Desclée, 1965) 140-324.

Levy, J., *Wörterbuch über Talmud und Mischna* (Berlin: B. Harz, 1924).

Lightfoot, R. H., *The Gospel Message of St. Mark* (Oxford: Clarendon, 1952).

Lindars, B., *New Testament Apologetic: The Doctrinal Significance of the Old Testament Quotations* (London: SCM, 1961).

Linnemann, E., *Gleichnisse Jesu: Einführung und Auslegung* (Göttingen: Vandenhoeck & Ruprecht, 1966).

Lövestam, E., *Son and Saviour: A Study of Acts 13,32-37: With an Appendix: 'Son of God' in the Synoptic Gospels* (ConNt 18; Lund: Gleerup, 1961).

Lövestam, E., *Spiritus blasphemia: Eine Studie zu Mk 3,28f par Mt 12,31f, Lk 12,10* (Scripta minora regiae societatis humaniorum litterarum lundensis, 1966-67/1; Lund: Gleerup, 1968).

Lohmeyer, E., *Das Evangelium des Markus: Nach dem Handexemplar des Verfassers durchgesehene Ausgabe mit Ergänzungsheft* (Meyer, 1/2; Göttingen: Vandenhoeck & Ruprecht, 1963).

Lohmeyer, E., *Kultus und Evangelium* (Göttingen: Vandenhoeck & Ruprecht, 1942).

Lohmeyer, E., *Gottesknecht und Davidssohn* (FRLANT 61; Göttingen: Vandenhoeck & Ruprecht, 1953).

Lohmeyer, E., "Vom Sinn der Gleichnisse Jesu," *Urchristliche Mystik: Neutestamentliche Studien* (Darmstadt: Wissenschaftliche Buchgesellschaft, 1958) 123-57.

Lohse, E., "Hosianna," *NovT* 6 (1963) 113-19.
Lohse, E., "*Huios Dauid*," *TWNT* 8, 482-92.
Lubac, H. de, *Aspects of Buddhism* (New York: Sheed and Ward, 1954).
Lundström, G., *The Kingdom of God in the Teaching of Jesus: A History of Interpretation from the Last Decades of the Nineteenth Century to the Present Day* (Edinburgh: Oliver and Boyd, 1963).
Luz, U., "Das Geheimnismotiv und die markinische Christologie," *ZNW* 56 (1965) 9-30.
Lyonnet, S., *Quaestiones in epistulam ad Romanos: Prima series* (2d ed.; Rome: Biblical Institute, 1962).

McLoughlin, S., "Les accords mineurs Mt-Lc contre Mc et le problème synoptique: Vers la théorie de deux sources," *ETL* 43 (1967) 17-40.
Mally, E. J., "The Gospel according to Mark," *JBC* 2, 21-61.
Manson, T. W., *The Teaching of Jesus* (Cambridge: Cambridge University, 1963).
Manson, T. W., *Ethics and the Gospel* (London: SCM, 1966).
Manson, W., *Jesus the Messiah* (London: Hodder and Stoughton, 1961).
Manson, W., *The Gospel of Luke* (The Moffat NT Commentary; London: Hodder and Stoughton, 1955).
Manson, W., "The Purpose of the Parables: A Re-examination of St. Mark IV. 10-12," *ExpT* 68 (1956-57), 132-35.
Marxsen, W., *Mark the Evangelist* (Nashville: Abingdon, 1969).
Marxsen, W., *The Lord's Supper as a Christological Problem* (*FBBS* 25; Philadelphia: Fortress, 1970).
Marxsen, W., *The Resurrection of Jesus of Nazareth* (London: SCM, 1970).
Marxsen, W., "Redaktionsgeschichtliche Erklärung der sogenannten Parabeltheorie des Markus," *Der Exeget als Theologe: Vorträge zum Neuen Testament* (Gütersloh: Mohn, 1968) 13-28.
Masson, C., *Les paraboles de Marc IV* (Cahiers théologiques de l'actualité protestante, 11; Neuchâtel: Delachaux & Niestlé, 1945).
Mastin, B. A., "The Date of the Triumphal Entry," *NTS* 16 (1969-70) 76-82.
Maurer, C., "Das Messiasgeheimnis des Markusevangeliums," *NTS* 14 (1967-68) 515-26.
Mauser, U. W., *Christ in the Wilderness* (*SBT* 39; London: SCM, 1963).
Meinertz, M., " 'Dieses Geschlecht' im Neuen Testament," *BZ* n.s. 1 (1957) 283-89.
Merk, A., *Novum Testamentum graece et latine* (9th ed.; Rome: Biblical Institute, 1964).
Meye, R. P., *Jesus and the Twelve: Discipleship and Revelation in Mark's Gospel* (Grand Rapids: Eerdmans, 1968).
Meye, R. P., "Mark 4,10: 'Those about Him with the Twelve,' " *SE* 2 (*TU* 87), 211-18.
Meye, R. P., "Messianic Secret and Messianic Didache in Mark's Gospel," *Oikonomia: Heilsgeschichte als Thema der Theologie: Festschrift für Oscar Cullmann* (ed. F. Christ; Hamburg-Bergstedt: H. Reich, 1967) 57-68.

Michaelis, W., "Die Davidssohnschaft Jesu als historisches und kerygmatisches Problem," *Der historische Jesus und der kerygmatische Christus: Beiträge zum Christusverständnis in Forschung und Verkündigung* (eds. H. Ristow and K. Matthiae; Berlin: Evangelische Verlagsanstalt, 1960) 317-30.

Minear, P. S., "The Needle's Eye: A Study in Form Criticism," *JBL* 61 (1942) 157-69.

Minette de Tillesse, G., *Le secret messianique dans l'évangile de Marc* (*LD* 47; Paris: Cerf, 1968).

Montefiore, C. G., *The Synoptic Gospels I-II* (New York: Ktav, 1968).

Montefiore, H. and H. E. W. Turner, *Thomas and the Evangelists* (*SBT* 35; London: SCM, 1962).

Müller, H.-P., "Die Verklärung Jesu," *ZNW* 51 (1960) 56-64.

Mussner, F., *Christ and the End of the World: A Biblical Study in Eschatology* (Contemporary Catechetics Series; South Bend: University of Notre Dame, 1965).

Mussner, F., "Gleichnisauslegung und Heilsgeschichte: Dargetan am Gleichnis von der selbstwachsenden Saat (Mk 4,26-29)," *TTZ* 64 (1955) 257-66.

Mussner, F., "Die Bedeutung von Mk 1,14f für die Reichsgottesverkündigung," *TTZ* 66 (1957) 257-75.

Mussner, F., "Gottesherrschaft und Sendung Jesu nach Mk 1,14f," *Praesentia Salutis: Gesammelte Studien zu Fragen und Themen des Neuen Testaments* (Düsseldorf: Patmos, 1967) 81-98.

Mussner, F., "1Q Hodajoth und das Gleichnis vom Senfkorn (Mk 4,30-32 Par)," *BZ* 4 (1960) 128-30.

Mussner, F., "'Evangelium' und 'Mitte des Evangeliums': Ein Beitrag zur Kontroverstheologie," *Gott in Welt: Festgabe für K. Rahner* (Freiburg: Herder, 1964), 1, 492-514; also in *Praesentia Salutis*, 159-77.

Nagel, W., "Der historische Jesus im Abendmahl der Kirche," *Der historische Jesus und der kerygmatische Christus*, 543-53.

Nauck, W., "Salt as a Metaphor in Instructions for Discipleship," *ST* 6 (1952) 165-78.

Nauck, W., "Die Bedeutung des leeren Grabes für den Glauben an den Auferstandenen," *ZNW* 47 (1956) 243-76.

Neirynck, F., "The Tradition of the Sayings of Jesus: Mark 9,33-50," *Concilium* 20 (1966) 62-74.

Nestle, E. and K. Aland, *Novum Testamentum graece* (Stuttgart: Württembergische Bibelanstalt, 1963).

Neuhäusler, E., *Anspruch und Antwort Gottes: Zur Lehre von den Weisungen innerhalb der synoptischen Jesusverkündigung* (Düsseldorf: Patmos, 1962).

Niederwimmer, K., "Johannes Markus und die Frage nach dem Verfasser des zweiten Evangeliums," *ZNW* 58 (1967) 172-88.

Nineham, D. E., *The Gospel of St Mark* (Pelican Gospel Commentaries; Harmondsworth: Penguin, 1963).

Nineham, D. E., "The Order of Events in St. Mark's Gospel—an Examination of Dr. Dodd's Hypothesis," *Studies in the Gospels: In Memory of R. H. Lightfoot* (ed. D. E. Nineham; Oxford: Blackwell, 1955) 223-39.

Nisin, A., *Histoire de Jésus* (Paris: Seuil, 1961).

Oepke, A., *"Eis," TDNT* 2, 420-34.

Oepke, A., *"Kalyptō,* etc.," *TDNT* 3, 556-92.

Oepke, A., *"Mesitēs,* etc.," *TDNT* 4, 598-624.

Oepke, A., *"Pais,* etc.," *TDNT* 5, 636-54.

O'Rourke, J. J., "A Note Concerning the Use of *eis* and *en* in Mark," *JBL* 85 (1966) 349-51.

Paschen, W., *Rein und Unrein: Untersuchung zur biblischen Wortgeschichte* (*StANT* 24; München: Kösel, 1970).

Peacock, H. F., "The Theology of the Gospel of Mark," *RevExp* 55 (1958) 393-99.

Peisker, C. H., "Konsekutives *hina* in Markus 4,12," *ZNW* 59 (1958) 126-27.

Percy, E., *Die Botschaft Jesu: Eine traditionskritische und exegetische Untersuchung* (Lund: Gleerup, 1953).

Perrin, N., *The Kingdom of God in the Teaching of Jesus* (NT Library; London: SCM, 1963).

Perrin, N., *Rediscovering the Teaching of Jesus* (NT Library; London: SCM, 1967).

Pesch, R., *Naherwartungen: Tradition und Redaktion in Mk 13* (Kommentare und Beiträge zum Alten und Neuen Testament; Düsseldorf: Patmos, 1968).

Pesch, R., "Anfang des Evangeliums Jesu Christi: Eine Studie zum Prolog des Markusevangeliums (Mk 1,1-15)," *Die Zeit Jesu: Festschrift für Heinrich Schlier* (eds. G. Bornkamm and K. Rahner; Freiburg: Herder, 1970) 108-44.

Pesch, W., *Der Lohngedanke in der Lehre Jesu verglichen mit der religiösen Lohnlehre des Spätjudentums* (Münchener theologische Studien 1, Hist. Abt., Bd. 7; München: Zink, 1955).

Pesch, W., "Die sogenannte Gemeindeordnung Mt 18," *BZ* 7 (1963) 220-35.

Plummer, A., *A Critical and Exegetical Commentary on the Gospel according to St. Luke* (*ICC*; Edinburgh: Clark, 1956).

Preisker, H., *"Makran,* etc.," *TDNT* 4, 372-74.

Quesnell, Q., *The Mind of Mark: Interpretation and Method through the Exegesis of Mark 6,52* (*AnBib* 38; Rome: Biblical Institute, 1969).

Rad, G. von, *Deuteronomy* (OT Library; Philadelphia: Westminster, 1966).

Rad, G. von, *Old Testament Theology* (2 vols.; Edinburgh: Oliver and Boyd, 1962-65).

Rawlinson, A. E. J., *St. Mark* (Westminster Commentaries; London: Methuen, 1956).

Rengstorf, K. H., *Das Evangelium nach Lukas* (*NTD* 3; Göttingen: Vandenhoeck & Ruprecht, 1949).

Rengstorf, K. H., *"Didaskō,* etc.," *TDNT* 2, 135-65.

Reploh, K.-G., *Markus—Lehrer der Gemeinde: Eine redaktionsgeschichtliche Studie zu den Jüngerperikopen des Markus-Evangeliums* (Stuttgarter biblische Monographien, 9; Stuttgart: Katholisches Bibelwerk, 1969).

Richardson, A., *An Introduction to the Theology of the New Testament* (London: SCM, 1961).

Riddle, W. D., "Mark 4,1-34: The Evolution of a Gospel Source," *JBL* 56 (1937) 77-90.

Riesenfeld, H., *Jésus transfiguré: L'arrière-plan du récit évangélique de la transfiguration de Notre-Seigneur* (Acta seminarii neotestamentici uppsaliensis, 16; Copenhagen: Munksgaard, 1947).

Rigaux, B., "Révélation des mystères et perfection à Qumrân et dans le Nouveau Testament," *NTS* 4 (1957-58) 237-62.

Rigaux, B., "La seconde venue de Jésus," *La venue du messie: messianisme et eschatologie* (*RechBib* 6; Bruges: Desclée de Brouwer, 1962) 173-216.

Ringgren H., *The Faith of Qumran: Theology of the Dead Sea Scrolls* (Philadelphia: Fortress, 1963).

Robinson, J. A. T., *Jesus and His Coming* (London: SCM, 1957).

Robinson, J. M., *The Problem of History in Mark* (*SBT* 21; London: SCM, 1962).

Robinson, J. M., *A New Quest of the Historical Jesus* (*SBT* 25; London: SCM, 1966).

Robinson, T. H., *The Gospel of Matthew* (Moffat NT Commentary; London: Hodder and Stoughton, 1951).

Roloff, J., "Das Markusevangelium als Geschichtsdarstellung," *EvT* 27 (1969) 73-93.

Romaniuk, K., "Repentez-vous, car le royaume des cieux est tout proche (Matt. IV. 17 par.)," *NTS* 12 (1965-66) 259-69.

Sabbe, M., "Le baptême de Jésus," *De Jésus aux évangiles: Tradition et rédaction dans les évangiles synoptiques* (ed. I. de la Potterie; Gembloux: Duculot, 1967) 184-211.

Sahlin, H., "Die Perikope vom gerasenischen Besessenen und der Plan des Markusevangeliums," *ST* 18 (1964) 159-72.

Sandmel, S., "Prolegomena to a Commentary on Mark," *JBR* 31 (1963) 294-300.

Schenke, L., *Auferstehungsverkündigung und leeres Grab: Eine traditionsgeschichtliche Untersuchung von Mk 16,1-8* (*SBS* 33; Stuttgart: Katholisches Bibelwerk, 1968).

Schierse, F. J., "Historische Kritik und theologische Exegese: Erläutert an Mk 9,1 par," *Scholastik* 29 (1954) 520-36.

Schille, G., "Das Leiden des Herrn: Die evangelische Passionstradition und ihr 'Sitz im Leben,'" *ZTK* 52 (1955) 161-205.

Schille, G., "Bemerkungen zur Formgeschichte des Evangeliums: Rahmen und Aufbau des Markus-Evangeliums," *NTS* 4 (1957-58) 1-24.

Schilling, F. A., "What Means the Saying about Receiving the Kingdom of God as a Little Child (*tēn basileian tou theou hōs paidion*)? Mk X. 15; Lk XVIII. 17," *ExpT* 77 (1964-65) 56-58.

Schlatter, A., *Das Evangelium des Lukas* (Stuttgart: Calwer, 1960).
Schmid, J., *The Gospel according to Mark* (RNT 1; Staten Island, N.Y.: Mercier, 1968).
Schmidt, K. L., *Der Rahmen der Geschichte Jesu: Literarkritische Untersuchung zur ältesten Jesusüberlieferung* (Darmstadt: Wissenschaftliche Buchgesellschaft, 1964).
Schmidt, K. L., "Basileus, etc.," *TDNT* 1, 564-93.
Schmidt, K. L. and M. A., "Pachynō, etc.," *TDNT* 5, 1022-31.
Schnackenburg, R., *God's Rule and Kingdom* (Freiburg/Montreal: Herder/Palm, 1963).
Schnackenburg, R., *The Moral Teaching of the New Testament* (Freiburg/Montreal: Herder/Palm, 1965).
Schnackenburg, R., *Das Evangelium nach Markus* (Geistliche Schriftlesung 2/1; Düsseldorf: Patmos, 1966).
Schnackenburg, R., *The Gospel according to St John I: Introduction and Commentary on Chapters 1-4* (Herder's Theological Commentary on the NT; Freiburg/Montreal: Herder/Palm, 1968).
Schnackenburg, R., "Mk 9,35-50," *Synoptische Studien: Festschrift für A. Wikenhauser* (München: Zink, 1953) 184-206.
Schnackenburg, R., "Die Vollkommenheit des Christen nach den Evangelien," *Geist und Leben* 32 (1959) 420-33.
Schnackenburg, R., "Kirche und Parousie," *Gott in Welt: Festgabe für K. Rahner* (Freiburg: Herder, 1964), 1, 551-87.
Schnackenburg, R., "Naherwartung," *LTK*² 7, 777-79.
Schneider, J., "Erchomai," *TDNT* 2, 666-84.
Schniewind, J., *Das Evangelium nach Markus* (München: Siebenstern Taschenbuch, 1968).
Schrage, W., *Das Verhältnis des Thomas-Evangeliums zur synoptischen Tradition und zu den koptischen Evangelienübersetzungen* (BZNW 29; Berlin: Töpelmann, 1964).
Schreiber, J., *Theologie des Vertrauens: Eine redaktionsgeschichtliche Untersuchung des Markusevangeliums* (Hamburg: Furche, 1967).
Schreiber, J., "Die Christologie des Markusevangeliums: Beobachtungen zur Theologie und Komposition des zweiten Evangeliums," *ZTK* 58 (1961) 154-83.
Schürmann, H., *Der Paschamahlbericht: Lk 22,(7-14.)15-18: 1. Teil einer quellenkritischen Untersuchung des lukanischen Abendmahlsberichtes Lk 22,7-38* (NTAbh 19/5; Münster: Aschendorff, 1953).
Schürmann, H., *Der Abendmahlsbericht Lucas 22,7-38 als Gottesdienstordnung, Gemeindeordnung, Lebensordnung* (Schriften zur Pädagogik und Katechetik, 9; Paderborn: Schöningh, 1963).
Schürmann, H., "Die Anfänge christlicher Osterfeier II," *TQ* 131 (1951) 414-25.
Schürmann, H., "Eschatologie und Liebesdienst in der Verkündigung Jesu," *Kaufet die Zeit aus: Beiträge zur christlichen Eschatologie: Festgabe für Th. Kampmann* (ed. H. Kirchhof; Paderborn: Schöningh, 1959) 39-71.

Schürmann, H., "Das hermeneutische Hauptproblem der Verkündigung Jesu," *Traditionsgeschichtliche Untersuchungen zu den synoptischen Evangelien* (Düsseldorf: Patmos, 1968) 13-35.

Schütz, R., "Apokalyptik, III. Altchristliche," *RGG*³ 1, 467-69.

Schulz, S., *Die Stunde der Botschaft: Einführung in die Theologie der vier Evangelisten* (Hamburg: Furche, 1967).

Schulz, S., "Markus und das Alte Testament," *ZTK* 58 (1961) 184-97.

Schulz, S., "Die Bedeutung des Markus für die Theologiegeschichte des Urchristentums," *SE* II, 135-45.

Schweizer, E., *The Good News According to Mark* (Richmond, Va.: John Knox, 1970).

Schweizer, E., "Das Herrenmahl im Neuen Testament: Ein Forschungsbericht," *TLZ* 79 (1954) 577-92; also in *Neotestamentica* (Zürich: Zwingli, 1963) 344-70.

Schweizer, E., Review of H. Schürmann, *Der Paschamahlbericht: Lk 22,(7-14.) 15-18*, *TLZ* 80 (1955) 156-57.

Schweizer, E., "Anmerkungen zur Theologie des Markus," *Neotestamentica*, 93-104.

Schweizer, E., "Mark's Contribution to the Quest of the Historical Jesus," *NTS* 10 (1963-64) 421-32.

Schweizer, E., "Die theologische Leistung des Markus," *EvT* 24 (1964) 337-55.

Schweizer, E., "Zur Frage des Messiasgeheimnisses bei Markus," *ZNW* 56 (1965) 1-8.

Schweizer, E., "*Huios*, Neues Testament," *TWNT* 8, 364-95.

Schweizer, E., *The Lord's Supper according to the New Testament* (*FBBS* 18; Philadelphia: Fortress, 1967).

Siegman, E. F., "Teaching in Parables," *CBQ* 23 (1961) 161-81.

Simpson, R. T., "The Major Agreements of Matthew and Luke against Mark," *NTS* 12 (1965-66) 273-84.

Sjöberg, E., *Der verborgene Menschensohn in den Evangelien* (Lund: Gleerup, 1955).

Sloyan, G. S., " 'Primitive' and 'Pauline' Concepts of the Eucharist," *CBQ* 23 (1961) 1-13.

Soden, H. von, *Griechisches Neues Testament* (Göttingen: Vandenhoeck & Ruprecht, 1913).

Soiron, T., *Die Logia Jesu* (*NTAbh* 6/4; Münster: Aschendorff, 1916).

Souček, J. B., "Salz der Erde und Licht der Welt: Zur Exegese von Matth. 5, 13-16," *TZ* 19 (1963) 169-79.

Stählin, G., "*Skandalon*, etc.," *TWNT* 7, 338-58.

Stewart, R. A., "The Parable Form in the Old Testament and the Rabbinic Literature," *EvQ* 36 (1964) 133-47.

Strack, H. L. and P. Billerbeck, *Kommentar zum Neuen Testament aus Talmud und Midrasch I-VI* (München: Beck, 1922-65).

Strecker, G. "Zur Messiasgeheimnistheorie im Markusevangelium," *SE* III, 87-104.

Strecker, G., "Die Leidens- und Auferstehungsvoraussagen im Markusevangelium," *ZTK* 64 (1967) 16-39.
Strecker, G., "Die historische und theologische Problematik der Jesusfrage," *EvT* 29 (1969) 453-76.
Streeter, B. H., *The Four Gospels: A Study of Origins* (London: Macmillan, 1956).
Suhl, A., *Die Funktion der alttestamentlichen Zitate und Anspielungen im Markusevangelium* (Gütersloh: Mohn, 1965).
Sundwall, J., *Die Zusammensetzung des Markusevangeliums* (Acta academiae aboensis, humaniora, 9/2; Åbo: Åbo Akademi, 1934).
Swete, H. B., *The Gospel according to St. Mark* (London: Macmillan, 1898).

Tasker, R. V. G., *The Greek New Testament (Being the Text Translated in the New English Bible)* (Oxford/Cambridge: Cambridge University, 1964).
Taylor, V., *The Gospel according to St. Mark* (2d ed.; London: Macmillan, 1966).
Thackeray, H. St. John, *The Septuagint and Jewish Worship: A Study in Origins* (2d ed.; London: Oxford, 1923).
Theissing, J., *Die Lehre Jesu von der ewigen Seligkeit* (Breslau: Müller & Seiffert, 1940).
Thurian, M., *L'eucharistie: Mémorial du Seigneur: Sacrifice d'action de grâce et d'intercession* (2d ed.; Collection communauté de Taizé; Neuchâtel: Delachaux et Niestlé, 1963).
Tillard, J. M. R., "L'Eucharistie, sacrement de l'espérance ecclésiale," *NRT* 83 (1961) 561-92.
Tischendorf, C., *Novum Testamentum graece* (Graz: Akademische Druck- und Verlagsanstalt, 1965).
Tödt, H. E., *The Son of Man in the Synoptic Tradition* (NT Library; London: SCM, 1965).
Trilling, W., *Das wahre Israel: Studien zur Theologie des Matthäus-Evangeliums* (3rd ed.; StANT 10; München: Kösel, 1964).
Trilling, W., *Christusverkündigung in den synoptischen Evangelien: Beispiele gattungsgemässer Auslegung* (Biblische Handbibliothek, 4; München: Kösel, 1969).
Trocmé, E., *La formation de l'évangile selon Marc* (Études d'histoire et de philosophie religieuses, 57; Paris: Presses Universitaires de France, 1963).
Trocmé, E., "Marc 9,1: prédiction ou réprimande?" *SE* II, 259-65.
Trocmé, E., "Pour un Jésus public: les évangélistes Marc et Jean aux prises avec l'intimisme de la tradition," *Oikonomia: Heilsgeschichte als Thema der Theologie. Festschrift für Oscar Cullmann* (ed. F. Christ; Hamburg-Bergstedt: H. Reich Evang. Verlag, 1967) 42-50.
Turner, C. H., "Marcan Usage: Notes, Critical and Exegetical, on the Second Gospel," *JTS* 25 (1923-24) 377-86; 26 (1924-25) 12-20, 145-56, 225-40, 337-46; 27 (1925-26) 58-62; 28 (1926-27) 9-30, 349-62; 29 (1927-28) 275-89, 346-61.
Turner, N., "The Style of St Mark's Eucharistic Words," *JTS* 8 (1957) 108-11.

Tyson, J. B., "The Blindness of the Disciples in Mark," *JBL* 80 (1961) 261-68.

Vaganay, L., "Le schématisme du discours communautaire à la lumière de la critique des sources," *RB* 60 (1953) 203-44.

Vermes, G., *The Dead Sea Scrolls in English* (Harmondsworth: Penguin, 1965).

Via, D. O., Jr., "Matthew on the Understandability of the Parables," *JBL* 84 (1965) 430-32.

Vielhauer, P., "Erwägungen zur Christologie des Markusevangeliums," *Zeit und Geschichte: Dankesgabe an Rudolf Bultmann* (Tübingen: Mohr, 1964) 155-69; also in *Aufsätze zum Neuen Testament* (Theologische Bücherei 31; München: Kaiser, 1965) 199-214.

Vielhauer, P., "Ein Weg zur neutestamentlichen Christologie? Prüfung der Thesen Ferdinand Hahn's," ibid., 141-98.

Vögtle, A., "Exegetische Erwägungen über das Wissen und Selbstbewusstsein Jesu." *Gott in Welt: Festgabe für Karl Rahner* (Freiburg: Herder, 1964) 1, 608-67.

Vogels, H., "Mk 14,25 und Parallelen," *Vom Wort des Lebens: Festschrift für Max Meinertz* (*NTAbh*, 1. Ergänzungsband; Münster: Aschendorff, 1951) 93-104.

Vogels, H. J., *Novum Testamentum graece et latine* (Düsseldorf: Schwann, 1922).

Vogt, E., " 'Mysteria' in textibus Qumran," *Bib* 37 (1956) 247-57.

Walter, N., "Zur Analyse von Mc 10,17-31," *ZNW* 53 (1962) 206-18.

Walter, N., "Tempelzerstörung und synoptische Apokalypse," *ZNW* 57 (1966) 38-49.

Walther, G., *Jesus, das Passalamm des neuen Bundes: Der Zentralgedanke des Herrenmahles* (Gütersloh: Bertelsmann, 1950).

Weeden, T. J., "The Heresy that Necessitated Mark's Gospel," *ZNW* 59 (1968) 145-59.

Weiser, A., *The Psalms* (OT Library; Philadelphia: Westminster, 1962).

Weiser, A., *Die Propheten: Hosea, Joel, Amos, Obadja, Micha* (ATD 24; Göttingen: Vandenhoeck & Ruprecht, 1967).

Weiss, J., "Zum reichen Jüngling Mk 10,13-27," *ZNW* 11 (1910) 79-83.

Weiss, K., "Ekklesiologie, Tradition und Redaktion in der Jüngerunterweisung, Mark. 8,27-10,52," *Der historische Jesus und der kerygmatische Christus: Beiträge zum Christusverständnis in Forschung und Verkündigung* (eds. H. Ristow and K. Matthiae; Berlin: Evangelische Verlagsanstalt, 1960) 414-38.

Wellhausen, J., *Das Evangelium Marci* (2d ed.; Berlin: Reimer, 1909).

Werner, M., *Der Einfluss paulinischer Theologie im Markusevangelium* (BZNW 1; Giessen: Töpelmann, 1923).

Westermann, C., "Sprache und Struktur der Prophetie Deuterojesajas," *Forschung am Alten Testament: Gesammelte Studien* (Theologische Bücherei, 24; München: Kaiser, 1964) 92-170.

Westermann, C., *Isaiah 40-66* (OT Library; Philadelphia: Westminster, 1969).

Westermann, C., "Vergegenwärtigung der Geschichte in den Psalmen," *Forschung am Alten Testament: Gesammelte Studien*, 306-335.
Wilder, A. N., *Eschatology and Ethics in the Teaching of Jesus* (New York: Harper, 1950).
Wilkens, W., "Die Redaktion des Gleichniskapitels Mark. 4 durch Matth.," *TZ* 20 (1964) 305-27.
Wilkens, W., "Zur Frage der literarischen Beziehung zwischen Matthäus und Lukas," *NovT* 8 (1966) 48-57.
Winandy, J., "Le logion de l'ignorance (Mc., XIII,32; Mt., XXIV,36)," *RB* 75 (1968) 63-79.
Windisch, H., *Der Sinn der Bergpredigt* (2d ed.; Untersuchungen zum Neuen Testament, 16; Leipzig: Hinrichs, 1937).
Windisch, H., "Die Verstockungsidee in Mc 4,12 und das kausale *hina* der späteren Koine," *ZNW* 26 (1927) 203-9.
Windisch, H., "Die Sprüche vom Eingehen in das Reich Gottes," *ZNW* 27 (1928) 163-92.
Winter, P., "Marginal Notes on the Trial of Jesus II," *ZNW* 50 (1959) 221-51.
Wrede, W., *Das Messiasgeheimnis in den Evangelien: Zugleich ein Beitrag zum Verständnis des Markusevangeliums* (Göttingen: Vandenhoeck & Ruprecht, 1963).

Zerwick, M., *Untersuchungen zum Markus-Stil: Ein Beitrag zur stilistischen Durcharbeitung des Neuen Testaments* (Scripta pontificii instituti biblici; Rome: Biblical Institute, 1937).
Zerwick, M., *Biblical Greek* (Scripta pontificii instituti biblici, 114; Rome: Biblical Institute, 1963).
Zimmerli, W., "Das Gotteswort des Ezechiel," *Gottes Offenbarung: Gesammelte Aufsätze zum Alten Testament* (Theologische Bücherei, 19; München: Kaiser, 1963) 133-47.
Zimmerli, W., "Der 'neue Exodus' in der Verkündigung der beiden grossen Exilspropheten," *Gottes Offenbarung*, 192-204.
Zimmermann, H., " 'Mit Feuer gesalzen werden'. Eine Studie zu Mk 9,49," *TQ* 139 (1959) 28-39.
Zirker, H., *Die kultische Vergegenwärtigung der Vergangenheit in den Psalmen* (*BBB* 20; Bonn: Hanstein, 1964).

INDEX OF MODERN AUTHORS

Aland, K., 3, 46, 106, 122, 123, 159, 183, 203
Albright, W. F., 10
Arndt, W. F., 26, 59, 78, 106, 115, 125, 162, 187
Asting, R., 7, 12, 13, 14, 22, 24

Baltensweiler, H., 113, 116, 117, 121, 122
Barrett, C. K., 66, 80, 206
Barth, G., 88
Bartsch, H.-W., 213
Beare, F. W., 16, 49, 93, 108, 127, 128, 137, 156, 159, 183, 201, 204, 237
Beasley-Murray, G. R., 110, 222, 223, 229
Begrich, J., 220, 221
Behm, J., 5, 148, 196
Benoit, P., 185, 186, 187, 193
Berkey, R. F., 16
Best, E., 8, 43, 159, 179, 243
Betz, J., 183, 194, 196, 199
Beyer, K., 176
Beyer, W., 66
Bieler, L., 95
Bieneck, J., 96
Black, M., 3, 15, 46, 82, 106, 107, 122, 123, 125, 126, 152, 156, 157, 159, 173, 176, 183, 203
Blank, J., 76
Blass, F., 33, 68, 126, 176, 187, 204
Blinzler, J., 241, 242, 243
Boobyer, G. H., 47, 52, 73, 74, 92, 101, 120, 206, 236, 238
Born, A. van den, 194
Bornkamm, G., 4, 25, 29, 44, 88, 92, 93, 132, 138, 141, 142, 148, 165, 177, 178, 179, 180, 184, 195, 196, 204, 207, 209, 210, 211, 212
Bosch, D., 12, 21
Bousset, W., 97, 239
Brandon, S. G. F., 70, 224, 228
Bratcher, R. G., 239
Brown, R. E., 76, 87, 92, 147, 215, 216, 217
Bruce, F. F., 92
Bultmann, R., 4, 5, 6, 8, 17, 26, 33, 35, 36, 42, 43, 47, 49, 52, 54, 56, 57, 58, 73, 74, 80, 81, 96, 100, 103, 107, 108, 118, 119, 123, 127, 128, 129, 137, 138, 139, 142, 148, 152, 153, 154, 155, 156, 157, 160, 162, 164, 165, 172, 173, 175, 177, 178, 183, 187, 195, 196, 201, 204, 205, 207, 210, 231, 232, 233, 238, 241
Burger, C., 11, 43, 44
Burkill, T. A., 17, 24, 38, 41, 42, 47, 53, 55, 68, 74, 92, 93, 96, 97, 101, 165, 184, 191, 208, 210, 235, 240
Burkitt, F. C., 36

Campbell, J. Y., 15, 210
Cerfaux, L., 51, 53, 82, 87, 92
Childs, B. S., 10
Clark, K. W., 15, 16, 210
Clark, N., 183, 184, 189, 192
Conzelmann, H., 25, 81, 88, 95, 179, 198, 207, 210, 211, 212, 224, 226, 227, 228, 229
Coutts, J., 53, 54, 84, 104
Cranfield, C. E. B., 17, 18, 20, 26, 33, 35, 38, 42, 49, 52, 60, 67, 76, 82, 92, 93, 100, 102, 120, 125, 128, 129, 131, 137, 148, 152, 153, 154, 156, 165, 169, 173, 175, 177, 179, 191, 196, 203, 209, 229, 234, 237, 238
Creed, J. M., 15, 129, 184, 210
Cullmann, O., 154, 175, 213

Dahl, N. A., 100, 108, 110, 111, 118, 119, 130, 131, 132
Dalman, G., 35, 139, 143, 187, 188, 192, 194, 196, 208, 226
Daube, D., 44, 178, 179
Davey, N., 128
Debrunner, A., 33, 68, 126, 176, 187, 204
Delling, G., 226
Delorme, J., 12, 15, 19, 20, 22, 49, 68, 84, 89, 92, 93
Descamps, A., 156, 157, 174
Dibelius, M., 35, 36, 78, 97, 195, 238, 241
Dockx, S., 184, 186, 191
Dodd, C. H., 15, 25, 27, 36, 76, 93, 110, 125, 127, 129, 205, 210
Duncan, G. S., 37

Dupont, J., 47, 49, 52, 67, 68, 73, 74, 107, 108, 109, 110, 113, 114, 117, 118, 119, 120, 121, 122, 123, 129, 132, 138, 139, 148, 150, 152, 153, 154, 194, 200

Ebeling, H. J., 51, 73, 74, 103
Elliger, K., 220, 221
Elliott, J. H., 222

Farrer, A., 24, 53
Fenton, J. C., 26
Fitzmyer, J. A., 39
Flender, H., 22
Friedrich, G., 9
Fuchs, E., 75, 76, 90, 100, 112, 113, 114, 117
Fuller, R. H., 61, 95, 140, 191, 193, 199, 209, 213, 219

Gardner, H., 24
Gealy, F. D., 47, 74
George, A., 100, 102
Gerhardsson, B., 100, 103
Gingrich, W. F., 26, 59, 78, 106, 115, 125, 162, 187
Glasson, T. F., 213
Gnilka, J., 47, 48, 49, 50, 52, 53, 55, 56, 63, 64, 65, 67, 68, 73, 74, 78, 79, 85, 89, 93, 94, 101, 102, 109, 111, 117, 118, 119, 129, 131, 213, 219
Goppelt, L., 183, 186, 196
Gould, E. P., 19, 21, 93, 115, 231
Grässer, E., 107, 108, 129, 132, 196, 209, 210, 211, 212, 213, 238
Grant, F. C., 76, 93, 124, 129
Gressmann, H., 123, 129
Grobel, K., 72
Grundmann, W., 18, 20, 21, 30, 36, 41, 54, 63, 71, 72, 80, 81, 92, 100, 119, 124, 127, 128, 137, 138, 144, 148, 149, 151, 159, 153, 154, 159, 172, 177, 178, 181, 185, 187, 205, 206, 207, 223, 224, 225, 226, 227, 230, 232, 234, 237, 238, 239, 241, 243
Güttgemanns, E., 27
Guthrie, D., 27

Haenchen, E., 5, 24, 29, 33, 36, 41, 44, 101, 125, 154, 180, 206, 231, 232, 233, 234, 243

Hahn, F., 4, 30, 35, 36, 38, 39, 40, 44, 103, 179, 185, 186, 187, 190, 191, 194, 196, 200, 222, 223, 232
Hamilton, N. Q., 7, 235
Harder, G., 107, 108, 109, 112, 116, 117, 122
Hartman, L., 194
Hartman, Lars, 223, 224, 230
Hatch, E., 150
Hauck, F., 75, 77, 82, 124, 127, 129
Hawkins, J. C., 32, 33, 34, 67, 128, 155, 162, 164, 185
Held, H. J., 88
Hermaniuk, M., 55, 68, 93
Higgins, A. J. B., 184, 193, 196, 199, 200
Hoskyns, E., 128
Howard, J. K., 117, 177
Huck, A., 203
Hunter, A. M., 100
Hutton, W. R., 15, 16

Iersel, B. M. F. van, 35, 38, 170

Jeremias, J., 47, 48, 49, 50, 52, 67, 68, 69, 73, 74, 75, 78, 79, 92, 93, 100, 101, 102, 103, 110, 111, 117, 119, 124, 125, 127, 128, 129, 130, 131, 137, 139, 149, 165, 176, 178, 179, 183, 184, 186, 187, 188, 189, 191, 192, 193, 194, 195, 196, 208, 212, 214, 241, 242
Johnson, A. R., 75, 77
Johnson, S. E., 17, 93, 101, 159
Jülicher, A., 47, 68, 108, 109, 110, 115, 123, 124, 128, 129, 133
Jüngel, E., 75, 76, 77, 84, 100, 107, 112, 113, 114, 129, 131, 132

Käsemann, E., 12, 27
Kahlefeld, H., 100, 113
Kaiser, O., 219, 220
Karnetzki, M., 132
Keck, L. E., 4, 5, 8, 9, 11, 19, 21, 22, 29, 63, 84, 95, 96, 224
Kee, H. C., 96, 97
Kilmartin, E. J., 183, 189, 190, 191, 199, 200
Kilpatrick, G. D., 18
Klostermann, E., 18, 34, 39, 42, 54, 93, 100, 123, 125, 129, 136, 137, 148, 159, 193, 195, 196, 204, 206, 243

Knox, W. L., 36, 44
Koch, R., 169
Kuby, A., 17, 81
Kümmel, W. G., 6, 15, 16, 35, 36, 107, 108, 117, 118, 129, 132, 139, 179, 185, 186, 191, 196, 198, 203, 205, 210, 215, 216
Kuhn, H.-W., 36
Kuss, O., 111, 118, 130, 132

Lacan, M.-F., 148
Ladd, G. E., 92, 100, 118, 119, 131, 140, 142, 179
Lagrange, M.-J., 7, 17, 18, 24, 26, 47, 54, 55, 60, 67, 68, 74, 76, 93, 100, 104, 117, 123, 124, 127, 129, 148, 177, 180, 194, 196, 203
Lambrecht, J., 4, 205, 206, 218, 222, 223, 224, 225, 226, 228, 229
Leaney, R. A. C., 130, 194
Lebeau, P., 185, 187, 189, 190, 191, 193, 194, 195, 196
Leenhardt, F.-J., 183, 186, 187, 189, 198, 199, 200
Légasse, S., 160, 161, 162, 164, 169, 170, 171
Léon-Dufour, X., 163
Levy, J., 152
Lietzmann, H., 203
Lightfoot, R. H., 1, 15, 17, 25, 26, 163, 239
Lindars, B., 42, 43, 44
Linnemann, E., 77, 100, 101
Lövestam, E., 37, 39, 66
Lohmeyer, E., 1, 5, 6, 7, 19, 20, 21, 26, 35, 38, 39, 41, 42, 44, 60, 63, 75, 76, 77, 82, 84, 93, 103, 107, 115, 116, 117, 124, 129, 137, 138, 140, 148, 151, 152, 153, 157, 160, 164, 174, 181, 189, 190, 196, 198, 199, 200, 226, 231, 232, 233, 234, 238, 239, 241, 243
Lohse, E., 35, 38, 39, 44
Lubac, H. de, 166
Lundström, G., 15, 100, 119, 140, 141, 149
Luz, U., 29, 95
Lyonnet, S., 169

McLoughlin, S., 87
Mally, E. J., 55

Manson, T. W., 47, 48, 68, 133, 140, 143, 149, 165, 177, 210, 212, 215, 216
Manson, W., 67, 84, 92, 127, 129
Marxsen, W., 4, 8, 9, 10, 11, 12, 13, 14, 17, 19, 20, 26, 27, 28, 31, 32, 47, 48, 49, 50, 52, 53, 63, 67, 68, 71, 74, 75, 77, 82, 196, 207, 222, 223, 224, 226, 227, 228, 230
Masson, C., 47, 49, 52, 73, 100, 101, 103, 124, 125, 129
Mastin, B. A., 36
Maurer, C., 61
Mauser, U. W., 19, 20, 237
Meinertz, M., 231
Merk, A., 3, 106, 122, 123, 159, 183, 203
Metzger, B. M., 3, 46, 106, 122, 123, 159, 183, 203
Meye, R. P., 63, 70, 89, 243
Michaelis, W., 38, 39
Minear, P. S., 164
Minette de Tillesse, G., 31, 33, 38, 42, 43, 50, 51, 52, 53, 59, 61, 62, 63, 64, 67, 71, 73, 75, 77, 82, 93, 95, 97, 98, 103, 162, 224, 225, 228, 229
Montefiore, C. G., 35, 36, 138, 179, 197
Montefiore, H., 125
Müller, H.-P., 238
Mussner, F., 4, 5, 6, 7, 9, 14, 17, 19, 20, 21, 22, 26, 28, 109, 117, 131, 231

Nagel, W., 198, 199
Nauck, W., 125, 241, 242
Neirynck, F., 155, 156, 173, 174
Nestle, E., 3, 46, 106, 122, 123, 159, 183, 203
Neuhäusler, E., 143, 148, 149, 164, 165, 175, 177, 178, 179
Niederwimmer, K., 32
Nineham, D. E., 5, 6, 17, 21, 27, 30, 40, 47, 52, 63, 74, 77, 80, 81, 100, 101, 115, 117, 128, 137, 138, 153, 154, 159, 164, 165, 175, 177, 180, 181, 223, 234, 236, 237, 243
Nisin, A., 215, 217

Oepke, A., 26, 96, 149
O'Rourke, J. J., 26

Paschen, W., 83

Peacock, H. F., 12
Peisker, C. H., 67
Percy, E., 107, 115, 119, 130, 136, 137, 138, 139, 144, 148, 153, 154, 162, 164, 165, 207, 210
Perrin, N., 5, 15, 119, 140, 141, 142, 148, 149, 165, 204, 205, 208, 209, 210, 213
Pesch, R., 4, 18, 19, 21, 24, 25, 42, 56, 71, 75, 144, 163, 202, 205, 206, 218, 222, 223, 224, 225, 226, 227, 228, 229, 230, 234, 237
Pesch, W., 148, 172
Plummer, A., 125, 128
Preisker, H., 180

Quesnell, Q., 64

Rad, G. von, 9, 10, 11, 76, 77, 80, 220, 221
Rawlinson, A. E. J., 93, 104, 117, 127, 129, 159
Redpath, H. E., 150
Rengstorf, K. H., 84, 129
Reploh, K.-G., 7, 21, 103, 104
Richardson, A., 154
Riddle, W. D., 51, 102
Riesenfeld, H., 26
Rigaux, B., 212, 215, 216, 217
Ringgren, H., 92
Robinson, J. A. T., 213, 214
Robinson, J. M., 7, 8, 17, 19, 21, 26, 53, 56, 59, 80, 81, 82, 101, 138, 139, 140, 210
Robinson, T. H., 139
Roloff, J., 8, 27, 28, 29, 30, 95, 98, 99, 232
Romaniuk, K., 5, 7

Sabbe, M., 19
Sahlin, H., 97
Sandmel, S., 70, 113
Schenke, L., 241, 242
Schierse, F. J., 209, 237
Schille, G., 17, 240, 241, 242
Schilling, F. A., 137
Schlatter, A., 83, 123
Schmid, J., 17, 35, 38, 39, 40, 41, 49, 82, 93, 100, 101, 103, 117, 124, 129, 137, 148, 156, 179, 224, 230, 243
Schmidt, K. L., 4, 5, 17, 19, 80, 86, 129, 154, 173
Schmidt, M. A., 80, 86
Schnackenburg, R., 15, 16, 29, 42, 71, 82, 92, 93, 100, 101, 103, 111, 117, 119, 121, 129, 131, 133, 138, 139, 140, 141, 143, 147, 148, 149, 155, 156, 165, 172, 174, 178, 179, 194, 199, 201, 208, 210, 212, 215, 218, 219, 230, 233, 235
Schneider, J., 38
Schniewind, J., 7, 22, 38, 39, 77, 82, 105, 129, 173, 179, 210, 215, 231, 243
Schrage, W., 125
Schreiber, J., 8, 19, 20, 24, 95, 96, 155
Schürmann, H., 34, 141, 184, 185, 187, 188, 189, 192, 193, 194, 195, 196, 197, 199, 200
Schütz, R., 210
Schulz, S., 4, 28, 29, 31, 67, 82, 102, 132, 237
Schweizer, E., 4, 12, 17, 18, 19, 20, 22, 25, 29, 30, 32, 35, 37, 39, 51, 54, 55, 57, 58, 67, 69, 71, 72, 73, 75, 80, 81, 84, 92, 93, 95, 99, 100, 102, 103, 110, 113, 137, 138, 148, 153, 154, 155, 156, 157, 160, 163, 164, 174, 175, 179, 181, 183, 192, 193, 194, 198, 200, 206, 210, 212, 224, 227, 230, 232, 234, 238, 239, 241, 243
Siegman, E. F., 67, 76, 90, 92, 93
Simpson, R. T., 19, 86
Sjöberg, E., 38, 43, 47, 63, 64, 67, 73, 78, 79
Sloyan, G. S., 193
Soden, H. von, 3, 46, 106, 107, 122, 123, 159, 183, 203
Soiron, T., 103
Souček, J. B., 175
Stählin, G., 174, 176
Stewart, R. A., 76
Strack H. L. and P. Billerbeck, 35, 38, 42, 124, 127, 139, 149, 154, 157, 165, 176, 189, 190, 196, 208, 209, 243
Strecker, G., 27, 28, 29, 40, 93, 206, 232, 234
Streeter, B. H., 122, 123, 128, 129
Suhl, A., 10, 11, 31, 37, 42, 44, 48, 51, 67, 68, 87, 108, 128, 133, 194, 195
Sundwall, J., 4, 5, 7, 19, 34, 137, 138, 155, 156, 157, 160, 173, 223, 225, 231
Swete, H. B., 89, 92, 93, 117

Tasker, R. V. G., 3, 46, 106, 122, 123, 159, 183, 203
Taylor, V., 17, 18, 19, 20, 26, 32, 33, 34, 37, 39, 40, 41, 49, 50, 52, 54, 57, 60, 63, 65, 67, 68, 75, 80, 81, 82, 83, 90, 93, 97, 100, 101, 105, 106, 107, 108, 110, 123, 124, 125, 126, 127, 129, 137, 138, 148, 153, 154, 155, 157, 163, 164, 170, 173, 174, 175, 176, 177, 179, 181, 187, 193, 196, 199, 203, 204, 209, 210, 223, 225, 230, 232, 237, 238, 241, 242, 243
Thackeray, H. St. John, 50
Theissing, J., 140, 152
Thurian, M., 191, 194, 196
Tillard, J. M. R., 198
Tischendorf, C., 3, 46, 47, 106, 122, 123, 183, 203
Tödt, H. E., 234
Trilling, W., 4, 5, 6, 7, 20, 21, 26, 140
Trocmé, E., 49, 51, 55, 61, 73, 74, 93, 215
Turner, C. H., 7, 26, 32, 60, 61, 82, 155, 156, 162
Turner, H. E. W., 125
Turner, N., 186, 188, 196, 232
Tyson, J. B., 70

Vaganay, L., 156, 173, 174
Vermes, G., 2
Via, D. O., Jr., 47, 101
Vielhauer, P., 8, 12, 29, 39, 41, 95, 97, 186, 239

Vögtle, A., 205, 207, 208, 209, 212, 215, 216, 217, 218, 235
Vogels, H., 184, 193, 196
Vogels, H. J., 3, 46, 106, 107, 122, 123, 159, 183, 203
Vogt, E., 87, 92

Walter, N., 148, 159, 160, 161, 162, 164, 224, 228, 229
Walther, G., 199
Weeden, T. J., 29, 70, 95, 103
Weiser, A., 38, 151, 220
Weiss, J., 163
Weiss, K., 30
Wellhausen, J., 8, 14, 34, 108, 155, 159, 196, 197
Werner, M., 14
Westermann, C., 10, 220, 221
Wikgren, A., 3, 46, 106, 122, 123, 159, 183, 203
Wilder, A. N., 141, 142
Wilkens, W., 31, 81, 86, 88, 89, 104
Winandy, J., 226, 229
Windisch, H., 68, 139, 141
Winter, P., 242
Wrede, W., 9, 93, 94, 95, 102

Zerwick, M., 6, 7, 18, 50, 79, 162, 164, 231
Zimmerli, W., 220, 221
Zimmermann, H., 173, 175
Zirker, H., 9, 10

INDEX OF SUBJECTS

Amazement, 26, 58-59, 80-81, 153, 161-63
Church, 27-31, 70-71, 89-92, 102, 104, 121, 130-35, 171, 195-96, 198-202, 211-12, 246-47
Controversies, 56-57, 61-62
Christian Service, 155-58, 171, 181, 244, 246
Crowd, 55-57, 65-67, 70-72
Demons, 56-57

Disciples, 29-30, 51-53, 60, 62-65, 88-89, 163-64, 174, 243
Enemies of Jesus, 56-62, 66, 71-72, 177
Eschatology, 15-25, 89-90, 101-2, 104-6, 109-14, 116-18, 130-35, 139-43, 158, 176, 186, 189-91, 206, 209-22
Exorcism, 56-57, 96-97
Fear, see Amazement
Following of Jesus, 63-64, 206, 234

Ignorance, see Understanding
Impenitence, 66-70, 79, 83-86, 89
Jerusalem, 40-41
John the Baptist, 18-20, 25, 45
Knowledge, see Understanding
Messiah, 37-39, 41, 64, 232-34
Messianic secret, 28-31, 38, 44, 61-62, 92-99, 103, 206, 222-23, 238, 245
Mysteries, 87-88, 145
Parable of the Sower and its Interpretation, 99-102, 120-21
Parousia, 207, 213-15, 224-31, 234-36, 237, 245

Proclamation, 12-14, 22, 25, 27-28, 228, 245, 247
Prophetic Perspective, 219-222
Resurrection of Jesus, 10-13, 89, 121, 237
Riches, 163-69, 171
Son of David, 35, 38-40, 42-44
Son of God, 23, 43, 99, 233
Teaching, 22, 57, 74-75, 83-85, 120-21, 179
Transfiguration, 23, 236-40
Understanding, 30, 42-43, 51, 56, 59-62, 64, 69, 76-83, 86-90, 91-92, 122, 145-47, 232-34, 238, 244
Vita Jesu, 27-28, 244

BIBLICAL INDEX

GENESIS	
15:1	78
31:54	189
49:11	36,37

EXODUS	
18:12	189
20:20-22	167
22:24	165
24:3-8	200
24:11	189

LEVITICUS	
1:2	16
1:3	16
1:5	16
1:10	16
1:13	16
9:5	16
9:7	16
9:8	16
10:4	16
10:5	16
16:1	16
19:18	178
21:17	16
25:5	117
25:11	117

NUMBERS	
6:3-5	194
6:5	194
19:2	37

DEUTERONOMY	
6:5	178
12:7	189
14:23	189
14:26	189
15:7-8	168
15:11	168
15:20	189
21:3	37
32:1	209

1 SAMUEL	
9:12-14	189
15:32	42

2 KINGS	
4:27	154
19:29	117
23:10	176

TOBIT	
3:15	150

7:17	150
10:12	150
10:13	150
11:14	150
12:1	150
14:3	150
14:4	150
14:8	150
14:10	150

JOB	
4:12	144
22:22	144
28:28	146
29:12-13	168
29:16	168

PSALMS	
49:5	76
78:1	151
78:2-4	151
103:12	128
118:25-26	38
119:9	151

PROVERBS	
1:2-4	145
1:3	144

1:4	147	51:17	145,147	**HOSEA**	
1:6	76	51:26	144	8:13	189
1:9	144				
2:1	144,145,152	**ISAIAH**		**JOEL**	
2:6	145,147	3:4	150	4:13	108
3:1	152	3:5	150		
4:10	144	5:1	220	**AMOS**	
Ch. 8	138	6:9	47	2:6-8	167
8:4-36	151	6:9-10	47,69	2:8	189
8:22-31	146	8:23–9:1	220		
9:9	144,145,147	9:2-6	220	**JONAH**	
10:8	144	10:19	150	4:11	149
16:17	144	11:1-8	220		
21:11	144	11:6-8	150	**ZEPHANIAH**	
23:10-11	168	25:6	189	1:7	189
24:22	144	42:9	190	3:2	144
30:1	144	43:18-19	190	3:7	144
		65:17	190		
QOHELETH				**HAGGAI**	
2:26	147	**JEREMIAH**		2:20-23	220
		2:30	144		
WISDOM		5:3	144	**ZECHARIAH**	
7:7	147	7:28	144	6:13	220
7:15	147	7:31	176	9:9	36,37
7:17	147	9:19	144	14:4	42
8:21	147	17:23	144		
9:4	147	19:5-6	176	**MATTHEW**	
9:17	147	22:13	168	1:23	219
12:7	145	22:17	168	3:10	114
19:14	145	30:31-33	200	3:12	114
		35:6-7	194	5:15	103
SIRACH				5:29-30	176
2:4	144	**EZEKIEL**		5:44	185
6:23	144	11:23	42	5:45	143
15:17	147	17:1	77	6:10	208
24:3-12	146	17:23	128	7:2	103
30:1-13	149,151	21:1-5	77	8:11	190
30:2-3	151	24:3	77	9:6	125
39:3	76	31:6	128	9:11-13	190
39:5-8	146			10:13-15	185
41:1	144	**DANIEL**		10:23	207
43:33	147	4:11	15	10:26	103
45:26	147	Ch. 7	228	10:40	152,185
47:15	76	7:1-8	228	11:6	119
47:17	77	7:13	208	11:11	140
50:12	144	Ch. 8	228	11:12	140
51:13-16	146	Ch. 11	228	11:25	152
51:16	144-45	Ch. 12	228	12:28	210

Ch. 13	88	1:4-5	18	2:15	55,65
13:2	125	1:7	18,19	2:15-17	61,65,190
13:8	200	1:7-8	20	2:16	62
13:11	86	1:8	18	2:16-17	57
13:13	54	1:9	18,26	2:17	57,58
13:15	54	1:9-11	23	2:18	62
13:17	192	1:10	18,126	2:18-22	57,201
13:19	88	1:11	23	2:19	57-58
13:23	88	1:14	3,4,6-7,	2:20	62,195,201
13:31	125		17-20,24-25	2:21-22	191,201
13:31-32	123	1:14-15	3-32	2:24	62
13:32	200	1:15	1-7,15-17,	2:24-28	57
13:51	88		21,23,25-26,	2:25	57-58
14:23	125		44,91,105,202,	2:26	53
14:33	93		235,247	2:27	57-58,204
15:35	125	1:16	26,163	2:28	62
16:12	88	1:16-20	14,22,164	Ch. 3	55,233
16:28	213	1:17	170	3:1-5	60
17:13	88	1:20	170	3:1-6	98
17:20	129	1:21	57	3:2-5	57
18:3	137,139,148	1:22	57-59,85	3:4	57,62
18:4	139,148,150	1:23	61	3:5	57
18:10	173	1:23-26	57	3:6	57,61,62,79
18:14	173	1:24	56	3:6-8	65
19:12	143	1:27	57-59,83,85,	3:7	65
19:28	191		190	3:7-12	8,96
20:26-27	155	1:33	55	3:7-10	185
21:9	35	1:34	56,61	3:8	41
21:31	140	1:35	24	3:11	83
22:2-8	190	1:38	57	3:11-12	56-57,61
22:9-13	190	1:38-39	18	3:12	56
23:11	155	1:45	18,55	3:13	54,58,126
23:13	140	Ch. 2	233	3:13-15	63
25:10	190	2:1	24	3:14	14,18
25:29	103	2:1-12	98	3:19	20
26:29	201	2:1-3:6	53,62	3:20	55,65
26:39	125	2:2	55,57,73	3:20-35	53-55
27:57	243	2:3-12	60-61	3:21	53-54
		2:5-10	61	3:22	41,53,67,72
MARK		2:6	33	3:23	52,54,57-58,
1:1	7,9,12,14,	2:6-11	57		72,75
	17-21,201	2:7	61-62	3:23-29	66,72
1:1-3	18-21	2:8	57	3:23-30	66,86
1:1-13	19	2:10	60-62,124	3:24	54
1:2-3	18-21	2:10-11	61	3:27	96
1:2-8	18	2:12	58-59	3:28-29	62,66,91,94
1:2–8:26	24	2:13	55,57,65	3:29	85
1:4	5	2:14	164,170	3:30	67,72,91,94

3:31	54		66-71,79,88,		126,128,133,200
3:31-35	54,122		105	4:33	51,73-75
3:32	54	4:13	50-53,60,64,	4:33-34	49,51,57,
3:33-34	57		69,75,78,80,		72-74
3:34	54,57		88,101,	4:34	73-75,78,134
3:35	122,185		120-21,202	4:35-41	93,126
Ch. 4	30,52,55,69,	4:14	202	4:35–5:43	96
	79,88,102	4:14-19	89	4:36	26
4:1	24,53,55,57,	4:14-20	49-50,63,	4:37	127
	74,124		75,77,82,101,	4:39	127
4:1-2	49,55,58,66,		102,120-21,133,	4:40	26,80-81
	74,134		202	4:40-41	93
4:1-13	69	4:15	88,246	4:41	58-59,80,93,
4:1-25	127	4:17	49,89,122		127
4:1-34	51	4:19	121,246	5:1-20	97
4:2	50,57,74-75,	4:20	88,101,102,	5:3	188
	83,85,124		122	5:8	83
4:3	81,103	4:21	50,57,102	5:15	58,59
4:3-4	124	4:21-22	93,103-5,	5:18-20	33,97
4:3-8	77		139	5:19	33,97
4:3-20	49	4:21-23	50	5:20	18,32,58,59
4:7	126	4:21-24	103	5:21	24,55,65
4:7-8	126	4:21-25	2,50,68,77,	5:24	55
4:8	126,200		102,105,204,	5:24-34	98
4:9	50,81,103		233	5:33	58-59
4:10	31,46,52,53,	4:22	103-4	5:36	26,58-59
	56,66,70-71,	4:23	81,103-4	5:42	58-59
	75,89,105,134,	4:24	50,57,81,102-4	6:2	57-59
	233	4:24-25	50,103-4,	6:6	32,57-58
4:10-11	54		202	6:7	53,58,83
4:10-12	2,46-106	4:25	103	6:9	80
4:10-13	50,51,75	4:26	50,57,106,108,	6:10	204
4:10-20	75		114-15,119,121,	6:12	18
4:11	1,9,22,29,		124	6:14-16	44
	47-51,53-55,	4:26-29	2,23,46,49,	6:20	58
	57,66-67,69,		77,106-22,247	6:30–10:52	24
	72-73,75-76,	4:27	115-18,121,122	6:31-52	96
	78-79,81,85-86,	4:27-28	114-15,121	6:34	57
	88-90,92-93,	4:28	106,114,116	6:36	32
	99,104,121,134,	4:29	34,107-8,114,	6:38	60
	149,155,163,188,		116-17,121	6:45-52	93
	235,247	4:30	50,57,75,	6:49	80
4:11-12	47-49,51,		122-23	6:50	58-59,80
	53,66,69,75,	4:30-32	2,23,46,77,	6:51	58-59,126
	77,80,86,93,		122-34,247	6:52	51,60,64,79,80
	101-5,204	4:30–5:43	126	6:53	124
4:11-25	93	4:31	122,124,126	6:53-56	96
4:12	47,54-55,	4:32	116,122-23,	6:56	32

Ch. 7	233		40-41,64,	9:18	64
7:1	33,41		89,99,155,	9:23	26
7:1-5	65		171,232-33,	9:28	64
7:2	33		238	9:28-29	63,93
7:2-3	38	8:28	44	9:30	24
7:3	55	8:30	232-33	9:30-32	40
7:3-4	82,83	8:31	40,57,62,81,	9:30-36	81
7:5-15	57		232,234,240	9:30-37	155
7:9	57,204	8:31-32	40	9:31	20,57
7:12	188	8:31–16:8	24	9:32	51,58-60,64,
7:14	54,58,65,79,	8:32	73,232-33		80,81
	80-81	8:33	64,89,232,	9:33	20,155,172-74
7:14-23	75		235,246	9:33-34	81,173,234
7:15	77,82,91	8:34	31,53-55,58,	9:33-35	156,158,175
7:17	52,75,77		71,231-32	9:33-37	89,155-57,
7:17-18	75	8:34-37	29,175		174-75
7:17-23	93	8:34-38	202,231,	9:33-41	173-74
7:18	50,60,64,77		233,240	9:33-50	63,156,
7:18-23	63,77	8:34-9:1	102,181,234		172-75
7:19	82,91,126	8:35	8,9,14,15,181,	9:34	20,64,155-57,
7:19-23	75		233		173
7:31	32	8:36	233	9:35	155-58,246
7:31-37	81	8:37	233	9:35-50	103
7:36	65	8:38	15,25,89,202,	9:36	137-38,156-57,
7:36-37	18		204-7,231,		172-73
7:37	58,59		233-35	9:36-37	137,139,
Ch. 8	231,233	8:38–9:1	25,153,206		156,158
8:1	54,58	Ch. 9	1,231	9:37	137,148,152,
8:1-21	69	9:1	2,15,23,25,33,		156-58,172-74
8:3	33,40		58,90,203-40,247	9:37-50	173
8:11-12	57	9:2	238	9:38	174
8:11-13	62,69	9:2-8	240	9:38-40	172
8:12	57	9:5	164	9:38-41	175
8:14-21	69,93	9:6	26,58-60,64,80	9:39	172
8:17	60,79,80-81		238	9:40	172,174
8:17-21	64,80	9:7	23	9:41	172,174
8:18	60,69,80	9:8	188	9:41-50	173
8:21	60,69,81,88,204	9:9	29,31,238	9:42	148,172,174,
8:22-26	81	9:9-13	63		176
8:23	32	9:10	64,192	9:42-47	172,175
8:26	32,65	9:11-13	88	9:42-48	174,175
8:27	20,32,231	9:12	50,195	9:42-50	172,173
8:27-29	15,233	9:12-13	20	9:43	173-74,176
8:27-30	24	9:13	24	9:43-48	172,175
8:27-33	232	9:13-16	172	9:44	172
8:27–9:1	204,206,	9:14	55,233	9:45	173,176
	232-33,245	9:15	26,58-59	9:46	**172**
8:27–10:52	24,29,30,	9:17	188		

9:47	2,136,171-72, 181	10:25	83,159-62,164, 168,171	11:4	33,42
9:48	172,174	10:26	58-59,159-64	11:4-7	33,42
9:49	172,174-75	10:26-27	159-60,169	11:5	33
9:49-50	175	10:27	159,161-64,170	11:7	34
9:50	172-75	10:28	14,153,164	11:9	37-41
Ch. 10	1,172	10:28-29	164	11:9-10	37-38,42,44
10:1	55,57,65	10:28-30	158,181,202	11:10	2,3,32-44,208
10:1-2	65	10:29	8,9,14	11:11	34,40-41
10:2	62,233	10:29-30	170	11:14	164
10:2-9	57	10:29-31	164	11:17	57
10:3	57	10:30	23,89,181,247	11:18	57-59,62,65,85
10:5	57-58	10:32	20,40,58,59,126	11:27	24,65
10:10	26,62	10:32-33	126	11:27-33	20,41,51
10:10-12	63,93	10:32-34	40,81	11:27–12:12	24
10:11	50	10:32-45	155	11:28	62,226
10:11-12	62,82	10:33	20,126	11:29	57
10:13	137,152,154, 172	10:35	155	11:31-33	62
		10:35-37	64	11:32	58-59,65
10:13-16	2,136-38	10:35-41	234	11:33	57,62
10:14	64,137-38, 151-55	10:35-45	81,155,157, 174	12:1	57-58,72,75
				12:1-9	61,66,86,91
10:14-15	136-58,181, 246-47	10:36	155	12:1-12	24,56
		10:37	156	12:9	62,66,85,91
10:15	23,136-39,143, 147-49,151-58, 160,170	10:38	60,195	12:10	61
		10:41	64	12:12	24,56-62,65, 75,80
		10:42	38,54,58,60,181		
10:16	137-38,152-56, 163,170-72	10:42-43	171,246	12:13	24,33,57,65
		10:43	155,246	12:13-17	57
10:17	162	10:43-44	155,157	12:13-37	178
10:17-22	160-64,170	10:43-45	155	12:13-44	24
10:19	162,170	10:44	155	12:14	57,62
10:21	153,162,170	10:45	155,171,246	12:15	57
10:21-27	2	10:46-52	43	12:16	57
10:22	153,161-64	10:52	20	12:17	57-59
10:23	83,159-64, 168, 170-71	Ch. 11	233	12:18	24,65
		11:1	16,32,40-42	12:18-27	57
10:23-25	136,153, 158-71,181, 246	11:1-2	34	12:19-23	62
		11:1-3	33,42	12:24	57-58
		11:1-7	34,40-41	12:28	24,177
		11:1-10	31	12:28-34	177-79
10:23-26	159	11:1-26	24	12:29	57-58,178
10:23-27	160	11:1-11	31,40,44	12:29-31	180
10:23-31	63	11:1–12:44	24	12:30	178
10:24	26,58-59,83, 151,159-65, 169-71	11:1–13:37	34	12:32	178
		11:1–16:8	24	12:32-33	178
		11:2	32	12:33	62,178
10:24-27	161	11:3	33,37	12:34	2,24,62,136, 177-82,188

12:35	43,57,164,178,179	13:26	90,205,206,208,235	14:47	33,34
12:35-36	24	13:28	60,75,83,224,228	14:48	164
12:35-37	43,44,178			14:49	57
12:37	65	13:28-29	224	14:50	64
12:37-38	43	13:28-31	223,230	14:53–15:5	24
12:38	57,85	13:28-32	224-25,229	14:61-64	62
12:43	43,54,58	13:29	60,83,224-25,227,229-30	14:62	25,57,102,208
12:44	188			14:62-64	61,80
Ch. 13	24,25,63,88,102,134,203,205,218,222-31	13:30	2,204,205,207,216,217,218,224,225,229-30,235,239	14:64	61
				14:66-72	64
				14:68	60
				14:69	188
13:1	225			14:70	188
13:1-2	225	13:32	216-18,222-24,227,229-30	14:71	60
13:1-4	222,226,230			15:1	20
13:1-5	225	13:33	60,223-24	15:5	58,188
13:2	225-26,228	13:33-36	222-23	15:8	126
13:3	26,42,70,223	13:33-37	3,224,230	15:10	20
13:3-4	225	13:34	2,223	15:11	65
13:4	224-26,228-30	13:35	60,223	15:15	20
13:5	50,224,228	13:35-36	223	15:35	33,34,65
13:5-6	228,246	13:37	71,202,223-24,246	15:39	12,23,41,99,239
13:5-13	226-27			15:42	241
13:5-22	229	14:1-2	65	15:42-47	241-42
13:5-23	133,228-29	14:1–16:8	24	15:43	2,203,240-43
13:5-27	228	14:2	59	15:44	58
13:5-37	225	14:3-8	14,195	15:44-45	241-42
13:7	3,89,222,226,229	14:6	64	15:46	242
13:7-8	228	14:9	9,14,18	15:47	241
13:8	227	14:11	20	16:5	58
13:9	20,224,245	14:12-13	34	16:6	58
13:9-11	9,13,247	14:12-25	24	16:7	29
13:9-13	89,228	14:18	20,188	16:8	58
13:10	3,14,18,108,133,201,226,247	14:21	20,188		
		14:22-23	184,187	**LUKE**	
		14:22-24	184	2:25	243
13:11	20,89	14:22-25	187,197	2:38	243
13:11-13	245	14:23	197	3:15	68
13:12	20	14:24	186-87,200-1	4:2	188
13:13	68,226,228	14:25	2,183-202	6:17-19	185
13:14	226-28,230	14:26	42	6:27-28	185
13:14-20	228	14:28	29	6:38	103
13:14-23	228	14:33	58	7:33-34	190
13:14-27	227	14:37	64	8:4-18	127
13:21-22	228,246	14:40	60,64	8:8	200
13:23	224,228-29	14:41	20	8:10	54,86
13:24	188,228-29	14:42	16,20	8:13	122
13:24-27	229,235	14:43	65	8:15	122

8:17	188	**JOHN**		**PHILIPPIANS**	
8:21	122,185	3:3	137,152,208	2:7	79
9:18	52	3:5	137,152	2:22	150
9:27	208	3:14	18	**COLOSSIANS**	
9:48	155	6:64	113	2:20-23	83
10:8-12	185	6:66	113	3:20	150
10:16	152,185	7:26	68		
10:19	188	8:52	209	**1 THESSA-**	
10:21	152	13:20	152	**LONIANS**	
10:30	128	16:30	26	1:5	79
11:20	210	19:38	243	2:5	79
11:30	18			4:13	211
11:33	103	**ACTS**		4:15-17	211
11:52	140	3:18	6		
12:2	103	4:25	35	**1 TIMOTHY**	
12:8	119	5:31	5	3:4	150
12:16	128	11:17-18	5	3:12	150
13:18	75,123	11:18	5		
13:18-19	122-23,127	20:21	5	**2 TIMOTHY**	
13:19	123,200	23:15	16	1:8	15
13:29	190	24:16	26	1:12	15
14:2	128			2:25	68
14:15-24	190	**ROMANS**			
14:16	128	1:3-4	11,13,39,43	**TITUS**	
15:2	190	1:4	205	1:6	150
15:11	128	1:16	15		
15:16	192			**HEBREWS**	
16:1	128	**1 CORIN-**		2:10	209
16:16	140	**THIANS**		7:19	16
16:19	128	2:9	18	9:17	68
17:6	129	4:4	26		
17:20	208,217-18	11:23-25	186	**JAMES**	
17:20-21	210	11:26	186	4:8	16
17:22	192	15:51	211	5:7-8	118
17:26	18				
18:24-27	160	**2 CORIN-**		**1 PETER**	
19:12	128	**THIANS**		1:14	150
19:26	103	5:17	191		
22:7-14	192	12:14	150	**2 PETER**	
22:15	184,191-92			3:13	191
22:15-18	183-88	**GALATIANS**			
22:16	184,188,191,194	4:4	6	**REVELATION**	
22:17	184,187,192	6:15	191	3:20	190
22:18	184,188,191,208			18:14	188
22:26	155	**EPHESIANS**		19:9	190
22:29	190	1:10	6	21:1-5	191
23:53	188	6:1	150	22:27	138
24:35	115				

EXTRABIBLICAL LITERATURE

1 ENOCH
5:8	145
25:5	189
27:1-3	176
37:4	145, 147
37:5	145
38:1	82
38:1-2	85
38:5-6	166
43:3	82
43:3-4	145
45:1	82
45:1-2	85
45:4	190
46:4-8	166
53:1-7	167
54:1-5	176
54:1-6	167
56:3-4	176
58:1	82
60:1	146
62:1-12	167
62:13-14	189
63:3	88
63:7	166
63:10	166
68:1	145
68:5	88
72:1	190
79:1	151
82:1	145
82:1-3	151
82:2	145
90:26-27	176
91:2	151
94:6-9	166
96:8	166
97:8-10	167
98:2	166
98:3	166
98:6	166
99:12-13	167
99:15-16	167
101:8	145
103:9-15	167
103:15	167

2 APOC. BAR.
14:13	144
29:4	189
29:8	189
44:12	190
51:3	144
70:2	118

TEST. LEVI
18:11	188

2 ESDRAS
4:28-29	118
4:35-40	118
5:10	146
5:26	209
7:14	144
7:36	176
7:74	118
7:102	86
7:114-15	86
8:4	145
8:41	86
8:52	189
10:38	146
14:25-26	146
14:44	146

1QH
5:36	88
6:14-17	131
8:4-9	131
8:6	88
8:11	88
10:27	88, 146
11:27-28	88, 146
12:11-13	88, 146
13:18-19	88, 146
14:8	88, 146
16:11-12	146
16:11-13	88
17:17	88, 146
18:27	88, 146

1QM
3:8-9	88
14:9	88
14:14	88
16:11	88
17:9	88

1QSa
2:11-22	190

1Q27
27:5	88
27:7	88

TARGUM MICAH
4:8	208
7:14	190